Feed Additives in Pig Feeding: 2nd Edition

Feed Additives in Pig Feeding: 2nd Edition

Guest Editor

Małgorzata Kasprowicz-Potocka

MDPI

Basel • Beijing • Wuhan • Barcelona • Belgrade • Novi Sad • Cluj • Manchester

Guest Editor
Małgorzata Kasprowicz-Potocka
Department of Animal Nutrition
Poznań University of Life Sciences
Poznań
Poland

Editorial Office
MDPI AG
Grosspeteranlage 5
4052 Basel, Switzerland

This is a reprint of the Special Issue, published open access by the journal *Animals* (ISSN 2076-2615), freely accessible at: https://www.mdpi.com/journal/animals/special_issues/428DQUO25U.

For citation purposes, cite each article independently as indicated on the article page online and as indicated below:

Lastname, A.A.; Lastname, B.B. Article Title. *Journal Name* **Year**, *Volume Number*, Page Range.

ISBN 978-3-7258-6532-1 (Hbk)
ISBN 978-3-7258-6533-8 (PDF)
https://doi.org/10.3390/books978-3-7258-6533-8

Contents

About the Editor vii

Dongfeng Sun, Wenfei Tong, Shaochen Han, Mengjun Wu, Peng Li, Youguo Li and Yunxiang Liang
Effects of Dietary Supplementation with *Lactobacillus reuteri* Postbiotics on Growth Performance, Intestinal Flora Structure and Plasma Metabolome of Weaned Piglets
Reprinted from: *Animals* **2025**, *15*, 204, https://doi.org/10.3390/ani15020204 1

Narathon Innamma and Kampon Kaeoket
Feeding Sows with Multi-Species Probiotics During Late Pregnancy and the Lactating Period Influences IgA Concentration in Colostrum and Subsequently Increases the Survival Rate of Piglets in Porcine Epidemic Diarrhea Outbreak Herd
Reprinted from: *Animals* **2025**, *15*, 103, https://doi.org/10.3390/ani15010103 17

Frank R. Dunshea, Ian McCauley and Robert J. Smits
Effects of Dietary Bromide, Magnesium and Tryptophan and Immunocastration on Growth Performance and Behaviour of Entire Male Pigs
Reprinted from: *Animals* **2024**, *14*, 3685, https://doi.org/10.3390/ani14243685 29

Sheng Huang, Li Yang, Li Wang, Yu Chen, Xiuliang Ding, Feiyun Yang, et al.
The Effects of Octapeptin Supplementation on Growth Performance, Serum Biochemistry, Serum Immunity, and Gut Microbiota in Weaned Piglets
Reprinted from: *Animals* **2024**, *14*, 2546, https://doi.org/10.3390/ani14172546 41

Min Ma, Yuriko Enomoto, Tomotsugu Takahashi, Kazuyuki Uchida, James K. Chambers, Yuki Goda, et al.
Study of the Effects of Condensed Tannin Additives on the Health and Growth Performance of Early-Weaned Piglets
Reprinted from: *Animals* **2024**, *14*, 2337, https://doi.org/10.3390/ani14162337 55

Jakavat Ruampatana, Junpen Suwimonteerabutr, Kunaporn Homyog, Wanwimon Mekboonsonglarp, Korntip Kanjanavaikoon, Wouter Van der Veken, et al.
Clostridium butyricum Probiotic Feed Additive: Modulation of Sow Milk Metabolomics and Mitigation of Pre-Weaning Piglet Diarrhea
Reprinted from: *Animals* **2024**, *14*, 2098, https://doi.org/10.3390/ani14142098 69

Hyuk-Jun Lee, Bu-Gil Choi, Young-Ho Joo, Chang-Hyun Baeg, Ji-Yoon Kim, Dong-Hyeon Kim, et al.
The Effects of Microbial Additive Supplementation on Growth Performance, Blood Metabolites, Fecal Microflora, and Carcass Characteristics of Growing–Finishing Pigs
Reprinted from: *Animals* **2024**, *14*, 1268, https://doi.org/10.3390/ani14091268 88

Prester C. John Okafor and Nitipong Homwong
Dietary 25 Hydroxyvitamin D3 Improved Serum Concentration Level and Alkaline Phosphatase Activity during Lactation but Had Meager Impact on Post-Farrowing Reproductive Performance in Sows
Reprinted from: *Animals* **2024**, *14*, 419, https://doi.org/10.3390/ani14030419 100

Alfredo Valverde Lucio, Ana Gonzalez-Martínez, Julio Gabriel Ortega and Evangelina Rodero Serrano
Effects of Alternative Cassava and Taro Feed on the Carcass and Meat Quality of Fattening Pigs Reared under Ecuadorian Backyard Systems
Reprinted from: *Animals* **2023**, *13*, 3086, https://doi.org/10.3390/ani13193086 113

Rumbidzai Blessing Nhara, Upenyu Marume and Carlos Wyson Tawanda Nantapo
Potential of Organic Acids, Essential Oils and Their Blends in Pig Diets as Alternatives to Antibiotic Growth Promoters
Reprinted from: *Animals* **2024**, *14*, 762, https://doi.org/10.3390/ani14050762 . . . . **127**

Paloma Islas-Fabila, Patricia Roldán-Santiago, Luis Alberto de la Cruz-Cruz, Ofelia Limón-Morales, Anna Dutro-Aceves, Héctor Orozco-Gregorio and Herlinda Bonilla-Jaime
Importance of Selected Nutrients and Additives in the Feed of Pregnant Sows for the Survival of Newborn Piglets
Reprinted from: *Animals* **2024**, *14*, 418, https://doi.org/10.3390/ani14030418 . . . . **143**

John O'Doherty, Alison Dowley, Eadaoin Conway and Torres Sweeney
Nutritional Strategies to Mitigate Post-Weaning Challenges in Pigs: A Focus on Glucans, Vitamin D, and Selenium
Reprinted from: *Animals* **2024**, *14*, 13, https://doi.org/10.3390/ani14010013 . . . . **161**

Hyunjun Choi and Sung Woo Kim
Characterization of β-Glucans from Cereal and Microbial Sources and Their Roles in Feeds for Intestinal Health and Growth of Nursery Pigs
Reprinted from: *Animals* **2023**, *13*, 2236, https://doi.org/10.3390/ani13132236 . . . . **177**

Mariah R. Jansen and Kwangwook Kim
Evaluating the Effects of Non-Nutritive Sweeteners on Pigs: A Systematic Review
Reprinted from: *Animals* **2024**, *14*, 3032, https://doi.org/10.3390/ ani14203032 . . . . **189**

About the Editor

Małgorzata Kasprowicz-Potocka

Małgorzata Kasprowicz-Potocka, Poznań University of Life Sciences (Department of Animal Nutrition on Faculty of Veterinary Medicine and Animal Sciences). For over 25 years, I have been scientifically working with animals. My accomplishments include around 60 research papers and over 250 popular science publications, as well as several patents. I have participated in 12 national and international research projects, including Horizon 2020 and Horizon Europe. My research interests focus on the use of feed additives (eubiotics, enzymes, and herbs) in pig nutrition and on methods of processing plant feeds to improve their nutritional value, as well as on the use of by-products from industry in animal feed. In pig research, I specialize in analyzing the digestibility of feeds and feed raw materials, studying nutrient retention, analyzing gases excreted by pigs, as well as evaluating the biochemical, histomorphometric, and metabolic parameters of the gastrointestinal tract in pigs.

animals

MDPI

Article

Effects of Dietary Supplementation with *Lactobacillus reuteri* Postbiotics on Growth Performance, Intestinal Flora Structure and Plasma Metabolome of Weaned Piglets

Dongfeng Sun [1], Wenfei Tong [2], Shaochen Han [2], Mengjun Wu [2], Peng Li [2], Youguo Li [1,*] and Yunxiang Liang [1,*]

1 National Key Laboratory of Agricultural Microbiology and College of Life Science and Technology, Huazhong Agricultural University, Wuhan 430070, China; sdf365@126.com
2 Hubei Key Laboratory of Animal Nutrition and Feed Science, Wuhan Polytechnic University, Wuhan 430023, China; tong1wenfei@163.com (W.T.); 17633859146@163.com (S.H.); wumengjun93@163.com (M.W.); lp1536698031@163.com (P.L.)
* Correspondence: youguoli@mail.hzau.edu.cn (Y.L.); liangyunxiang@mail.hzau.edu.cn (Y.L.)

Simple Summary

Intestinal health is related to the healthy and efficient breeding of piglets, which needs to be focused on in the post-antibiotic era. Microecological agents play an important role in improving the intestinal health of piglets; however, many of the mechanisms have not been characterized. In the present study, we present an updated report of *Lactobacillus reuteri* postbiotics on the growth performance, intestinal flora structure and plasma metabolome of weaned piglets. Our outcomes demonstrate that *Lactobacillus reuteri* postbiotics improve the antioxidant function and reduce the mortality of piglets by regulating the structure of intestinal flora and upregulating the content of coenzyme Q10 in serum. Our findings provide an important theoretical basis for the application of *Lactobacillus reuteri* postbiotics in piglet production and provide new data for the healthy and efficient breeding of piglets.

Abstract

Probiotics and their postbiotics have the potential to improve the health and growth performance of piglets, which has brought them widespread attention in the post-antibiotic era. In the present study, the effects of dietary supplementation of *Lactobacillus reuteri* postbiotics on the growth performance, intestinal flora structure and plasma metabolome of weaned piglets were investigated. A total of 816 healthy male piglets with uniform weight were divided into two treatment groups: piglets in the control (CTR) group were fed with a basic diet, and the ones in the LAC group were fed with the basic diet supplemented with 500 mg/kg *Lactobacillus reuteri* postbiotics. There were six replicates in each group and 68 piglets in each replicate. The animal trial lasted for 30 days. The feces and blood of piglets were collected for investigation, and the growth performance during the trial was counted. Our outcomes show that dietary supplementation with *Lactobacillus reuteri* postbiotics had no effect on the growth performance of piglets; however, it reduced the mortality rate of piglets by 6.37%. The levels of total superoxide dismutase in the serum, propionic acid and butyric acid in the feces were elevated, and the content of malondialdehyde in the serum was decreased with *Lactobacillus reuteri* postbiotics-treated piglets ($p < 0.05$). The fecal flora sequencing results show that the relative abundance of Firmicutes and monoglobus was upregulated, and the relative abundance of Bacteroides was downregulated with *Lactobacillus reuteri* postbiotics-treated piglets ($p < 0.05$). In addition, the levels of propionic acid and butyric acid in the feces were positively correlated with the relative abundance of Firmicutes and negatively correlated with the relative abundance of Bacteroides ($p < 0.05$). The plasma metabolome results show that dietary supplementation with *Lactobacillus*

 https://doi.org/10.3390/ani15020204

reuteri postbiotics raised the level of coenzyme Q10 in the serum, and the abundance of coenzyme Q10 was positively correlated with the relative abundance of Firmicutes and the level of total superoxide dismutase in the serum. In conclusion, dietary supplementation with *Lactobacillus reuteri* postbiotics contributed to improving the antioxidant function and reducing the mortality of piglets by regulating the structure of intestinal flora and upregulating the content of coenzyme Q10 in serum.

Keywords: biochemical profiling; microbial diversity; probiotic-derived metabolites; swine

1. Introduction

Previously, antibiotics were allowed to be added to feed due to their beneficial effect on intestinal health in livestock and poultry, as this would help improve growth performance [1]. However, due to the problem of bacterial resistance caused by antibiotics, this could have a negative impact on the health of people. Therefore, the use of antibiotics in feed is no longer permitted. Although livestock products are safer, the intestinal health problems of livestock and poultry are prominent without antibiotics in their diet, and this is particularly evident in young animals [2]. Weaned piglets, for example, are equipped with underdeveloped digestive and immune systems, which make them extremely sensitive to the external environment, such as abnormal environmental changes, pathogen infection, transportation and other stress [3]. The development of safe and effective alternatives to antibiotics in feed and the enhancement of piglets' ability to respond to environmental stress is one of the key topics in the current research on livestock.

It is well known that antibiotics in feed can regulate the structure of intestinal flora, such as inhibiting the proliferation of harmful bacteria and elevating the abundance of beneficial bacteria, thus improving the intestinal physiological function of livestock [4], while developing antibiotic substitutes helps to regulate the intestinal flora structure and improve the intestinal function of livestock in the post-antibiotic era [5]. Hence, we should pay more attention to the intestinal flora of livestock reared on feed with and without the use of antibiotics. Studies have demonstrated that probiotics and their postbiotics are able to improve the growth performance and health of piglets by improving the structure of intestinal flora and regulating the immune and antioxidant functions of piglets [5,6]. Among them, *Lactobacillus reuteri* is a lactobacillus that has been reported to colonize the intestines of almost all vertebrates and mammals [7]. Therefore, *Lactobacillus reuteri* is safe for animals and humans, which is the main reason why it is widely studied and used in medicine and food. A study demonstrated that *Lactobacillus reuteri* could metabolize and produce bioactive substances, such as short-chain fatty acids and indole-3-acetaldehyde, which contribute to improving intestinal health and growth performance [8,9]. However, the application of postbiotics developed based on the metabolites of *Lactobacillus reuteri* and its bacterial composition in pig production is rarely reported.

We believed it would be interesting to study the effects of *Lactobacillus reuteri* postbiotics on the growth performance and intestinal flora structure of piglets. In the present study, *Lactobacillus reuteri* postbiotics were used to conduct experiments in large populations of piglets and harvest samples based on non-invasive sampling methods to investigate the effects of dietary supplementation of *Lactobacillus reuteri* postbiotics on the growth performance, intestinal flora structure and plasma metabolome of weaned piglets. The purpose of this study was to provide a theoretical basis for the application of *Lactobacillus reuteri* postbiotics in the healthy and efficient breeding of piglets.

2. Materials and Methods

2.1. Experiment Design and Animal Management

A single-factor experimental design was implemented to perform a 30-day animal trial in the breeding farm of New Hope in Zhenyuan, Guizhou, China. The 30-day use period was based on our experience with over 2000 broiler chickens and our results with over 20 weaned piglets (unpublished data). The 30-day animal trial period was chosen for the following reasons: our preliminary experimental results demonstrated that 30 days of *Lactobacillus reuteri* postbiotics supplementation was sufficient to improve the health of piglets. In addition, considering that the cost of *Lactobacillus reuteri* postbiotics per ton of feed addition might be higher than CNY 30, long-term use would increase the breeding cost, and the experimental protocol and management of the pigs were approved by the Animal Care and Use Committee of New Hope Liuhe Co., Ltd., (Chengdu, China). The code for ethical inspection was IAS 2023-32. A total of 816 healthy male weaned piglets (those pigs with the same genetic background of Durc× Landrace× Yorkshire were weaned at 21 days of age after birth and started to be kept in corrals on a diet of feed for a 3-day acclimation period. With uniform body weight (7.06 ± 0.73 kg), they were randomly divided into the control group (CTR) and *Lactobacillus reuteri* postbiotics group (LAC). There were 6 replicates in each treatment and 68 piglets in each replicate. Piglets in the control group were fed a basic diet, and the diet formula was formulated according to the NRC 2012 nutritional requirements standard for piglets, and the formula and nutritional composition are shown in Table 1. Piglets in the LAC group were fed with the basic diet supplemented with 500 mg/kg *Lactobacillus reuteri* postbiotics purchased from Hubei Lan Good Microbial Technology Co. Ltd., Yichang, Hubei, China. The control group diet was correspondingly supplemented with an equal volume of a *Lactobacillus reuteri* postbiotics carrier (wheat bran and zeolite powder were mixed according to a 1:1 mass ratio). During the animal trial, piglets in each group were fed the same batch of feed without any change, all piglets were free to feed and water, and the routine immunization procedures were carried out according to the requirements of the farm. Briefly, all piglets were raised in a room, with each of the 68 piglets housed individually in a concrete enclosure without bedding (20 × 40 m^2), and the room temperature was maintained at 25 ± 2 °C, and a 24-hour lighting schedule was implemented. The number of piglet deaths during the animal trial was recorded, and all piglets were weighed at the end of the trial, and the feed consumption was measured. At the same time, two pigs from each replicate group were selected to collect blood from the anterior vena cava for the detection of antioxidant-related indexes and metabolome. Fresh feces were collected at the end of the animal trial to investigate the intestinal flora structure and levels of short-chain fatty acids.

Table 1. Feed formulation and nutritional value on dry matter basis.

Feed Formulation			Nutrient Level [3]		
Ingredients	**CTR**	**LAC**	**Item**	**CTR**	**LAC**
Corn	31.10	31.10	NE, kcal	2570	2570
Soybean meal	11.00	11.00	DE, kcal	3509	3509
Extruded soybean	10.00	10.00	CP, %	17.70	17.70
Wheat germ	4.00	4.00	EE, %	5.79	5.79
Extruded corn	20.00	20.00	CF, %	2.82	2.82
Whey powder	7.50	7.50	NDF, %	8.09	8.09
Fermented soybean meal	5.00	5.00	Ca, %	0.62	0.62
White sugar	2.00	2.00	total P, %	0.60	0.60
Wheat bran	1.70	1.70	Ash, %	4.77	4.77

Table 1. *Cont.*

Feed Formulation			Nutrient Level [3]		
Ingredients	**CTR**	**LAC**	**Item**	**CTR**	**LAC**
Coconut oil powder	1.25	1.25			
Soybean oil	1.00	1.00			
$Ca(H_2PO_4)_2$	0.89	0.89			
L-lysine	0.66	0.66			
Calcium formate	0.50	0.50			
Acidifying agent	0.50	0.50			
Stone powder	0.40	0.40			
L-valine	0.36	0.36			
L-threonine	0.31	0.31			
DL-methionine	0.31	0.31			
Montmorillonite	0.45	0.45			
L-tryptophan	0.27	0.27			
Trace element premix [1]	0.25	0.25			
Zinc oxide	0.18	0.18			
NaCl, 98.5%	0.18	0.18			
Choline chloride, 60%	0.08	0.08			
Vitamin premix [2]	0.05	0.05			
Phytase, 20,000 IU	0.03	0.03			
Sandoquine	0.02	0.02			
Sweetening agent	0.01	0.01			
Lactobacillus reuteri postbiotics	0.00	0.05			

[1] In the trace mineral elements, the content of copper was 40,000 mg/kg, the content of iron was 75,000 mg/kg, the content of zinc was 30,000 mg/kg, and the content of manganese was 35,000 mg/kg. [2] The main elements in vitamin premix were as follows: VA: 20 million IU/kg, VD_3: 10 million IU/kg, VE: 100,000 mg/kg, VK_3: 10,000 mg/kg, VB_1: 5000 mg/kg, VB_2: 12,000 mg/kg, VB_6: 4000 mg/kg and Niacinamide: 60,000 mg/kg. [3] Calculation level.

2.2. *Growth Performance*

The average daily gain (ADG) is expressed as the difference between the weight at the end and the weight at the start divided by the number of days in the trial. The average daily feed intake (ADFI) is characterized by the amount of feed consumed during the animal trial period divided by the number of days tested. The ratio of ADFI to ADG indicates the feed conversion efficiency (FCR), and the ratio of the number of piglets that died during the trial to the number of piglets at the start of the trial is indicative of piglet mortality.

2.3. *Blood Biochemical Indexes and Antioxidant Related Parameters*

Blood samples were collected from the anterior vena cava into either heparinized, and then they was centrifuged for 10 min at 3500 r/min and 4 °C to harvest the plasma. A Hitachi 7060 Automatic Biochemical Analyzer (Hitachi, Japan) was used to measure the blood biochemical indexes shown in Table 2. The contents of the antioxidant enzymes and peroxide products, for instance, superoxide dismutase (SOD), malondialdehyde (MDA), myeloperoxidase (MPO), and glutathione peroxidase (GSH-px), were investigated according to the steps of the kit purchased from Nanjing Jiancheng Bioengineering Institute, Nanjing, China.

Table 2. Results of serum biochemical index.

Item	CTR	LAC	p-Value
TB, μmol/L	0.30 ± 0.13	0.31 ± 0.12	0.979
TP, g/L	61.13 ± 2.90	57.91 ± 1.00	0.317
ALB, g/L	34.57 ± 1.74	32.34 ± 0.88	0.285
GLB, g/L	26.56 ± 1.49	25.57 ± 1.05	0.595
AST, U/L	73.20 ± 17.14	59.51 ± 3.09	0.451
ALT, U/L	44.70 ± 3.53	43.44 ± 3.58	0.806
ALP, U/L	320.50 ± 18.21	315.70 ± 14.14	0.837
TC, mmol/L	2.38 ± 0.09	2.31 ± 0.10	0.637
TG, mmol/L	0.49 ± 0.03	0.43 ± 0.05	0.263
GLU, mmol/L	5.51 ± 0.25	5.42 ± 0.21	0.796
CA, mmol/L	3.20 ± 0.13	3.06 ± 0.07	0.368
P, mmol/L	2.99 ± 0.26	3.24 ± 0.13	0.420
CREA, μmol/L	72.22 ± 4.14	77.73 ± 4.91	0.400
HDL, mmol/L	0.87 ± 0.04	0.85 ± 0.05	0.813
LDL, mmol/L	1.19 ± 0.05	1.11 ± 0.06	0.274
GGT, U/L	53.13 ± 3.65	59.23 ± 3.02	0.213
CK, U/L	2261.18 ± 230.19	2049.52 ± 238.95	0.532
LDH, mmol/L	1116.80 ± 139.45	991.04 ± 85.45	0.443

TB = total bilirubin, TP = total protein, ALB = albumin, GLB = globulin, AST = glutamic oxalacetic transaminase, ALT = glutamic–pyruvic transaminase, ALP = alkaline phosphatase, TC = total cholesterol, TG = triglyceride, GLU = glucose, CA = calcium, P = phosphorus, CREA = creatinine, HDL = high-density lipoprotein, LDL = low-density lipoprotein, GGT = glutamyltranspeptidase, CK = creatine kinase, and LDH = lactate dehydrogenase.

2.4. Levels of Short-Chain Fatty Acids in Feces

Fresh feces were collected and placed in a sterile EP tube and frozen with liquid nitrogen, then transferred into a −80 °C refrigerator for storage. The contents of short-chain fatty acids in feces were determined according to the method described by Li et al. [10]. Briefly, 0.5 g of feces was weighed and placed in a clean EP tube containing 1.5 mL of ultra-pure water, and then it was mixed well and left to rest for 30 min. Next, 1 mL of supernatant was collected after centrifugation at 4 °C and 15,000 r/min for 20 min. The supernatant was transferred into a new 2 mL EP tube containing 0.2 mL 25% metaphosphates solution, which was mixed well and centrifuged again under the same conditions as above to collect the supernatant, and then the supernatant was passed through a 0.22-micron filter membrane, and the filtrate was taken to be measured. A gas chromatography analyzer (Agilent GC-MS 7890B) was used to detect the levels of acetic acid, propionic acid and butyric acid. The chromatographic column was an Agilent DB-FFAP, the chromatographic column parameter was 30 m × 0.25 mm × 0.25 μm, and the carrier gas was helium with purity greater than 99.99%. In the process of gas phase implementation, a sample size of 1 μL was required for treatment at 50 °C for 1 min, followed by heating at 10 °C/min to 200 °C. The inlet temperature was 250 °C, and the flow rate was 1.0 mL/min. EI was used as the ionization mode, the electron energy was −70 eV, and the shunt ratio was controlled at 2:1. Additionally, the temperature of the ion source and the transmission line were maintained at 280 °C, 0.954 kV was a must as the electron multiplier voltage, and 150 °C was needed in the four-stage rod temperature. These test conditions guaranteed a scanning range of 33–200 m/z. Finally, data were analyzed with the agilent mass hunter.

2.5. 16S Sequencing and Analysis

The fresh feces were harvested at the end of the animal trial and frozen with liquid nitrogen, then transferred into a −80 °C refrigerator for storage. The 16S sequencing and analysis were implemented according to the method described by Li et al. [11]. Briefly, the bacterial DNA was extracted, its concentration and purity were measured, and the V3–V4 region was amplified using a universal primer and named 515 F and 806 R of the 16S rDNA gene. After the mixing and purification of PCR products were performed, the sequencing library was formed by terminal repair, the addition of the A-tail, and the addition of sequencing joints. An Illumina HiSeq2500 PE250 platform (Illumina, San Diego, CA, USA) was used to perform sequencing at Novogene Bioinformatics Technology Co., Ltd., Beijing, China. The raw data were spliced and filtered to obtain clean data, and then the final ASVs were obtained through DADA2 based on the clean data, and species annotation was made based on the ASVs, and the relative abundance, Alpha diversity calculation, Venn diagram and petal diagram were analyzed. LEFse software (Version 1.0) was used to analyze the dominant flora between groups, and R (Version 2.15.3) was used to perform a T-test analysis between the two groups to obtain differential flora. Finally, PICRUST2 functional prediction analysis was used to implement the functional prediction of differences between groups at Level 3 to obtain valuable information for the study. The original data of 16S sequencing have been uploaded to the NCBI database with the login number PRJNA1195387.

2.6. Plasma Metabolome

Blood samples were collected from the anterior vena cava into either heparinized, and then they were centrifuged for 10 min at 3500 r/min and 4 °C to harvest the plasma. A liquid chromatography–tandem mass spectrometry (LC-MS/MS) (ultra-high liquid chromatograph, LC-30A, Shimadzu, Japan, and mass spectrometer: TripleTOF 6600+, SCIEX, Foster City, CA, USA) platform was used to investigate the plasma metabolome. Specifically, 50 μL plasma was added into a clean EP tube containing 300 μL of 20% acetonitrile–methanol internal standard extraction solution and mixed for 3 min and centrifuged at 4 °C and 12,000 r/min for 10 min to obtain 200 μL of the supernatant. The supernatant was placed in a −20 °C refrigerator for 30 min, and then 180 μL of the supernatant was obtained by centrifugation at 4 °C and 12,000 r/min for 3 min and analyzed with the supernatant. Under chromatographic conditions, the Water ACQUITR Premier HSS T3 column (1.8 μm and 2.1 mm × 100 mm) was used, and mobile phase A was a 0.1% formic acid aqueous solution and mobile phase B was a 0.1% formic acid acetonitrile solution. The column temperature was 40 °C, the flow rate was 0.4 mL/min, and the sample size was 4 μL. The mass spectrum conditions of the AB TripleTOF 6600 are shown in Table 3. The samples were extracted and tested by Wuhan Meiwei Metabolic Biotechnology Co., Ltd. Wuhan, China. The original data format of the mass spectrometry machine needed to be converted to mzXML, and XCMS sequencing was used to perform peak extraction, alignment, filtration and retention time correction and then identify metabolites based on the mtDNA method with the Meiwei metabolic database and integrated public database. Finally, a list of all differential metabolites was obtained based on the criterion that the CV value of QC (quality control) samples was less than 0.3. R (Version 2.15.3) was used for PCA and OPLS-DA analysis, and Volcano maps with differential metabolite bar plots were drawn based on a VIP greater than 1. The top 50 metabolites with the largest VIP (variable importance in projection) values were used to map the chord, and the typical differential metabolites of interest were used for independent sample T-test analysis.

Table 3. The mass spectrum conditions of AB TripleTOF 6600.

Item	ESI+	ESI−
Duration, min	10	10
Ion Spray Voltage, V	5000	−4000
Temperature, °C	550	450
Ion Source Gas 1, psi	50	50
Ion Source Gas 2, psi	60	60
Curtain Gas, psi	35	35
Declustering Potential, V	60	−60
MS1 Collision Energy, V	10	−10
MS2 Collision Energy, V	30	−30
Collision Energy Spread, V	15	15
MS1 TOF Masses, Da	50–1000	50–1000
MS2 TOF Masses, Da	25–1000	25–1000
MS1 Accumulation Time, s	0.2	0.2
MS2 Accumulation Time, s	0.04	0.04
Candidate Ions	18	18

2.7. Statistical Analysis

All data were analyzed by one-way ANOVA procedure in the SPSS 23.0 software (SPSS Inc. Chicago, IL, USA) and expressed as mean ± SD, and the independent sample T-test was used to analyze the differences between groups. In addition, Pearson correlation analysis in the SPSS 23.0 software (SPSS Inc., Chicago, IL, USA) was used to investigate correlations between different flora and observed indicators, as well as between different metabolites and different bacteria or observed indicators. A value of $p < 0.05$ was taken to indicate statistical significance. If necessary, the first author can be contacted by email regarding the original data of the full text.

3. Results

3.1. Effects of Lactobacillus reuteri Postbiotics on Growth Performance and Blood Biochemical Indices of Weaned Piglets

Compared with the control group, dietary supplementation with *Lactobacillus reuteri* postbiotics had no significant effect on the growth performance of piglets. However, it is worth mentioning that although we did not observe a statistical difference, *Lactobacillus reuteri* postbiotics treatment reduced the mortality of weaned piglets by 6.37% (Figure 1A). In addition, dietary supplementation with *Lactobacillus reuteri* postbiotics did not cause abnormal changes in the blood biochemical indices (Table 2).

3.2. Effects of Lactobacillus reuteri Postbiotics on Antioxidant-Related Parameters in Plasma and Levels of Short-Chain Fatty Acids in Feces

Dietary supplementation with *Lactobacillus reuteri* postbiotics tended to decrease the level of MDA ($p = 0.068$) and significantly raised the content of SOD ($p < 0.05$). Additionally, the levels of propionic acid and butyric acid in the feces were elevated in the diet treated with *Lactobacillus reuteri* postbiotics ($p < 0.05$) (Figure 1B,C).

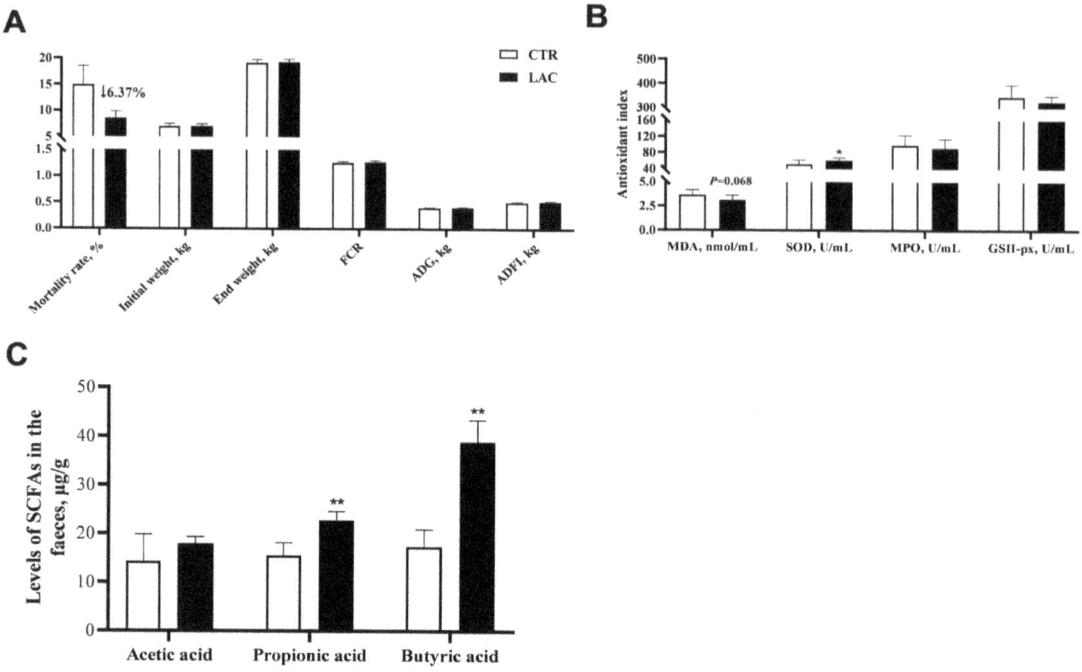

Figure 1. Effects of *Lactobacillus reuteri* postbiotics on growth performance, plasma antioxidant parameters and fecal short-chain fatty acid content of piglets. Among them, results of growth performance are arranged in (**A**), the antioxidant parameters in plasma are shown in (**B**), and levels of short-chain fatty acids are arranged in (**C**). FCR represents feed conversion efficiency, ADFI represents average daily feed intake, ADG represents average daily gain, MDA represents malondialdehyde, SOD represents superoxide dismutase, MPO represents myeloperoxidase, and GSH-px represents glutathione peroxidase. * $0.01 < p < 0.05$, and ** $0.001 < p < 0.01$.

3.3. Effects of *Lactobacillus reuteri* Postbiotics on Intestinal Flora Structure of Feces

Dietary supplementation with *Lactobacillus reuteri* postbiotics had no effect on the α-diversity of intestinal flora, and the number of OTUs unique to the LAC group was 168, and the number of OTUs shared with the control group was 960. Additionally, dietary supplementation with *Lactobacillus reuteri* postbiotics upregulated the relative abundance of Firmicutes and downregulated the relative abundance of Bacteroidetes (Figure 2). The outcomes based on the LEFse analysis show that Firmicutes and *Lachnospiraceae-NK3A20* were the dominant bacteria in the LAC group, and Bacteroidetes was the dominant bacteria in the CTR group (Figure 3A,B). The T-test analysis further confirmed that *Lactobacillus reuteri* postbiotics contributed to upregulating the relative abundance of Firmicutes and downregulating the relative abundance of Bacteroidetes (Figure 3C,D).

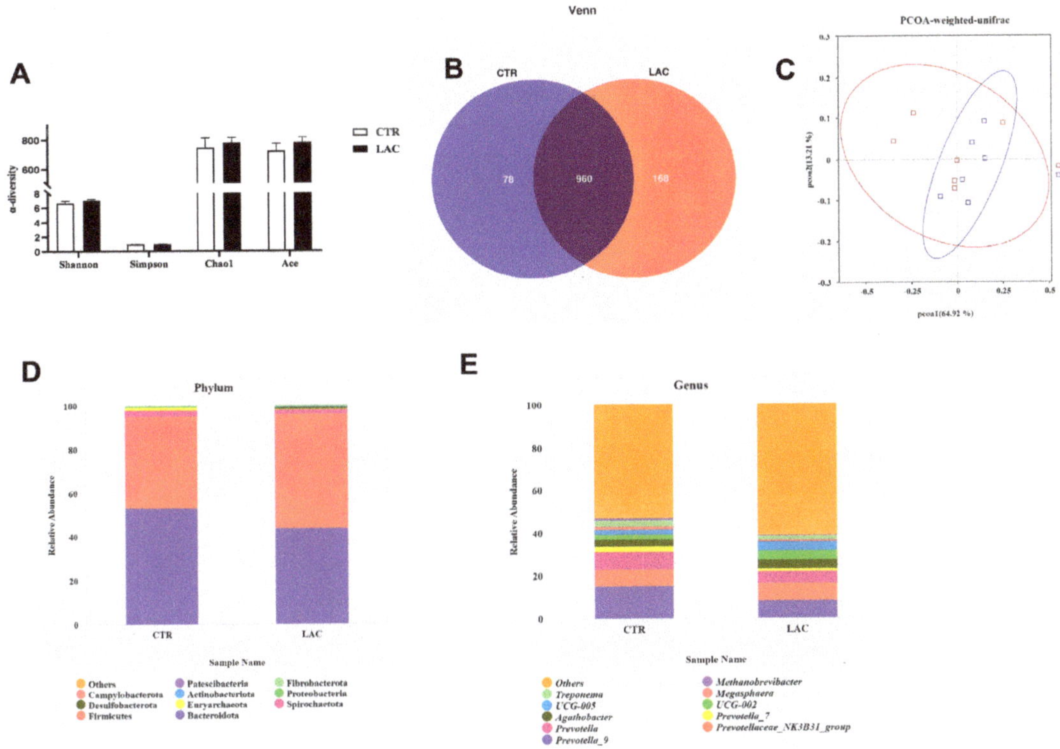

Figure 2. Effects of *Lactobacillus reuteri* postbiotics on fecal flora structure of piglets. Among them, results of α-diversity are arranged in (**A**), the Venn and PCOA results are shown in (**B**) and (**C**), respectively. (**D**) and (**E**) show the relative abundance of the top ten bacteria at phyla level and genus level, respectively.

The outcomes of the correlation analysis show that the contents of propionic acid and butyric acid in the feces were positively correlated with the relative abundance of Firmicutes and negatively correlated with the relative abundance of Bacteroidetes ($p < 0.05$). The functional prediction results demonstrate that *Lactobacillus reuteri* postbiotics enhanced bacterial signal transduction, bacteria motility proteins, secretory system, lipid metabolism and other functions (Figure 4).

3.4. Effects of *Lactobacillus reuteri* Postbiotics on Plasma Metabolome

The metabolite composition in the plasma of the LAC and CTR piglets was inconsistent. Specifically, dietary supplementation with *Lactobacillus reuteri* postbiotics raised 47 metabolites and decreased 86 metabolites. The typical upregulated metabolites were 1-hexadecanoyl-2-(9Z,12Z-octadecadienoyl)-sn-glycero-3-phosphocholine (MW0012968), 1-oleoyl-2-palmitoyl-sn-glycero-3-phosphocholine (MW0057016), coenzyme Q10 (MW0048971), PE-NMe2 (18:1(9Z)/18:1(9Z)) (MW0060366) and 1,2-distearoyl-sn-glycero-3-phosphocholine (MW0011927). The typical downregulated metabolites were 5-Hydroxy-6-methoxy-3-methyl-2-octaprenyl-1,4-benzoquinone (MW0142519), 3-hexanoyl-NBD cholesterol (MW0014037), glucocerebrosides (MW0053661), 1-palmitoyl-3-adrenoyl-sn-glycerol (MW0049772) and PC(14:1(9Z)/P-18:1(11Z)) (MW0056845) (Figure 5).

The differences in metabolites were mainly concentrated in glycerophospholipids and sphingolipids, as well as ketones and hormones. The outcomes of the correlation analysis show these upregulated metabolite levels were positively correlated with the relative abundance of Firmicutes and the content of SOD in plasma and negatively correlated with the relative abundance of Bacteroidetes. This was the opposite in the metabolites that were downregulated ($p < 0.05$). The two different metabolites we focused on were coenzyme Q10 and 3-hexanoyl-NBD cholesterol, and dietary supplementation with *Lactobacillus reuteri* postbiotics elevated the level of coenzyme Q10 and decreased the level of 3-hexanoyl-NBD cholesterol in the plasma of the piglets ($p < 0.05$) (Figure 6).

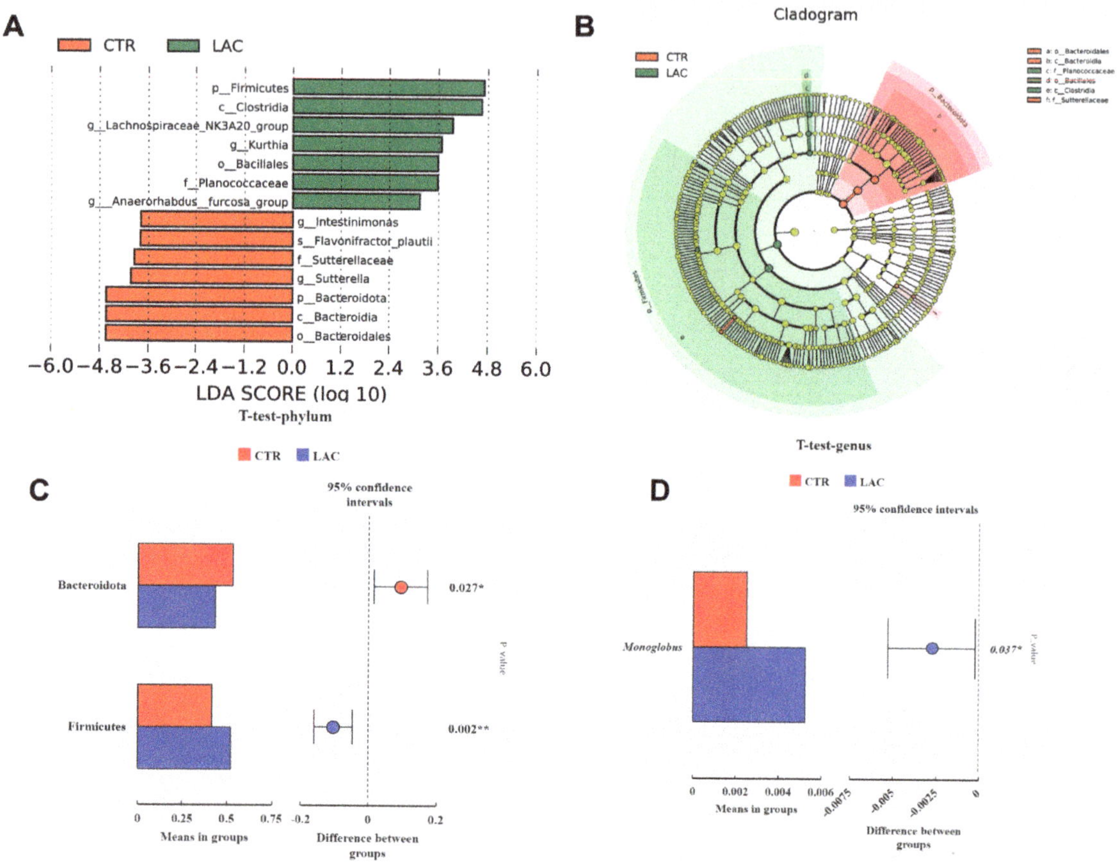

Figure 3. Results of flora structure difference between CTR and LAC group based on LEFse and T-test analyses. Among them, results based on LEFse analysis are shown in (**A**,**B**), and outcomes based on T-test analysis are arranged in (**C**,**D**). * $0.01 < p < 0.05$, and ** $0.001 < p < 0.01$.

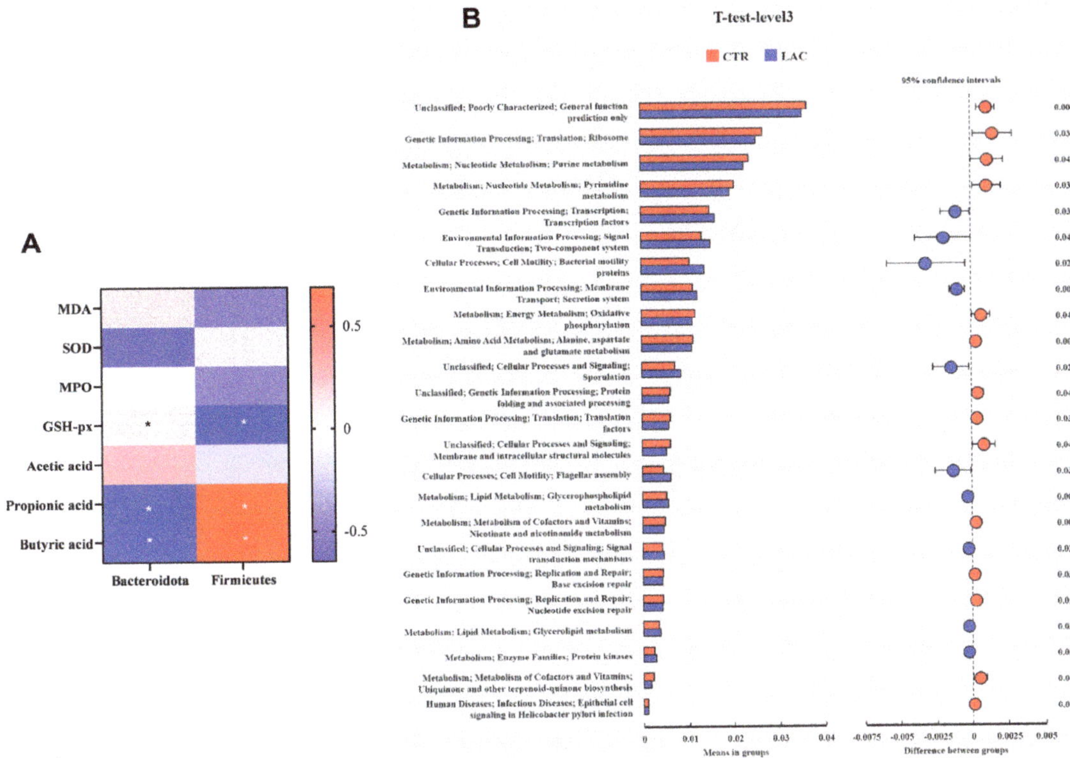

Figure 4. Results of correlation analysis and functional prediction. Among them, results of correlation analysis are arranged in (**A**), and (**B**) shows outcomes of functional prediction. MDA represents malondialdehyde, SOD represents superoxide dismutase, MPO represents myeloperoxidase, GSH-px represents glutathione peroxidase. * 0.01 < p < 0.05, and ** 0.001 < p < 0.01.

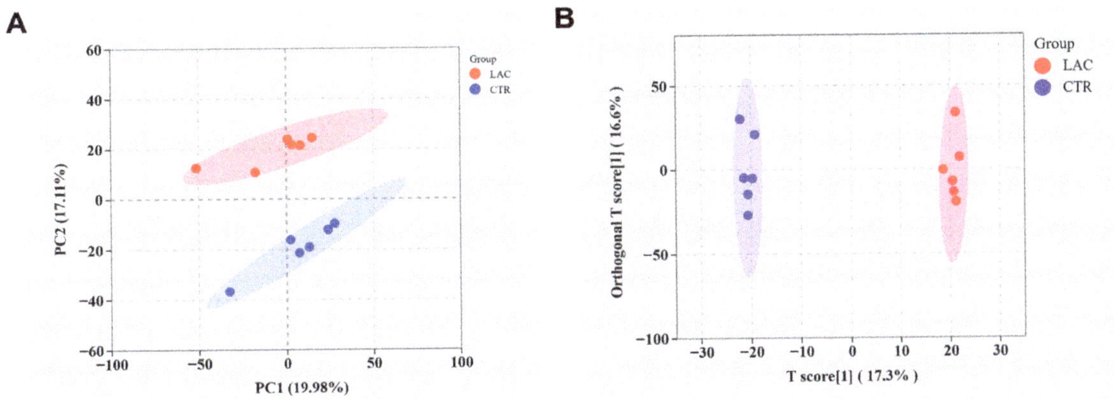

Figure 5. *Cont.*

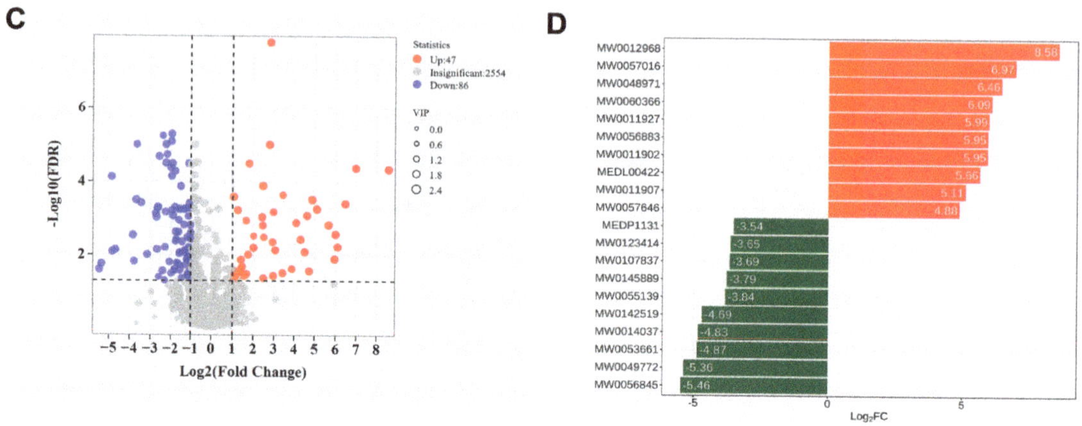

Figure 5. Effects of Lactobacillus reuteri postbiotics on serum metabolome. Among them, results of PCA and OPLS-DA analyses are arranged in (**A**,**B**), the volcanic maps are shown in (**C**), and typical differential metabolite bars are arranged in (**D**). 1-hexadecanoyl-2-(9Z,12Z-octadecadienoyl)-sn-glycero-3-phosphocholine = MW0012968, 1-oleoyl-2-palmitoyl-sn-glycero-3-phosphocholine = MW0057016, coenzyme Q10 = MW0048971, PE-NMe2 (18:1(9Z)/18:1(9Z)) = MW0060366, and 1,2-distearoyl-sn-glycero-3-phosphocholine = MW0011927. 5-Hydroxy-6-methoxy-3-methyl-2-octaprenyl-1,4-benzoquinone = MW0142519, 3-hexanoyl-NBD cholesterol= MW0014037, glucocerebrosides = MW0053661, and 1-palmitoyl-3-adrenoyl-sn-glycerol = MW0049772, PC (14:1(9Z)/P-18:1(11Z)) = MW0056845. Red represents upregulation, and green represents downregulation.

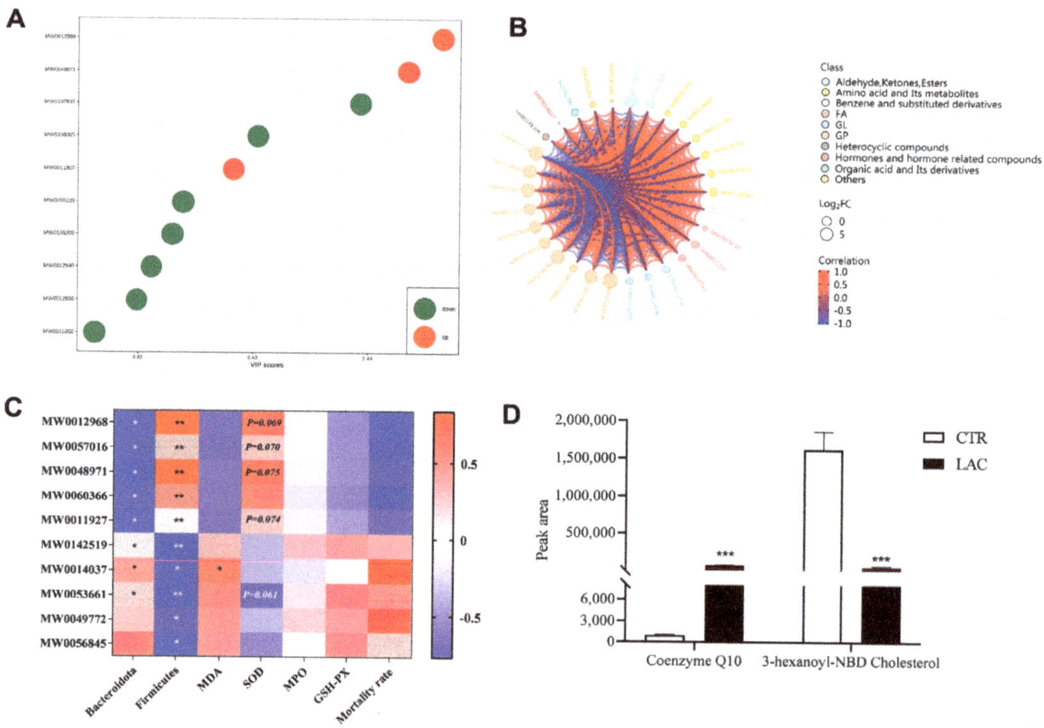

Figure 6. Analysis results of typical differential metabolites and their association with the indicators detected in the present study. Among them, VIP value diagram of differential metabolites and correlation

chord diagram of metabolites are arranged in (**A**) and (**B**), respectively. Results of typical differential metabolites and their association with the indicators detected in the present study are shown in (**C**), and the results of peak area regarding coenzyme Q10 and 3-hexanoyl-NBD cholesterol are arranged in (**D**). MDA represents malondialdehyde, SOD represents superoxide dismutase, MPO represents myeloperoxidase, and GSH-px represents glutathione peroxidase. * $0.01 < p < 0.05$, ** $0.001 < p < 0.01$, and *** $p < 0.001$.

4. Discussion

The weaning stage needs to be paid more attention in the healthy and efficient breeding process of pigs since the immune system and digestive system of the gastrointestinal tract of weaned piglets are not fully developed, which makes the piglets less resistant to exogenous stressors [3]. In the breeding process of piglets, pathogen infection, abnormal environmental stimulation, anti-nutrient factors in feed and other exogenous stressors enhance the negative effects of weaning stress on piglets [12,13]. These are the key reasons for the poor growth performance and high mortality of piglets at the weaning stage. As we observed in the present study, the mortality rate of piglets was close to 15%. Hence, it is a top priority for farmers or pig enterprises to reduce mortality and improve growth performance. Studies have demonstrated that dietary supplementation with *Lactobacillus reuteri* improved the intestinal flora structure, reduced the diarrhea rate, and then improved the growth performance of piglets [14–17]. Paradoxically, the beneficial effects of *Lactobacillus reuteri* were not always stable and might be closely related to the source of the strain and its ability to colonize the animal. Postbiotics, which are non-living active ingredients that contribute to improving the unstable efficacy of probiotics, have been extensively studied in recent years [16,17]. In the present study, although dietary supplementation with *Lactobacillus reuteri* postbiotics had no effect on the growth performance of weaned piglets, it was interesting to note that mortality was reduced by 6.37%. We suggest this will likely be highly welcomed by pig producers. In addition, the standard deviation of piglet mortality values in the LAC treatment group was smaller. This indicates that the *Lactobacillus reuteri* postbiotics intervention results in piglets showing better uniformity, which is also of interest to farmers. Subsequent experimental investigations were carried out to explain the reason for the beneficial effects of *Lactobacillus reuteri* postbiotics on the piglets.

Levels of biochemical markers in the blood could characterize the health of the body, such as elevated levels of glutamic oxalacetic transaminase and glutamic pyruvic transaminase, which indicate that the liver could be damaged [18]. Any exogenous factors might cause abnormal changes in the levels of biochemical indexes in blood, which would be a burden on the body. This justifies why we should not blindly take supplements [19]. In the present study, dietary supplementation with *Lactobacillus reuteri* postbiotics had no effect on the levels of biochemical indexes in the blood. This indicates that *Lactobacillus reuteri* postbiotics are clearly safe for piglets. With stimulated stressors, oxygen free radicals and peroxide products were expressed in large quantities in the piglets, which caused an imbalance between the oxidation and antioxidant systems. In order to maintain the balance of this system, the antioxidant system was activated, and then some antioxidant enzymes, such as SOD and GSH-px, were expressed to remove peroxide products, for instance, MDA or oxygen free radicals [20]. In the present study, dietary supplementation with *Lactobacillus reuteri* postbiotics raised the level of SOD and decreased the content of MDA. These outcomes illuminate that *Lactobacillus reuteri* postbiotics have the potential to relieve the oxidative stress of weaned piglets, and this might be one of the reasons why *Lactobacillus reuteri* postbiotics reduced the mortality of piglets in our present study.

A study demonstrated that *Lactobacillus reuteri* could metabolize to produce short-chain fatty acids, which helped to improve the antioxidant function and health status of piglets [21]. In this study, we also found that *Lactobacillus reuteri* postbiotics upregulated the contents of propionic acid and butyric acid in the feces of piglets. Although we did not observe the relevant indicators of intestinal function, the upregulation of short-chain fatty acids caused by *Lactobacillus reuteri* postbiotics intervention might be helpful to the intestinal health of piglets, which might be more evidence that dietary supplementation with *Lactobacillus reuteri* postbiotics contributed to reducing the mortality of piglets. We were willing to attribute that *Lactobacillus reuteri* postbiotics raised the levels of fecal short-chain fatty acids, improving the intestinal flora structure. The outcomes of the intestinal flora structure in the present study also confirmed our hypothesis. As we observed, dietary supplementation with *Lactobacillus reuteri* postbiotics upregulated the relative abundance of Firmicutes and downregulated the relative abundance of Bacteroidetes, which also showed a significant correlation with the changing trend of short-chain fatty acids. This study demonstrates that most of the bacteria in Firmicutes have the ability to metabolize and produce short-chain fatty acids, and the increased ratio of Firmicutes to Bacteroides might be more beneficial to the intestinal health of piglets [22]. Our outcomes suggest the potential of *Lactobacillus reuteri* postbiotics in improving intestinal health in piglets and that further research is worth pursuing in this area.

A study demonstrated that *Lachnospiraceae-NK3A20* might contribute to the metabolism of amino acids and glycerophospholipids, enhancing the antioxidant function of the body [23]. *Monoglubos*, a bacterium that degrades pectin to produce short-chain fatty acids, might be useful in maintaining immune homeostasis [24]. In the present study, *Lactobacillus reuteri* postbiotics upregulated the relative abundance of *Lachnospiraceae-NK3A20* and *Monoglubos* and enhanced the amino acid and lipid metabolism functions. In the present study, the correlation analysis between the outcomes of bacteria sequencing and serum metabolites shows that the changes in Firmicutes and Bacteroidetes induced by *Lactobacillus reuteri* postbiotics were significantly correlated with major differential metabolites. The correlation analysis further shows that these different metabolites were significantly correlated with the level of SOD in the plasma. These results are logically consistent, and they all point to the hypothesis that *Lactobacillus reuteri* postbiotics might improve the health of the body by regulating the structure of intestinal flora.

In addition to playing a key role in mitochondrial oxidative phosphorylation, coenzyme Q10 also acts as a fat-soluble antioxidant, plays an important role in fatty acid, pyrimidine and lysosome metabolism, and directly mediates the expression of many genes, including those associated with inflammation [25]. In the present study, dietary supplementation with *Lactobacillus reuteri* postbiotics raised the level of coenzyme Q10 in the plasma, and its abundance was positively correlated with the level of SOD in the plasma. This illuminates that the beneficial effect of *Lactobacillus reuteri* postbiotics on antioxidant function might be related to its upregulation of coenzyme Q10. Based on all the outcomes in the present study, we conclude that *Lactobacillus reuteri* postbiotics might improve the survival rate of piglets by reshaping the structure of intestinal flora and plasma metabolome. Although a detailed investigation of the regulatory mechanism of *Lactobacillus reuteri* postbiotics on intestinal health was beyond the scope of this work, we acknowledge that *Lactobacillus reuteri* postbiotics have great application potential in healthy piglet breeding.

5. Conclusions

Dietary supplementation with *Lactobacillus reuteri* postbiotics improved the survival rate of piglets by regulating the structure of intestinal flora and reshaping plasma metabolome, especially upregulating the content of coenzyme Q10.

Author Contributions: Y.L. (Yunxiang Liang) designed the study; D.S. wrote the manuscript; W.T., S.H. and M.W. collected and analyzed the experimental results; P.L. acquired funding; and P.L., Y.L. (Youguo Li) and Y.L. (Yunxiang Liang) participated in the revision of the paper. All authors contributed to the data interpretation and approved the final version of the manuscript. All authors have read and agreed to the published version of the manuscript.

Funding: This work was supported by the Knowledge Innovation Special Dawn Project of Wuhan city (2023020201020461), the Natural Science Foundation of Hubei Province (2023AFB354), the National Natural Science Foundation of China (32402801), the Hubei Science and Technology Talent Service Enterprise Project (2024DJC041), the Open Project of Hubei Key Laboratory of Animal Nutrition and Feed Science (202304) and the Open Project of Hubei Key Laboratory of Animal Nutrition and Feed Science (202402).

Institutional Review Board Statement: The experimental protocol and management of the pigs were approved by the Animal Care and Use Committee of New Hope Liuhe Co., Ltd. (Chengdu, China). The code for ethical inspection is IAS 2023-32.

Informed Consent Statement: Not applicable.

Data Availability Statement: If necessary, the first author can be contacted by email regarding the original data of the full text.

Conflicts of Interest: The authors declare no conflict of interest.

References

1. Wang, L.; Wang, C.; Peng, Y.; Zhang, Y.; Liu, Y.; Liu, Y.; Yin, Y. Research progress on anti-stress nutrition strategies in swine. *Anim. Nut.* **2023**, *13*, 342–360. [CrossRef] [PubMed]
2. Wu-Wu, J.; Guadamuz-Mayorga, C.; Oviedo-Cerdas, D.; Zamora, W.J. Antibiotic resistance and food safety: Perspectives on new technologies and molecules for microbial control in the food industry. *Antibiotics* **2023**, *12*, 550. [CrossRef] [PubMed]
3. Tang, X.; Xiong, K.; Fang, R.; Li, M. Weaning stress and intestinal health of piglets: A review. *Front. Immunol.* **2022**, *13*, 1042778. [CrossRef] [PubMed]
4. Zhao, J.; Zhang, Z.; Zhang, S.; Page, G.; Jaworski, N.W. The role of lactose in weanling pig nutrition: A literature and meta-analysis review. *J. Anim. Sci. Biotechnol.* **2021**, *12*, 10. [CrossRef]
5. Ali, M.S.; Lee, E.B.; Hsu, W.H.; Suk, K.; Sayem, S.; Ullah, H.; Lee, S.J.; Park, S.C. Probiotics and postbiotics as an alternative to antibiotics: An emphasis on pigs. *Pathogens* **2023**, *12*, 874. [CrossRef] [PubMed]
6. Liu, H.Y.; Zhu, C.; Zhu, M.; Yuan, L.; Li, S.; Gu, F.; Hu, P.; Chen, S.; Cai, D. Alternatives to antibiotics in pig production: Looking through the lens of immunophysiology. *Stress Biol.* **2024**, *4*, 1. [CrossRef]
7. Yang, B.; Liu, C.; Huang, Y.; Wu, Q.; Xiong, Y.; Yang, X.; Hu, S.; Jiang, Z.; Wang, L.; Yi, H. The responses of Lactobacillus reuteri LR1 or antibiotic on intestinal barrier function and microbiota in the cecum of pigs. *Front. Microbiol.* **2022**, *13*, 877297. [CrossRef] [PubMed]
8. Wang, W.; Ganzle, M. Toward rational selection criteria for selection of probiotics in pigs. *Adv. Appl. Microbiol.* **2019**, *107*, 83–112. [PubMed]
9. Xing, J.H.; Niu, T.M.; Zou, B.S.; Yang, G.L.; Shi, C.W.; Yan, Q.S.; Sun, M.J.; Yu, T.; Zhang, S.M.; Feng, X.Z.; et al. Gut microbiota-derived LCA mediates the protective effect of PEDV infection in piglets. *Microbiome* **2024**, *12*, 20. [CrossRef] [PubMed]
10. Li, P.; Zhao, S.; Teng, Y.; Han, S.; Yang, Y.; Wu, M.; Guo, S.; Ding, B.; Xiao, L.; Yi, D. Dietary supplementary with ellagic acid improves the intestinal barrier function and flora structure of broiler chicken challenged with *E. coli* K88. *Poult. Sci.* **2024**, *103*, 104429. [CrossRef] [PubMed]
11. Li, P.; Gao, M.; Fu, J.; Yan, S.; Liu, Y.; Mahmood, T.; Lv, Z.; Guo, Y. Dietary soya saponin improves the lipid metabolism and intestinal health of laying hens. *Poult. Sci.* **2022**, *101*, 101663. [CrossRef] [PubMed]
12. Habich, M.; Pawlinski, B.; Sady, M.; Siewruk, K.; Zielenkiewicz, P.; Gajewski, Z.; Szczesny, P. Stress-induced phosphaturia in weaned piglets. *Animals* **2020**, *10*, 2220. [CrossRef] [PubMed]
13. Hu, J.; Ma, L.; Nie, Y.; Chen, J.; Zheng, W.; Wang, X.; Xie, C.; Zheng, Z.; Wang, Z.; Yang, T.; et al. A microbiota-derived bacteriocin targets the host to confer diarrhea resistance in early-weaned piglets. *Cell Host Microbe* **2018**, *24*, 817–832. [CrossRef] [PubMed]
14. Yang, J.; Wang, J.; Shang, P.; Liu, Z.; Zhang, B.; Yang, D.; Zhang, H. Translocation of probiotics via gut-lung axis enhanced pulmonary immunity of weaned piglets exposed to low concentrations of ammonia. *Ecotoxicol. Environ. Saf.* **2024**, *284*, 116821. [CrossRef] [PubMed]

15. Dell'Anno, M.; Callegari, M.L.; Reggi, S.; Caprarulo, V.; Giromini, C.; Spalletta, A.; Coranelli, S.; Sgoifo, R.C.; Rossi, L. *Lactobacillus plantarum* and *Lactobacillus reuteri* as functional feed additives to prevent diarrhoea in weaned piglets. *Animals* **2021**, *11*, 1766. [CrossRef]
16. Wang, G.; Wang, X.; Ma, Y.; Cai, S.; Yang, L.; Fan, Y.; Zeng, X.; Qiao, S. *Lactobacillus reuteri* improves the development and maturation of fecal microbiota in piglets through mother-to-infant microbe and metabolite vertical transmission. *Microbiome* **2022**, *10*, 211. [CrossRef] [PubMed]
17. Yang, J.; Wang, C.; Huang, K.; Zhang, M.; Wang, J.; Pan, X. Compound *Lactobacillus* sp. administration ameliorates stress and body growth through gut microbiota optimization on weaning piglets. *Appl. Microbiol. Biotechnol.* **2020**, *104*, 6749–6765. [CrossRef]
18. Arzu, Y.M.; Vural, Y.S.; Bolu, S.; Danis, A. The relationship between blood biochemical parameters and oral health in children with obesity/overweight. *Br. Dent. J.* **2023**, *235*, 968–972. [CrossRef] [PubMed]
19. Sardana, K.; Sachdeva, S. Role of nutritional supplements in selected dermatological disorders: A review. *J. Cosmet. Dermatol.* **2022**, *21*, 85–98. [CrossRef] [PubMed]
20. Demirci-Cekic, S.; Ozkan, G.; Avan, A.N.; Uzunboy, S.; Capanoglu, E.; Apak, R. Biomarkers of oxidative stress and antioxidant defense. *J. Pharm. Biomed. Anal.* **2022**, *209*, 114477. [CrossRef] [PubMed]
21. Tang, J.; Zhao, M.; Yang, W.; Chen, H.; Dong, Y.; He, Q.; Miao, X.; Zhang, J. Effect of composite probiotics on antioxidant capacity, gut barrier functions, and fecal microbiome of weaned piglets and sows. *Animals* **2024**, *14*, 1359. [CrossRef] [PubMed]
22. Tang, Q.; Shen, D.; Dai, P.; Liu, J.; Zhang, M.; Deng, K.; Li, C. Pectin alleviates the pulmonary inflammatory response induced by PM (2.5) from a pig house by modulating intestinal microbiota. *Ecotoxicol. Environ. Saf.* **2023**, *261*, 115099. [CrossRef] [PubMed]
23. Liu, Y.; Liu, C.; Wu, H.; Meng, Q.; Zhou, Z. Small intestine microbiome and metabolome of high and low residual feed intake angus heifers. *Front. Microbiol.* **2022**, *13*, 862151. [CrossRef] [PubMed]
24. Verhaar, B.; Hendriksen, H.; de Leeuw, F.A.; Doorduijn, A.S.; van Leeuwenstijn, M.; Teunissen, C.E.; Barkhof, F.; Scheltens, P.; Kraaij, R.; van Duijn, C.M.; et al. Gut microbiota composition is related to AD pathology. *Front. Immunol.* **2021**, *12*, 794519. [CrossRef] [PubMed]
25. Hargreaves, I.; Heaton, R.A.; Mantle, D. Disorders of human coenzyme Q10 metabolism: An overview. *Int. J. Mol. Sci.* **2020**, *21*, 6695. [CrossRef]

animals

MDPI

Article

Feeding Sows with Multi-Species Probiotics During Late Pregnancy and the Lactating Period Influences IgA Concentration in Colostrum and Subsequently Increases the Survival Rate of Piglets in Porcine Epidemic Diarrhea Outbreak Herd

Narathon Innamma and Kampon Kaeoket *

Department of Clinical Sciences and Public Health, Faculty of Veterinary Science, Mahidol University, Phuttamonthon, Nakhon Pathom 73170, Thailand; narathon.inn@student.mahidol.edu
* Correspondence: kampon.kae@mahidol.edu

Simple Summary

Increasing the immunoglobulin A (IgA) potential of sow colostrum protects newborn piglets against infection during the pre- and post-weaning periods. Feeding pigs with multi-species probiotics (5 g/sow/day) via top dressing from 4 weeks before farrowing until weaning increases the IgA levels in colostrum during the first 6 h after farrowing. This subsequently improved the piglets' weaning weight and reduced the pre-weaning mortality rate in an outbreak breeder herd with porcine epidemic diarrhea (PED).

Abstract

The aim of the present study was to investigate the potential of a multi-species probiotics product to promote IgA-containing-colostrum production in sows during 24 h of lactation and subsequently promote piglet growth and diminish the pre-weaning mortality rate in a porcine epidemic diarrhea (PED)-infected herd. Sows in the 12th week of pregnancy (n = 20) with an average parity number of 2.4 $\pm$ 1.4 were divided into two groups: untreated control and probiotic-supplemented groups (treatment). They received a treatment composed of basal feed with a probiotic (5 g/sow/day) via top dressing from the 12th week of pregnancy until weaning. Colostrum samples were collected at 3, 6, 12, 24, and 48 h, and the Immunoglobulin A (IgA) levels were measured by using an ELISA kit. The weaning weight and pre-weaning mortality rate were recorded. There was a significantly higher level of IgA in the treatment group than there was in the control one ($p < 0.001$). In the treatment group, the highest level of IgA was found at 6 h (26.22 mg/mL), and the lowest level was found at 48 h (4.51 mg/mL). In the control group, the highest level of IgA level was found at 3 h (16.16 mg/mL), and the lowest level was found at 3 h (3.41 mg/mL). The treatment-administered pigs had a significantly higher ($p < 0.05$) weaning weight (5.90 kg/pig) when compared with that of the control pigs (3.90 kg/pig). A lower pre-weaning mortality rate (24.90%) was found in the treatment group when compared with that of the control pigs (53.60%) ($p < 0.05$). In conclusion, multi-species probiotic supplementation in sows increases the IgA level in colostrum during the first 6 h after farrowing, subsequently improving the weaning weight and reducing the pre-weaning mortality rate during porcine epidemic diarrhea (PED) outbreaks.

Keywords: sow colostrum; immunoglobulin A; porcine epidemic diarrhea; pre-weaning mortality rate; probiotics

 https://doi.org/10.3390/ani15010103

1. Introduction

Porcine epidemic diarrhea (PED) virus is an important cause of diarrhea in sows and piglets. Particularly in piglets, this porcine coronavirus causes severe damage to the small intestine by decreasing the proportion of crypt and villi from 1:7 to 1:3, subsequently driving the malabsorption of nutrients and electrolytes [1]. This is the primary cause of the high mortality rate (i.e., 20–50%) and growth retardation in piglets in infected herds, leading to economic losses in the pig industry worldwide. In the United States, the PEDV outbreak in 2013–2014 caused the pig population to decrease by approximately 3.2% compared to the previous year [2–4]. Nowadays, the most practical method is to provide passive immunity to piglets via colostrum from immunized sows. There are two well-documented techniques to immunize sows: vaccination of gilts at replacement unit and before farrowing and feeding the sows fresh small intestines from sick piglets with PED [4,5]. Using these techniques, sow colostrum will develop passive immunity or maternally derived antibodies (MDAs), especially Immunoglobulin G (IgG) and Immunoglobulin A (IgA) specific to PEDV [6,7]. In practice, in order to promote piglet health, every single piglet must intake colostrum from its mother as soon as possible after birth, since it is enriched with IgG and IgA. This is absorbed via the small intestine only for the first 24 h of life, the so-called "gut closure" period [6]; thus, to obtain the highest quantity, every single piglet should intake colostrum during the first 2–6 h of life. This is because IgG and IgA production in sow colostrum reaches the maximal level at about 2 h and this is maintained for 6 h; thereafter, the amounts of IgG and IgA decrease dramatically [8,9]. In normal sows, the average value of IgA in colostrum at birth is 23.8 mg/mL, and this decreases to 7.85 mg/mL at 6 h and to 4.59 mg/mL at 24 h after the onset of farrowing [10]. In addition, IgA localized at the surface of small intestines of piglets causes a mucosal barrier/mucosal immune response in order to prevent pathogen attachment [11]. Therefore, in practice, during an outbreak of PED, if a pig farmer can find a strategy to promote high IgA production in colostrum, this certainly guarantees that all the piglets keep their guts healthy.

Probiotic bacteria, the friendly bacteria of the gut, have multiple and various influences on the host, e.g., different organisms can influence the intestinal luminal environment, epithelial and mucosal barrier functions, and the mucosal immune system [12]. In a previous study, a multi-strain probiotic (10^{11} CFU/g of *Lactobacillus acidophilus*, *Bifidobacterium bifidum*, *Lactobacillus delbrueckii*, *Bacillus subtilis*, and *Lactobacillus rhamnosus*) had an overall positive effect on feed intake, the feed conversion ratio, and the IgG and IgG antibodies of Japanese quails and decreased the amount of *Esherichia coli* (*E. coli*) [13]. In pig production, probiotics increase the milk yield of lactating sows, enhance the welfare of pregnant sows, improve the growth of nursery pigs, and control diarrhea-causing pathogens in pig farms [14,15]. BACTOSAC-P™ (KMP Biotech Co. Ltd., Thailand) is a commercial multi-species probiotic product that contains 1.0×10^7 CFU/g of seven probiotic bacteria; these are *Lactobacillus acidophilus*, *Lactobacillus plantarum*, *Streptococcus faecium*, *Pediococcus pentosaceus*, *Bacillus subtilis*, *Bacillus licheniformis*, and *Saccharomyces cerevisiae*. The components of BACTOSAC-P™ are harvested from nature in Southeast Asia, and each dose has an equal concentration of the probiotics. BACTOSAC-P™ contains the selected probiotic bacteria, which can produce lipase enzymes to digest lipids in the intestines of hosts [16]. With these properties in mind, we hypothesize that BACTOSAC-P™ may facilitate an increase in IgA yield in colostrum in a sow herd during PED infection.

The aim of the present study was to investigate the potential of BACTOSAC-P™ to promote Immunoglobulin A-containing-colostrum production in sows during 24 h of lactation and subsequently promote piglet growth and diminish the pre-weaning mortality rate in a porcine epidemic diarrhea virus (PEDV)-infected herd.

2. Materials and Methods

2.1. Ethics Statement

This research project was approved by the Faculty of Veterinary Science—Institute Animal Care and Use Committee (FVS-IACUC-Protocol No. MUVS-2018-06-26), Mahidol University, Thailand.

2.2. Animals

This experiment was performed on 20 sows (primiparous and multiparous) in a pig farm located in Saraburi, Thailand. The animals were reared in a continuous farrowing system. The pregnant sows were housed in individual gestation crates. During lactation, the pigs remained in the individual stalls. This pig breeding farm was confirmed to be contaminated by PEDV in the middle of May 2019 by using antigen screening test kits (Bionote, Gyeonggi-do, Republic of Korea) and the PCR technique, and the experiment was carried out in the middle of June 2019. Altogether, 20 sows from the PED-infected herd, with an average parity number of 2.4 ± 1.4, were included in this study. At 12 weeks of pregnancy, all the sows were fed the minced small intestines from PED-infected piglets following the farm's protocol. The sows' gestation and lactation basal diets were formulated to meet the nutrient requirements of swine recommended by National Research Council [17]. The formulation and nutrient specifications of the gestating and lactating sows' diets are shown in Table 1. The sows were fed during the first, second, third, and fourth months of pregnancy approximately 2.0–2.5, 3.0–3.5, and 3.5 kg feed per sow per day, respectively. However, a week before parturition, they were fed 3 kg feed per sow per day. After parturition, the sows were fed 3.0 kg of feed per day, and the amount of feed offered to the sows was increased by 0.5 kg per day until they were fed ad libitum (6.0–7.0 kg) from week one of lactation until weaning according to the farm's protocol. Thereafter, they were randomly divided into 2 groups (i.e., control and treatment), with 10 in each group. They were kept in the same farrowing house equipped with an evaporative cooling system and given normal feed (control group) or a mixture of BACTOSAC-P™ (5 g/sow/day), which is the concentration recommended by the supplier (K.M.P.BIOTECH CO., LTD., Thailand), via top dressing on normal feed (treatment) from 4 weeks before farrowing until weaning (i.e., lactation period of 24 days). On day 113 of pregnancy, in order to take good care of the piglets, all the sows were induced via the injection of 5 mg Dinoprost into the perivulva area either at the 3 or 9 o'clock position as previously described [18,19]. All the sows were then farrowed at 24 h after induction between 7 and 10 am.

Table 1. Ingredient composition of gestation and lactation diets given on an as-fed basis.

Type of Diet	Gestation Diet	Lactation Diet
Ingredient composition (%)		
Broken rice	31.50	28.35
Soybean meal (44% CP)	16.00	10.00
Full fat soybean meal (36% CP)	-	16.00
Rice bran	45.10	36.20
Fish meal (60% CP)	3.00	2.00
Rice bran oil	-	2.50
Dicalcium phosphate (18% P)	2.00	2.50
Limestone	1.60	1.50
DL-Methionine	0.08	0.05
L-Lysine	0.12	0.15
Salt	0.35	0.50
Premix	0.25	0.25
Total	**100.00**	**100.00**

Table 1. *Cont.*

Type of Diet	Gestation Diet	Lactation Diet
Nutrient composition (%Dry matter Basis)		
Crude Protein	16.69	18.01
Metabolizable Energy (kcal/kg)	2995	3235
Calcium	1.08	1.02
Phosphorus	0.42	0.44
Methionine + Cystine	0.68	0.69
Lysine	0.85	0.96

2.3. Colostrum and Milk Sample Collection

The time when the first piglet was born was designated as 0 h. Colostrum and milk samples were collected from each sow in each group, considering the convenience of sample collection and the safety of both the animals and the personnel. A portion of colostrum were collected from all the teats as a pool sample (approximately 50 mL) at 3, 6, 12, and 24, and a portion of milk was collected from all the teats as a pool sample (approximately 50 mL) at 48 h, and then kept in sealed container and stored at −20 °C until analysis [9,20].

2.4. Measurement of Pig Immunoglobulin A (IgA) Level

The Immunoglobulin A (IgA) level was measured by using a pig IgA ELISA kit (Koma Biotech Inc., Seoul, Republic of Korea). Analysis was performed according to the manufacturer's instructions. Briefly, all the samples were diluted 100 times before measurement. The ELISA procedure involved the following steps: (1) coating: 100 µL of the diluted coating antibody was added to each well, which was then covered and incubated at 4 °C overnight. (2) Washing: the wells were washed four times with 300 µL of washing solution and the excess liquid was removed after the last wash. (3) Blocking: 200 µL of blocking solution was added per well and incubated at room temperature for 1 h; then it was washed again, as in step 2. (4) Reaction: 100 µL of standard, blank, or sample was added to each well in duplicate and incubated for 1 h at room temperature. (5) Detection: 100 µL of diluted detection antibody was added to each well and incubated for 1 h at room temperature, followed by washing. (6) Color Development: 100 µL of TMB or pink-ONE TMB solution was added to each well and the color was allowed to develop. (7) Stop Reaction: 100 µL of stop solution was added to each well and the absorbance was measured at 450 nm using a microplate reader. The results were recorded in ng/mL, and then the data were calculated in mg/mL [9].

2.5. Measurement of Backfat Thickness of Sows

The sows' backfat thickness was measured by using a digital backfat indicator (Renco Corp., Minneapolis, MN, USA). The average value from both sides of the P2 position (6.5 cm away from body midline at the last rib level) was used as the backfat thickness (mm) [21]. The sows' backfat thickness was measured at 4 weeks before farrowing and at weaning (day 24 of lactation).

2.6. Production Performance Parameters

Sow reproductive performance measurements included total number of piglets born, born alive, stillborn, mummified, and weaned per litter. Within 12 h of birth, the litter birth weights of these piglets were weighed and recorded. All piglets were processed according to the standard operating procedure established by the farm within 24 to 48 h of birth. Piglet processing included tail docking, needle teeth clipping, administering injectable iron, and castration of male piglets. Incidence of stillborn and mummified piglets was recorded at birth. Any pigs that died shortly before or during parturition, due to asphyxia or dystocia, were classified as stillborn. Piglets were monitored daily for instances of morbidity and mortality. Any dead piglets were recorded by date of death. Pre-weaning mortality in piglets per litter was calculated as a percentage, based on the number of piglets that die between birth and weaning, by the following formula:

$$\text{Preweaning Mortality per litter (\%)} = (\text{Number of piglet deaths before weaning in each litter/Total number of piglets born alive in each litter}) \times 100$$

One day before weaning, individual piglet body weights were determined and recorded to calculate total weight gain during the pre-weaning period. Piglets were weaned at about 24 $\pm$ 1.0 day of age.

2.7. Statistical Analysis

All data were tested for normality prior to analysis by examination of histograms and normal distribution plots using the Shapiro–Wilk Test. The IgA levels in the groups were analyzed at different timepoints by using IBM SPSS Statistics for Windows, version 26.0 (SPSS Inc., Chicago, IL, USA). ANOVA was used to analyze the pig IgA levels across the different timepoints in the groups, and we compared the means by using Duncan's multiple range test for a stepwise comparison approach, which can be more effective for identifying subtle differences in means. A T-test was used to analyze the production performance parameters between the groups. A statistically significant difference was defined as $p \leq 0.05$.

3. Results

According to the clinical findings, none of the sows in either of the groups showed clinical signs of PED when they were kept in the farrowing house. However, their piglets started showing clinical signs of PED at 4 days old. The pig IgA levels in colostrum in the control and treatment groups are shown in Figure 1. In the treatment group, the highest level of IgA was found in the sows fed BACTOSAC-P™ at 6 h (26.22 $\pm$ 7.09 mg/mL), the lowest level was found at 24 h (11.87 $\pm$ 11.58 mg/mL) ($p < 0.001$), and the IgA level of 4.51 $\pm$ 2.84 mg/mL in sow milk was found. In the control group, the highest level of IgA was found at 3 h (16.16 $\pm$ 2.50 mg/mL), and the lowest level was found at 24 h (3.41 $\pm$ 2.44 mg/mL) ($p < 0.001$) and the IgA level of 3.41 $\pm$ 2.44 mg/mL was found in sow milk. According to the comparison between the total IgA levels in colostrum and milk across the different timepoints within and between the groups in Figure 2, the pig IgA levels in both the groups from 3 to 48 h were higher in the treatment group than in the control group, especially at 6 h ($p = 0.10$).

The backfat thicknesses of both sow groups are shown in Table 2. There was no significant difference in terms of the backfat thickness of both the sow groups at the start of the experiment (1 month before farrowing). However, there was a significant difference in the backfat thickness of the sows 3 weeks after farrowing. The treatment-administered sows showed significantly thicker backfat (11.70 $\pm$ 0.14 mm) than those in the control group (11.13 $\pm$ 0.17 mm) ($p < 0.05$).

The production performance parameters are shown in Table 3. There was no significant difference in terms of the number of total piglets born per litter, the number of piglets born alive per litter, and the litter birth weight. However, the pre-weaning mortality rate was two times higher in the control group (53.6%) than that in the treatment group (24.9%). The same case was also found for the number of piglets weaned per litter. In addition, a significantly higher weaning weight (5.9 kg) was found in the treatment group than that in the control (3.9 kg) ($p < 0.05$).

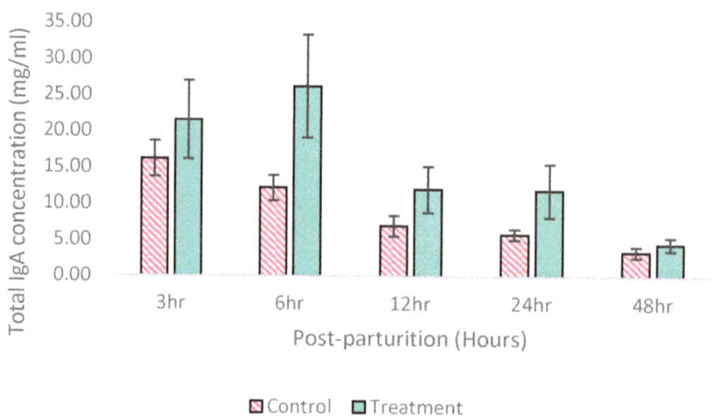

Figure 1. Immunoglobulin A (IgA) levels (means ± SE) in sow colostrum at 3, 6, 12, 24, and IgA in sow milk at 48 h after farrowing in control and the group of sows supplemented with probiotics; Treatment group.

		Control				Treatment				
		6hr	12hr	24hr	48hr	3hr	6hr	12hr	24hr	48hr
Control	3hr	0.058	0.010	0.004	0.000	0.378	0.198	0.318	0.346	0.000
	6hr		0.043	0.007	0.000	0.118	0.071	0.962	0.938	0.001
	12hr			0.374	0.033	0.018	0.016	0.164	0.230	0.164
	24hr				0.026	0.010	0.010	0.072	0.122	0.278
	48hr					0.004	0.005	0.017	0.037	0.364
Treatment	3hr						0.505	0.038	0.023	0.009
	6hr							0.018	0.014	0.009
	12hr								0.943	0.028
	24hr									0.045

Figure 2. p-value of the total IgA levels in colostrum and milk across the different timepoints within and between the groups.

Table 2. Backfat thickness in sows supplemented with probiotics (means ± SE).

Parameter	Control	Treatment	p-Value
Backfat thickness at 1 month before farrowing	11.98 ± 0.15	11.83 ± 0.21	0.478
Backfat thickness at 3 weeks after farrowing	11.13 ± 0.17 ^{a}	11.70 ± 0.14 ^{b}	0.047

Different superscript lower-case letters indicate significant difference within rows ($p < 0.05$).

Table 3. Productive performance in sows supplemented with probiotics (means ± SE).

Parameter	Control	Treatment	p-Value
Parity	3.0 ± 0.5	2.4 ± 0.5	0.625
Number of total born/L	13.1 ± 1.2	12.7 ± 0.9	0.552
Number born alive/L	11.3 ± 0.9	11.1 ± 0.9	0.867
Litter birth weight (kg)	15.5 ± 1.1	16.1 ± 1.2	0.830
Pre-weaning mortality rate per litter (%)	53.6 ± 11.3 ^{a}	24.9 ± 9.4 ^{b}	0.026
Number of weaned piglets/L	5.1 ± 1.2 ^{a}	8.1 ± 1.0 ^{b}	0.015
Weaning weight (kg)	3.9 ± 0.9 ^{a}	5.9 ± 0.3 ^{b}	0.032
Weaning weight gain (kg)	2.9 ± 0.7	4.5 ± 0.3	0.057

Different superscript lower-case letters indicate significant difference within rows ($p < 0.05$).

4. Discussion

PEDV remains a significant health issue, causing high mortality in pre-weaning piglets and economic losses [1–3]. Co-infection porcine deltacoronavirus with PDCoV exacerbates disease severity by increasing viral shedding, disrupting intestinal structure, and elevating infection levels. Limited vaccine efficacy, likely due to immunization challenges and misdiagnosis, underscores the need for new antiviral strategies against PEDV [22,23]. Therefore, it is necessary to explore new antiviral strategies to reduce the infectivity of the pandemic strain of PEDV among pigs.

Recent literature data highlight the beneficial effects of administering probiotics to pigs, such as the regulation of the intestinal microflora, the inhibition of pathogens in the gastrointestinal tract, improved intestinal barrier function, and the enhancement of mucosal immunity. The supplementation of probiotic Lactobacillus fermentum I5007 in newborn piglets can regulate the formation of gut microflora and reduce the number of enteropathogenic *Escherichia* spp. and *Clostridium* spp. in the gastrointestinal tract [24]. Lactic acid bacteria have shown antiviral effects. The cell-free supernatant (CFS) refers to the liquid obtained after separating cells from a culture medium through processes like centrifugation or filtration. CFS consists of the substances secreted by the cells during cultivation, including proteins, enzymes, metabolites, and bioactive compounds [25]. The cell-free supernatant (CFS) of the *Lactobacillus* spp. probiotic and live *Lactobacillus plantarum* and *Pediococcus* spp. showed protective effects against the pandemic strain of PEDV [26,27]. The CSF of lactobacilli could reduce the viral infectivity of Vero cells, and the live Lactobacillus plantarum strain 25F reduced the cytopathic effect of PEDV. One possible mechanism may be the blocking of viral adsorption into the host cells by CFS metabolites, for example, organic acids, organic compounds (diacetyl), short-chain fatty acids, and antimicrobial peptides [28,29]. Lactic acid bacteria enable the transformation of complex nutrients, such as plant cell wall components (pectin, cellulose, and hemicellulose), into simple sugars that ferment into short-chain fatty acids (SCFAs), mainly acetate, propionate, and butyrate [30]. Microbial-derived short-chain fatty acids are crucial for protecting the intestinal barrier and regulating the immune system to respond to viral infection [31]. Several experiments demonstrated that adding short- and medium-chain fatty acids (MCFAs) can effectively combat viral infection [32–37]. Selected MCFAs, mainly caprylic, capric, and lauric acids, and a related monoglyceride, glycerol monolaurate (GML), can inhibit African swine fever virus when administered as liquid or feed [38]. In the case of PEDV, butyrate provides protection from PEDV infection in the intestinal epithelial cells. One possible mechanism may be the activation of the innate immune response by GPR43. A previous study suggested a strategy involving the inhibitory effect of G protein-coupled receptors against PEDV infection [39].

Lactic acid bacteria (LAB) probiotics have been shown to enhance the intestinal barrier. The administration of *Lactobacillus delbrueckii* (LAB) to piglets during the suckling period can improve their intestinal morphology and barrier function [40]. A study found that LAB administration can cause greater jejunal and ileal villus heights and a better villus–crypt ratio when compared to those of untreated piglets. In addition, the administration of LAB slightly increases mRNA expression for occlusion and the quantity of ZO-1 in the jejuna and ilia of piglets during the suckling period, which indicates that LAB can improve intestinal barrier integrity [41]. These data suggest that LAB promotes the gut health of piglets during weaning, and this might increase their body weight and improve their health status. Probiotic bacteria stimulate the immune systems of pigs by triggering gut-associated lymphoid tissue (or GALT)-related activities. This occurs by increasing the quantities of T lymphocyte cells in the intestinal mucosa and Immunoglobulin A (IgA), which represents one type of antibody production, bringing out disease resistance in hosts [42].

Increasing the immune potential of sow colostrum protects newborn piglets against infection during the pre- and post-weaning periods. These results clearly show that probiotic-containing bacteria, BACTOSAC-P™, have the potential to promote IgA-containing colostrum and milk production in sows for 48 h during lactation and also have a beneficial effect on piglet growth and reduce the pre-weaning mortality rate during PED outbreaks in breeding herds. Considering the level of IgA in the pig colostrum and milk, the sows fed BACTOSAC-P™ (5 g/sow/day) produced more IgA during the first 48 h of lactation, particularly at 2–6 h. Similar results have been obtained in previous studies [43–45]. The reason for there being a higher IgA level in the pig colostrum in the treatment group might be that probiotic bacteria, friendly bacteria in the gut, influence the intestinal luminal environment, epithelial and mucosal barrier function, and the mucosal immune system [12]. In practice, farmers should ensure that all piglets intake colostrum from their mother before the gut closure phenomenon, which usually occurs at 24–36 h [46,47]. Nevertheless, the IgA level in colostrum reached its peak at about 6 h after farrowing, and thereafter gradually declined. In our experiment, the level of IgA during the first 6 h after farrowing is two times higher compared with that of the control; consequently, the piglets in the treatment group consumed more IgA-containing colostrum than the control group did. IgA may localize at the surface of the small intestines of piglets, performing a mucosal barrier/mucosal immune response in order to prevent pathogen attachment [11]. This mechanism may, at least in part, explain the higher weaning weight and lower pre-weaning mortality rate of the piglets from the sows fed BACTOSAC-P™. The factors that are associated with piglet colostrum consumption include the number of piglets weaned per litter and the initial and gained weaning weights of the piglets. Once the piglets started to suckle, maternal passive immunity was transferred from the sows to their piglets via colostrum [48]. In this work, probiotic supplementation in sows significantly enhanced the concentration of IgA via colostrum to the piglets and improved the survival and growth performance of the piglets. Similar results have been obtained in previous studies [49].

The gut microbiota is a community of microorganisms that include bacteria, archaea, fungi, and viruses that live in intestinal environments. The gut microbiota impacts the performance and health status of the host. Previous research shows that the initial colonization of the gut microbiota in piglets occurs at least within the immediate prenatal period [50]. Some studies indicate that the sows' microbiota and the rearing environment influence gut microbiota functionality and the composition of their offsprings. Exposure of piglets to sows' vaginas and their pen environment, exposure of sows and piglets to antibiotics, dietary treatments, and the length of time between sow microbiota modulation cause some differences between the sows' microbiota and that of their progeny. They also influence their immune system development, growth, and survival [51–53]. The importance of colostrum and milk in relation to gut microbiota development and piglet health has been documented [54]. Receiving colostrum and milk in the early lives of piglets is important for intestinal microbiota colonization and immune system development. In our study, we found that the sows supplemented with probiotics were healthy and could produce sufficient quantities of high-quality colostrum and milk for newborn piglets. In addition, the fact that the sows were constantly supplemented with probiotics may cause the transfer and shedding of beneficial microorganisms into the environment and their piglet's intestines [55,56]. These factors can support initial intestinal microbiota colonization in order to improve piglets' health and survival.

The probiotic-supplemented sows had thicker backfat compared to that of the control sows. Backfat thickness is a significant parameter for female pigs, which is associated with reproductive performance, for example, puberty attainment, the total piglets born (TB), the farrowing rate, and the period from weaning to the first interval. Moreover, backfat is a significant source of hormones related to puberty attainment, such as leptin, insulin-like growth factor-I (IGF-I), and progesterone (P4) [21].

In a previous study [57], PED infection caused the deterioration in the growth performance of piglets at the suckling period. In this study, the piglets of the sows that were fed via intestinal feedback to defend against PED and supplemented with probiotics had a greater average weight at weaning and a lower mortality rate than did those that were farrowed from the sows subjected to intestinal feedback, but no probiotic supplementation. This agrees with an earlier report stating that probiotics increase the milk yield of lactating sows, improve the growth rate in nursery pigs, and control diarrhea-causing pathogens in pig farms [14]. The probiotic blend used in this study, which includes species that produce lipase enzymes [16], was selected based on its potential to enhance gut health through multiple mechanisms. Lipase-producing probiotics have been shown to improve nutrient digestion and absorption, which are particularly relevant in sows and piglets that are susceptible to post-weaning diarrhea. Enhanced lipid digestion may alleviate the burden on the digestive system, thus reducing intestinal stress and improving overall gut health.

This study primarily focused on evaluating the key immune and reproductive parameters in sows and piglets, specifically IgA levels and reproductive performance. However, we acknowledge that the scope of the measured indicators was limited. While these outcomes provide important insights into the effects of multi-species probiotics, a more comprehensive assessment, including additional biomarkers, such as intestinal morphology, digestive enzyme activity, and microbial composition, could further strengthen our understanding of how probiotics influence the prevention of PED. These additional measures would allow us to evaluate the broader physiological impact of probiotics, particularly in relation to gut health, which is closely linked to the severity of PED. We recognize that this is a limitation of this study and recommend that future research incorporates a wider range of indicators to build upon these initial findings. While this study provides valuable insights into the effects of BACTOSAC-P™ on the reproductive performance and immunological response of lactating sows, it is important to acknowledge that this research primarily serves as initial verification of the probiotic product. The scope of this study was intentionally focused on evaluating the fundamental outcomes, such as the IgA levels and the piglet performance; however, we recognize that a more comprehensive exploration of the underlying mechanisms and broader physiological effects would strengthen these conclusions.

Future studies should aim to investigate additional biomarkers of gut health, including microbial composition, intestinal morphology, and cytokine profiles, to provide a more holistic understanding of how BACTOSAC-P™ interacts with the host. Moreover, exploring the dose–response relationships and the long-term effects of probiotic supplementation across the different stages of sow reproduction could offer deeper insights into the product's efficacy and mode of action. By expanding the scope of investigation, subsequent research will be better positioned to elucidate the synergistic or antagonistic interactions among multi-species probiotics and uncover their precise roles in improving sow and piglet health.

5. Conclusions

Feeding pigs multi-species probiotics (5 g/sow/day) via top dressing from 4 weeks before farrowing until weaning increases immunoglobulin A (IgA) levels in colostrum during the first 6 h post-farrowing. This enhanced immune response contributes to improved weaning weight and a lower pre-weaning mortality rate in breeder herds during porcine epidemic diarrhea outbreaks. These findings highlight the practical value of using hprobiotics to enhance sow and piglet health in the face of significant disease challenges.

Author Contributions: N.I.: conceptualization; methodology; data collection and analysis; writing—original draft. K.K.: conceptualization; methodology; data analysis; writing—review and editing; supervision; funding acquisition. All authors have read and agreed to the published version of the manuscript.

Funding: This research was supported by Mahidol University, the National Research Council of Thailand (NRCT, grant numbers 1059, 3728, and 17553), and K.M.P. Biotech (Thailand).

Institutional Review Board Statement: This research project was approved by the Faculty of Veterinary Science—Institute Animal Care and Use Committee (FVS-IACUC-Protocol No. MUVS-2018-06-26).

Informed Consent Statement: Not applicable.

Data Availability Statement: Data are contained within the article.

Acknowledgments: All laboratory facilities in this study were supported by the Faculty of Veterinary Science, Mahidol University. The authors would like to thank Sujin Sukchai and Nattaya Banglap for technical help during this experiment.

Conflicts of Interest: We clarify that there are no conflicts of interest due to any financial, personal, or other relationships with other people or organizations related to the material discussed in this manuscript.

References

1. Jung, K.; Wang, Q.; Scheuer, K.A.; Lu, Z.; Zhang, Y.; Saif, L.J. Pathology of US porcine epidemic diarrhea virus strain PC21A in gnotobiotic pigs. *Emerg. Infect. Dis.* **2014**, *20*, 662–665. [CrossRef] [PubMed]
2. Song, D.; Moon, H.; Kang, B. Porcine epidemic diarrhea: A review of current epidemiology and available vaccines. *Clin. Exp. Vaccine Res.* **2015**, *4*, 166–176. [CrossRef] [PubMed]
3. Jung, K.; Saif, L.J.; Wang, Q. Porcine epidemic diarrhea virus (PEDV): An update on etiology, transmission, pathogenesis, and prevention and control. *Virus Res.* **2020**, *286*, 198045. [CrossRef] [PubMed]
4. Lei, J.; Miao, Y.; Bi, W.; Xiang, C.; Li, W.; Zhang, R.; Li, Q.; Yang, Z. Porcine epidemic diarrhea virus: Etiology, epidemiology, antigenicity, and control strategies in China. *Animals* **2024**, *14*, 294. [CrossRef] [PubMed]
5. Lee, C. Porcine epidemic diarrhea virus: An emerging and re-emerging epizootic swine virus. *Virol. J.* **2016**, *12*, 193. [CrossRef]
6. Tucker, B.S.; Craig, J.R.; Morrison, R.S.; Smits, R.J.; Kirkwood, R.N. Piglet viability: A review of identification and pre-weaning management strategies. *Animals* **2021**, *11*, 2902. [CrossRef]
7. Song, X.; Qian, J.; Wang, C.; Wang, D.; Zhou, J.; Zhao, Y.; Wang, W.; Li, J.; Guo, R.; Li, Y.; et al. Correlation between the IgG/IgA antibody response against PEDV structural protein and virus neutralization. *Microbiol. Spectr.* **2023**, *11*, e0523322. [CrossRef]
8. Songsujaritkul, W.; Thepprechasakul, S.; Sukchai, S.; Luengyosluechakul, T.; Boonprakob, R.; Innamma, N.; Thitisak, P.; Kaeoket, K. Rising of IgA in sow colostrum after supplemented feed with probiotic Bactosac during late gestation. In Proceeding of the 25th International Pig Veterinary Society Congress, Chongqing, China, 11–14 June 2018; p. 970. Available online: https://www.theipvs.com/wp-content/uploads/2019/09/IPVS-2018.PROCEEDINGS.CHINA_.VOLUME-II.pdf (accessed on 20 June 2018).
9. Innamma, N.; Ngamwongsatit, N.; Kaeoket, K. The effects of using multi-species probiotics in late-pregnant and lactating sows on milk quality and quantity, fecal microflora, and performance of their offspring. *Vet. World* **2023**, *16*, 2055–2062. [CrossRef]
10. Markowska-Daniel, I.; Pomorska-Mól, M.; Pejsak, Z. Dynamic changes of immunoglobulin concentrations in pig colostrum and serum around parturition. *Pol. J. Vet. Sci.* **2010**, *13*, 21–27.
11. Porter, P. Transfer of immunoglobulins IgG, IgA and IgM to lacteal secretions in the parturient sow and their absorption by the neonatal piglet. *Biochim. Biophys. Acta* **1969**, *181*, 381–392. [CrossRef]
12. Ng, S.C.; Hart, A.L.; Kamm, M.A.; Stagg, A.J.; Knight, S.C. Mechanisms of action of probiotics: Recent advances. *Inflamm. Bowel Dis.* **2009**, *15*, 300–310. [CrossRef] [PubMed]
13. Akbarimehr, M.; Solka, M.; Nosrati, M.; Rasouli, B.; Vahabi-Asil, O.; Seidavi, A.; Hosseintabar-Ghasemabad, B.; Marchewka, J.; Sztandarski, P.; Jaszczyk, A.; et al. Multi-strain probiotic improved feed conversion ratio and selected health indicators in Japanese quail. *Anim. Sci. Pap. Rep.* **2023**, *41*, 345–358. [CrossRef]
14. Kenny, M.; Smidt, H.; Mengheri, E.; Miller, B. Probiotics—Do they have a role in the pig industry? *Animal* **2011**, *5*, 462–470. [CrossRef] [PubMed]
15. Pereira, M.M.C.; Andretta, I.; Franceschi, C.H.; Kipper, M.; Mariani, A.; Stefanello, T.; Carvalho, C.; Vieira, J.; Moura Rocha, L.; Ribeiro, A.M.L. Effects of multistrain probiotic supplementation on sows' emotional and cognitive states and progeny welfare. *Animals* **2024**, *14*, 847. [CrossRef]
16. Innamma, N.; Banglarp, N.; Poonsuk, K.; Thitisak, P.; Hankla, S.; Kaeoket, K. Evaluation and selection of *Bacillus* species based on enzyme production for probiotics candidate in Bactosac, Nutrition & Herd Health Management. In Proceedings of the 25th IPVS Congress, Chongqing, China, 11–14 June 2018.
17. National Research Council. *Nutrient Requirements of Swine*; National Academy Press: Washington, DC, USA, 2012.

18. Kaeoket, K. The Effect of Dose and Route of Administration of R-cloprostenol on the Parturient Response of Sows. *Reprod. Domest. Anim.* **2006**, *41*, 472–476. [CrossRef]
19. Kaeoket, K.; Chanapiwat, P. Inducing Farrowing in Sows by PGF2 alpha and Its Analogues. *Thai J. Vet. Med.* **2011**, *41* (Suppl. S2011), 31–37.
20. Nuntapaitoon, M.; Juthamanee, P.; Theil, P.K.; Tummaruk, P. Impact of sow parity on yield and composition of colostrum and milk in Danish Landrace × Yorkshire crossbred sows. *Prev. Vet. Med.* **2020**, *181*, 105085. [CrossRef]
21. Roongsitthichai, A.; Tummaruk, P. Importance of backfat thickness to reproductive performance in female pigs. *Thai J. Vet. Med.* **2014**, *44*, 171–178. [CrossRef]
22. Saeng-Chuto, K.; Madapong, A.; Kaeoket, K.; Piñeyro, P.E.; Tantituvanont, A.; Nilubol, D. Co-infection of porcine deltacoronavirus and porcine epidemic diarrhea virus induces early TRAF6-mediated NF-κB and IRF7 signaling pathways through TLRs. *Sci. Rep.* **2022**, *12*, 19443. [CrossRef]
23. Saeng-Chuto, K.; Madapong, A.; Kaeoket, K.; Piñeyro, P.E.; Tantituvanont, A.; Nilubol, D. Coinfection of porcine deltacoronavirus and porcine epidemic diarrhea virus increases disease severity, cell trophism and earlier upregulation of IFN-α and IL12. *Sci. Rep.* **2021**, *11*, 3040. [CrossRef]
24. Liu, H.; Zhang, J.; Zhang, S.; Yang, F.; Thacker, P.A.; Zhang, G.; Qiao, S.; Ma, X. Oral administration of Lactobacillus fermentum I5007 favors intestinal development and alters the intestinal microbiota in formula-fed piglets. *J. Agric. Food Chem.* **2014**, *62*, 860–866. [CrossRef] [PubMed]
25. Ong, L.; Henriksson, A.; Shah, N.P. Development of probiotic Cheddar cheese containing Lactobacillus acidophilus, Lb. casei, Lb. paracasei and Bifidobacterium spp. and the influence of these bacteria on proteolytic patterns and production of organic acid. *Int. Dairy J.* **2006**, *16*, 446–456. [CrossRef]
26. Sirichokchatchawan, W.; Temeeyasen, G.; Nilubol, D.; Prapasarakul, N. Protective effects of cell-free supernatant and live lactic acid bacteria isolated from thai pigs against a pandemic strain of porcine epidemic diarrhea virus. *Probiotics Antimicrob. Proteins* **2018**, *10*, 383–390. [CrossRef] [PubMed]
27. Huang, S.; Yu, Q.; Xie, L.; Ran, L.; Wang, K.; Yang, Y.; Gan, L.; Song, Z. Inhibitory effects of Lactobacillus plantarum metabolites on porcine epidemic diarrhea virus replication. *Res. Vet. Sci.* **2021**, *139*, 32–42. [CrossRef] [PubMed]
28. Vieco-Saiz, N.; Belguesmia, Y.; Raspoet, R.; Auclair, E.; Gancel, F.; Kempf, I.; Drider, D. Benefits and inputs from lactic acid bacteria and their bacteriocins as alternatives to antibiotic growth promoters during food-animal production. *Front. Microbiol.* **2019**, *10*, 57. [CrossRef] [PubMed]
29. Kaewchomphunuch, T.; Charoenpichitnunt, T.; Thongbaiyai, V.; Ngamwongsatit, N.; Kaeoket, K. Cell-free culture supernatants of Lactobacillus spp. and Pediococcus spp. inhibit growth of pathogenic Escherichia coli isolated from pigs in Thailand. *BMC Vet. Res.* **2022**, *18*, 60. [CrossRef]
30. Markowiak-Kopeć, P.; Śliżewska, K. The effect of probiotics on the production of short-chain fatty acids by human intestinal microbiome. *Nutrients* **2020**, *12*, 1107. [CrossRef]
31. Pascoal, L.B.; Rodrigues, P.B.; Genaro, L.M.; Gomes, A.B.D.S.P.; Toledo-Teixeira, D.A.; Parise, P.L.; Bispo-Dos-Santos, K.; Simeoni, C.L.; Guimarães, P.V.; Buscaratti, L.I.; et al. Microbiota-derived short-chain fatty acids do not interfere with SARS-CoV-2 infection of human colonic samples. *Gut Microbes* **2021**, *13*, e1874740. [CrossRef]
32. Jackman, J.A.; Boyd, R.D.; Elrod, C.C. Medium-chain fatty acids and monoglycerides as feed additives for pig production: Towards gut health improvement and feed pathogen mitigation. *J. Anim. Sci. Biotechnol.* **2020**, *11*, 44. [CrossRef]
33. Lerner, A.B.; Cochrane, R.A.; Gebhardt, J.T.; Dritz, S.S.; Jones, C.K.; DeRouchey, J.M.; Tokach, M.D.; Goodband, R.D.; Bai, J.; Porter, E.; et al. Effects of medium chain fatty acids as a mitigation or prevention strategy against porcine epidemic diarrhea virus in swine feed. *J. Anim. Sci.* **2020**, *98*, skaa159. [CrossRef]
34. Machado, M.G.; Sencio, V.; Trottein, F. Short-chain fatty acids as a potential treatment for infections: A closer look at the lungs. *Infect. Immun.* **2021**, *89*, IAI0018821. [CrossRef] [PubMed]
35. Roopashree, P.G.; Shetty, S.S.; Suchetha Kumari, N. Effect of medium chain fatty acid in human health and disease. *J. Funct. Foods* **2021**, *87*, 104724. [CrossRef]
36. Tran, H.T.T.; Truong, A.D.; Ly, D.V.; Hoang, T.V.; Chu, N.T.; Nguyen, H.T.; Dang, A.T.K.; De Vos, M.; Lannoo, K.; Bruggeman, G.; et al. The potential anti-African swine fever virus effects of medium chain fatty acids on In Vitro feed model: An evaluation study using epidemic ASFV strain circulating in Vietnam. *Open Vet. J.* **2021**, *11*, 346–355. [CrossRef] [PubMed]
37. Feng, C.; Jin, C.; Liu, K.; Yang, Z. Microbiota-derived short chain fatty acids: Their role and mechanisms in viral infections. *Biomed. Pharmacother.* **2023**, *160*, 114414. [CrossRef]
38. Jackman, J.A.; Hakobyan, A.; Zakaryan, H.; Elrod, C.C. Inhibition of African swine fever virus in liquid and feed by medium-chain fatty acids and glycerol monolaurate. *J. Anim. Sci. Biotechnol.* **2020**, *11*, 114. [CrossRef]
39. He, H.; Fan, X.; Shen, H.; Gou, H.; Zhang, C.; Liu, Z.; Zhang, B.; Wuri, N.; Zhang, J.; Liao, M.; et al. Butyrate limits the replication of porcine epidemic diarrhea virus in intestine epithelial cells by enhancing GPR43-mediated IFN-III production. *Front. Microbiol.* **2023**, *14*, 1091807. [CrossRef]

40. Li, Y.; Hou, S.; Peng, W.; Lin, Q.; Chen, F.; Yang, L.; Li, F.; Huang, X. Oral administration of lactobacillus delbrueckii during the suckling phase improves antioxidant activities and immune responses after the weaning event in a piglet model. *Oxidative Med. Cell. Longev.* **2019**, *2019*, 6919803. [CrossRef]
41. Chen, F.; Wang, H.; Chen, J.; Liu, Y.; Wen, W.; Li, Y.; Huang, X. *Lactobacillus delbrueckii* ameliorates intestinal integrity and antioxidant ability in weaned piglets after a lipopolysaccharide challenge. *Oxidative Med. Cell. Longev.* **2020**, *2020*, 6028606. [CrossRef]
42. Clancy, R. Immunobiotics and the probiotic evolution. *FEMS Immunol. Med. Microbiol.* **2003**, *38*, 9–12. [CrossRef]
43. Vitiñi, E.; Alvarez, S.; Medina, M.; Medici, M.; De Budeguer, M.V.; Perdigón, G. Gut mucosal immunostimulation by lactic acid bacteria. *Biocell* **2000**, *24*, 223–232.
44. Laskowska, E.; Jarosz, Ł.; Grądzki, Z. Effect of multi-microbial probiotic formulation bokashi on pro- and anti-inflammatory cytokines profile in the serum, colostrum and milk of sows, and in a culture of polymorphonuclear cells isolated from colostrum. *Probiotics Antimicrob. Proteins* **2019**, *11*, 220–232. [CrossRef] [PubMed]
45. Valeur, N.; Engel, P.; Carbajal, N.; Connolly, E.; Ladefoged, K. Colonization and immunomodulation by Lactobacillus reuteri ATCC 55730 in the human gastrointestinal tract. *Appl. Environ. Microbiol.* **2004**, *70*, 1176–1181. [CrossRef] [PubMed]
46. Weström, B.R.; Svendsen, J.; Ohlsson, B.G.; Tagesson, C.; Karlsson, B.W. Intestinal transmission of macromolecules (BSA and FITC-labelled dextrans) in the neonatal pig. Influence of age of piglet and molecular weight of markers. *Biol. Neonate* **1984**, *46*, 20–26. [CrossRef] [PubMed]
47. Wittum, T.E.; Perino, L.J. Passive immune status at postpartum hour 24 and long-term health and performance of calves. *Am. J. Vet. Res.* **1995**, *56*, 1149–1154. [CrossRef]
48. Chattha, K.S.; Vlasova, A.N.; Kandasamy, S.; Esseili, M.A.; Siegismund, C.; Rajashekara, G.; Saif, L.J. Probiotics and colostrum/milk differentially affect neonatal humoral immune responses to oral rotavirus vaccine. *Vaccine* **2013**, *31*, 1916–1923. [CrossRef]
49. Tsukahara, T.; Inatomi, T.; Otomaru, K.; Amatatsu, M.; Romero-Pérez, G.A.; Inoue, R. Probiotic supplementation improves reproductive performance of unvaccinated farmed sows infected with porcine epidemic diarrhea virus. *Anim. Sci. J.* **2018**, *89*, 1144–1151. [CrossRef]
50. Nowland, T.L.; Kirkwood, R.N.; Torok, V.A.; Plush, K.J.; Barton, M.D. Characterisation of early microbial colonisers within the spiral colon of pre- and post-natal piglets. *Life* **2021**, *11*, 312. [CrossRef]
51. Ma, C.; Gao, Q.; Zhang, W.; Zhu, Q.; Tang, W.; Blachier, F.; Ding, H.; Kong, X. Supplementing synbiotic in sows' diets modifies beneficially blood parameters and colonic microbiota composition and metabolic activity in suckling piglets. *Front. Vet. Sci.* **2020**, *7*, 575685. [CrossRef]
52. Beaumont, M.; Cauquil, L.; Bertide, A.; Ahn, I.; Barilly, C.; Gil, L.; Canlet, C.; Zemb, O.; Pascal, G.; Samson, A.; et al. Gut microbiota-derived metabolite signature in suckling and weaned piglets. *J. Proteome Res.* **2021**, *20*, 982–994. [CrossRef]
53. Liu, B.; Zhu, X.; Cui, Y.; Wang, W.; Liu, H.; Li, Z.; Guo, Z.; Ma, S.; Li, D.; Wang, C.; et al. Consumption of dietary fiber from different sources during pregnancy alters sow gut microbiota and improves performance and reduces inflammation in sows and piglets. *MSystems* **2021**, *6*, 10–1128. [CrossRef]
54. Vieira Borba, V.; Sharif, K.; Shoenfeld, Y. Breastfeeding and autoimmunity: Programing health from the beginning. *Am. J. Reprod. Immunol.* **2018**, *79*, e12778. [CrossRef] [PubMed]
55. Liao, S.F.; Nyachoti, M. Using probiotics to improve swine gut health and nutrient utilization. *Anim. Nutr.* **2017**, *3*, 331–343. [CrossRef] [PubMed]
56. Chen, W.; Ma, J.; Jiang, Y.; Deng, L.; Lv, N.; Gao, J.; Cheng, J.; Liang, J.B.; Wang, Y.; Lan, T.; et al. Selective maternal seeding and rearing environment from birth to weaning shape the developing piglet gut microbiome. *Front. Microbiol.* **2022**, *13*, 795101. [CrossRef] [PubMed]
57. Olanratmanee, E.O.; Kunavongkrit, A.; Tummaruk, P. Impact of porcine epidemic diarrhea virus infection at different periods of pregnancy on subsequent reproductive performance in gilts and sows. *Anim. Reprod. Sci.* **2010**, *122*, 42–51. [CrossRef]

animals

MDPI

Article

Effects of Dietary Bromide, Magnesium and Tryptophan and Immunocastration on Growth Performance and Behaviour of Entire Male Pigs

Frank R. Dunshea [1,2,3,*], Ian McCauley [3,*] and Robert J. Smits [4,5]

1. School of Agriculture, Food and Ecosystem Sciences, The University of Melbourne, Parkville, Melbourne, VIC 3010, Australia
2. Faculty of Biological Sciences, The University of Leeds, Leeds LS2 9JT, UK
3. Department of Primary Industries, 600 Sneydes Road, Werribee, VIC 3030, Australia
4. Tri Advice Pty Ltd., 8 Clubbe Cr., Macgregor, ACT 2615, Australia; robsmits@triadvice.au
5. Rivalea Australia, Corowa, NSW 2646, Australia

* Correspondence: fdunshea@unimelb.edu.au (F.R.D.); ian.mccauley@gmail.com (I.M.); Tel.: +61-412332805 (F.R.D.)

Simple Summary: The growth performance of group-housed boars is well below that of their individually housed contemporaries and much of this difference in performance can be ameliorated by immunocastration. The improvements in performance after immunocastration are, at least in part, attributable to an increase in time spent feeding and a reduction in aggressive and sexual activities. However, a negative aspect is the increase in carcass fat. In-feed dietary additives such as bromide, magnesium and tryptophan offer another means to improve the performance of entire male pigs, although the effects do not seem to be as pronounced as immunocastration. Entire male pigs appear to be less motivated to feed than immunocastrates and less inclined to enter the feeder. Therefore, it may be important to ensure that feeder spaces are not limiting entire male pigs. Dietary sedatives may modify the behaviour of group-housed entire pigs and improve growth performance.

Abstract: The growth of boars may be inhibited because of aggressive and/or sexual activity. Dietary Br, Mg and tryptophan (Trp) as well as immunocastration may reduce these behaviours. In Experiment 1, 200 boars and 40 barrows were allocated to six groups of four pens of 10 pigs per treatment. Control and immunocastrate (Improvac-vaccinated at 13 and 17 weeks, Imp) boars and barrows were fed a finisher ration while the others were fed diets supplemented with Mg (5 g Mg proteinate/kg), Br (140 mg NaBr/kg) and Trp (5 g Trp/kg). In experiment 2, 300 boars were stratified by weight and within three weight classes allocated to two pens of ten pigs per treatment. Control and Imp boars were fed a finisher ration while the other diets were supplemented with Br, Trp or both Br and Trp. In Experiment 1, average daily gain (ADG) was not affected by diet but the Imp boars had higher ADG than controls. Feed intake (FI) tended to be higher in all treatments compared to controls except for the Trp group. In Experiment 2, Imp boars had higher ADG and FI than other treatments while Br+Trp boars had higher ADG and FI than controls. These data suggest that immunocastration and dietary Trp and Br show promise for improving performance in group-housed boars.

Keywords: magnesium; bromide; tryptophan; boar; immunocastration

1. Introduction

Consumer preference is for pork from gilts or barrows rather than boars, and historically, male pigs have been castrated soon after birth in most parts of the world. However, in other regions such as Australasia, the United Kingdom and South Africa, male pigs have been kept entire, which has been purported to decrease the cost of production because of the better growth performance of boars compared to barrows. While this is certainly the case in male pigs experimentally housed in individual pens, the differences are not as pronounced when male pigs are housed in groups under commercial conditions [1–3]. For example, Suster et al. [2] found that over the final 4 weeks before slaughter, individually penned boars deposited 200 g/day more lean tissue than barrows. In contrast, there was no difference in group-penned animals. These differences are attributed to aggressive and sexual interactions between group-housed boars, which can be reduced by immunocastration [3,4].

Concerns about pig welfare issues surrounding castration have resulted in castration without anaesthesia being banned in some EU countries, with others likely to follow suit. For example, in 2002 the Norwegian parliament decided to ban castration from 1 January 2009, and until implementation of the ban, all castrations had to be performed under analgesia and by a veterinarian [5]. In 2010, 33 key stakeholders in the pork supply chain, including scientists, veterinarians and animal welfare organisations, voluntarily signed the European Declaration on Alternatives to Surgical Castration of Pigs. This agreement sought to end the practice of surgical castration of pigs without pain relief by 2012 and gradually phase out surgical castration entirely across the EU and European Free Trade Association (EFTA) countries by 2018 [6]. While this goal has not been completely achieved, some EU countries still desire to cease castration completely. In Australia, the Model Code of Practice for the Welfare of Pigs [7] recommends that alternative options that minimise or alleviate pain from elective husbandry procedures or the avoidance of their use should be adopted where possible. Therefore, the issues relating to sexual and aggressive activities of group-housed boars will continue to be a problem, particularly with regard to heavy-weight pigs.

Several researchers have attempted to modify the behaviour of group-housed pigs using dietary additives. For example, dietary tryptophan may raise brain serotonin and modify aggressive [8] or sleeping [9] behaviour in pigs. Furthermore, dietary magnesium supplementation has been demonstrated to reduce plasma catecholamine concentrations and the incidence of meat quality defects in negatively handled pigs [10]. Potassium bromide is a dietary neuroleptic that has been shown to decrease sexual and aggressive activities without altering the growth rate in growing bulls [11]. Also, dietary bromide has been found to increase [12] or have no effect [13] on growth in pigs. In the former study, there was an inhibition of sexual function or activity, which was reversed upon removal of the bromide from the diet. Therefore, it is possible that one or the other of these dietary treatments may be used to ameliorate the performance-detracting behaviours of group-housed boars. The aim of the present studies was to determine the growth performance of group-housed entire boars supplemented with dietary tryptophan, bromide and magnesium.

2. Materials and Methods

Both experiments were conducted at the Research and Development Unit (RDU) at Bunge Meat Industries (now Rivalea Australia Pty Ltd.) in Corowa, New South Wales, Australia. All procedures were approved by the Institutional Animal Ethics committee.

2.1. Experiment 1

2.1.1. Growth Performance

The study involved 240 male pigs (in two replicates) comprising 200 entire boars and 40 contemporary castrate (surgical castration at 2 weeks of age) pigs. Pigs were allocated to treatment at 13 weeks of age and placed in 12 pens of 20 pigs (2 pens per subsequent treatment) in the RDU. The pigs destined to become immunocastrates were given the first dose of an immunocastration vaccine (Improvac, Zoetis Animal Health, Parkville, VIC, Australia) at 13 weeks of age. All pigs received a standard pelleted wheat and lupin based grower ration containing 14.0 MJ DE and 201 g crude protein per kg ad libitum until 17 weeks of age. Feed consumed and liveweight gain per pen of pigs were determined over the period from 13 to 17 weeks of age. At 17 weeks of age, each group of 20 pigs were divided into groups of 10 pigs (pre-determined at 13 weeks of age) and moved into pens in the finisher shed in the BMI RDU. Immunocastrated pigs were given their second dose of Improvac at 17 weeks of age and dietary treatments began. Control boars, immunocastrate and surgical castrate boars were offered a commercial pelleted wheat- and lupin-based finisher ration containing 13.3 MJ DE and 164 g crude protein per kg. The three other dietary treatments offered to entire boars were finisher diet supplemented with magnesium (5 g magnesium proteinate/kg, Mg; Lienert, Roseworthy South Australia 5371, Australia), bromide (140 mg sodium bromide/kg, Br; CSA Scientific, Port Adelaide, South Australia, 5015, Australia) and tryptophan (5 g tryptophan/kg, Trp; Kemin Industries, Killara NSW 2071, Australia). All diets were offered ad libitum and feed intake and liveweight were determined on a per-pen basis weekly. Pigs were slaughtered at 22 weeks of age and slaughter weight, P2 fat, leg fat and dressing percentage were recorded.

2.1.2. Statistics

Growth performance over the late grower period between 13 and 17 weeks of age were analysed by ANOVA with sex group (Boar, Improvac or Castrate) as the main factors. All analyses were conducted using pen as the experimental unit. Growth performance over the finisher period between 17 and 22 weeks of age was analysed by ANOVA (Genstat for Windows 23rd Edition. VSN International, Hemel Hempstead, UK) with sex or diet group (control, Mg, Br or Trp boars, Improvac or castrate) as the main factors and replicate as a blocking factor. All analyses were conducted using pen as the experimental unit. Due to a mechanical failure at the abattoir, carcass weight was only obtained for the first replicate and so only these data have been used in the analyses of carcass weight and dressing out rate.

2.2. *Experiment 2*

2.2.1. Growth Performance

The second experiment involved 300 male pigs (in two replicates) that were weighed and allocated to treatment at 13 weeks of age to evaluate the most promising treatments from Experiment 1, Br and Trp. Pigs were stratified by weight into three 33.3 percentiles and randomly allocated to one of five treatments. The pigs destined to become immunocastrates (60 pigs) were given the first dose of Improvac at 13 weeks of age. All pigs received a standard grower ration ad libitum until 17 weeks of age. At 17 weeks of age, pigs were placed in groups of 10 pigs of each weight × treatment group (pre-determined at 13 weeks of age) and moved into pens in the finisher shed. Immunocastrate pigs were given their second dose of Improvac at 17 weeks of age and dietary treatments began. Control boars and immunocastrate boars were offered a commercial finisher ration containing 13.3 MJ DE and 164 g crude protein per kg. The three other dietary treatments offered to entire boars were the finisher diet supplemented with bromide (140 mg bromide chloride/kg, Br), tryptophan (5 g tryptophan/kg, Trp) or both bromide and tryptophan at the dose levels. All diets were offered ad libitum, and feed intake and liveweight were determined on a per-pen basis weekly. Pigs were slaughtered at 22 weeks of age, and slaughter weight, P2 fat, leg fat and dressing percentage were recorded.

2.2.2. Behavioural Measures

Direct measures of pig behaviour were taken on two occasions at 18 and 22 weeks of age. A total of 15 pens of pigs were observed by three trained people in each 50 min session. Each of the 3 observers spent 10 min of each 50 min session recording the behaviour of one group of 10 pigs at a time. At the end of 10 min, the observer moved to the next listed pen. Therefore, each observer recorded 5 pens of pigs over the 50 min session in the following order. Each day, every group of 5 pens was observed 3 times. Aggressive acts were defined as any incident involving two pigs where one or both pigs perform vigorous biting/slashing/pushing actions against the other, directed at any part of the body. Where a fight occurred between 2 pigs (i.e., there was reciprocal aggression), a bout criterion interval of 5 s was chosen to separate one bout from another. Where more than 2 pigs were involved, each pig was counted as having a separate bout. A mount was defined as the occurrence of a pig riding on the back of another pig, which may be standing, sitting or lying. As for aggressive activities, a 5 s bout length criteria was used to count a new act of mountings.

2.2.3. Statistics

Growth performance over the finisher period between 17 and 22 weeks of age was analysed by ANOVA (Genstat for Windows 23rd Edition. VSN International, Hemel Hempstead, UK) with sex or diet group (control, Br, Trp, Br+Trp and Improvac-treated boars) and weight group (heavy, medium and light) as the main factors and replicate as a blocking factor. All analyses were conducted using pen as the experimental unit. Behavioural data were analysed by restricted maximum likelihood (REML) analysis, with the main effects being sex or diet group (control, Br, Trp, Br+Trp and Improvac-treated boars), weight group (heavy, medium and light) and week (18 or 22 weeks) as the main factors, and replicate, observer and sequence as blocking factors.

3. Results

3.1. *Experiment 1*

Castrate pigs were 1.7 kg heavier (p = 0.017) than contemporary boars at 13 weeks of age (Table 1). There was no effect of sex on daily gain between 13 and 17 weeks of age. Consequently, surgical castrates tended to maintain their weight advantage at 17 weeks of age (+2.3 kg, p = 0.098). Surgical castrates ate 20% (p = 0.005) more feed and used feed 16% (p < 0.001) less efficiently than entire male pigs over the period from 13 to 17 weeks of age. There was no effect of primary vaccination with Improvac on any aspect of growth performance until secondary vaccination. Over 17 to 22 weeks, the Improvac-treated boars grew more quickly than all other classes of pigs (Table 2). In particular, the immunocastrates grew 26% (+199 g/d) faster than the control boars. While the sedatives had no significant effects on daily gain, all group means were numerically greater (+4 to +9%) than that of the control boars, as was the case for the surgical castrates (+12%).

Table 1. Effect of sex on growth performance over the late grower phase between 13 and 17 weeks of age in Experiment 1 [1].

	Boar	Improvac	Castrate	LSD [2]	p-Value
Liveweight (kg)					
13 week	44.5	43.9	46.1	1.10	0.017
17 week	66.4	65.8	68.7	2.31	0.098
Growth performance					
Daily gain (g/d)	782	781	807	70.2	0.71
Feed intake (g/d)	1867	1823	2233	190.7	0.005
FCR (g/g)	2.39	2.34	2.77	0.139	<0.001

[1] Improvac injections were given at 13 and 17 weeks of age. [2] Least significant difference (p = 0.05) between boars or Improvac-treated boars and castrate pigs. For least significant difference between boars and Improvac-treated boars, multiply by 1.265.

Feed intake of the castrates was higher than any other groups over the first 2 weeks of the finishing period (Table 2). There was no effect of any dietary additives on feed intake of entire boars over the latter part of the finishing phase. Immunocastrates increased their feed intake over the latter part of the finishing period to a similar level as the surgical castrates.

There was no effect of sex or dietary additives on feed conversion efficiency (FCR) over the first 2 weeks of the finishing period (Table 2). However, over the latter part of the finishing phase the FCR of the surgical castrate pigs was 21% higher than that of the control boars. Over the entire finishing period, the FCR of the Trp boars was 10% lower than that of the control boars, whereas the FCR of the surgical castrates was 17% higher than that of the control boars. The FCR of the Improvac-treated boars and the Br and Mg boars was not different from that of the control boars.

Table 2. Effect of sex and dietary additives on growth performance over the finisher phase between 17 and 22 weeks of age [1].

	Boars				Improvac	Castrate	LSD [2]	*p*-Value
	Control	Mg	Br	Trp				
Liveweight (kg)								
17 week	66.1	66.5	67.8	64.5	65.1	68.5	2.91	0.093
19 week	76.0	78.4	78.8	75.4	77.9	81.1	3.08	0.037
22 week	93.7	95.2	97.9	94.1	99.8	99.3	4.66	0.044
Daily gain (g/d)								
17–19 weeks	709	851	790	776	915	900	175.8	0.162
19–22 weeks	826	782	889	879	1025	852	211.1	0.284
17–22 weeks	778	806	849	834	977	869	105.8	0.017
Feed intake (g/d)								
17–19 weeks	1876	2188	2209	1862	2187	2583	199.9	<0.001
19–22 weeks	2410	2381	2354	2314	3106	3076	323.8	<0.001
17–22 weeks	2201	2351	2335	2137	2738	2880	320.1	<0.001
FCR (g/g)								
17–19 weeks	2.67	2.66	2.85	2.43	2.41	2.87	0.518	0.308
19–22 weeks	2.99	3.07	2.67	2.63	3.15	3.62	0.487	0.006
17–22 weeks	2.84	2.95	2.77	2.56	2.95	3.32	0.196	<0.001

[1] Improvac injections were given at 13 and 17 weeks of age. [2] Least significant difference ($p = 0.05$) between treatment groups.

Carcass weight was significantly increased in the castrates, the Improvac-treated boars and the boars fed diets containing Br (Table 3). In the surgical castrates and Br boars, this resulted from increased live weight (Table 2) and dressing rate (Table 3), whereas for the Improvac-treated boars, the increased carcass weight resulted from increased live weight. Dietary Mg and Trp also increased the dressing out rate. Surgical castrates had higher P2 (+5 mm) and leg fat (+4.8 mm) than the control boars, whereas there was no significant effect of any dietary additives or Improvac on either P2 or leg fat.

Table 3. Effect of sex and dietary additives over the finisher phase between 17 and 22 weeks of age on carcass characterisitcs at slaughter [1].

	Boars				Improvac	Castrate	LSD [2]	*p*-Value
	Control	Mg	Br	Trp				
Carcass weight (kg)	69.0	71.3	74.1	71.0	73.7	76.8	4.62	0.053
Dressing (g/kg)	751	761	761	760	755	773	8.70	0.009
P2 back fat (mm)	10.6	11.0	11.1	10.3	11.7	15.6	1.36	<0.001
Leg fat (mm)	13.7	12.6	13.5	12.9	15.1	18.5	2.32	<0.001

[1] Improvac injections were given at 13 and 17 weeks of age. [2] Least significant difference ($p = 0.05$) between treatment groups.

3.2. Experiment 2

Growth data are presented in Table 4. There were no effects of treatment on liveweight at 17 weeks of age, demonstrating that prior injection with a single priming dose of Improvac had no effect on growth performance of boars. As planned, there were clear differences in the initial liveweight of pigs classed as heavy, medium and light (approximately 8 kg between each class of pig). Over the period from 17 to 22 weeks, the Improvac-treated boars grew more quickly than all other classes of pigs. In particular, the immunocastrates grew 19% (+153 g/d) faster than the control boars. However, there was an interaction between treatment and weight such that the growth response was greatest in the medium-weight class of pigs treated with Improvac and least in the light pigs (Figure 1). While there were no significant individual effects of either bromide (+1.8%) or tryptophan (+2.2%) treatments on daily gain, pigs treated with both compounds grew significantly faster (10.3%) than controls. Importantly, there was an interaction between treatment and weight class such that this effect was most pronounced in the heavy pigs where pigs from all treatment groups grew faster than the control boars (Figure 1).

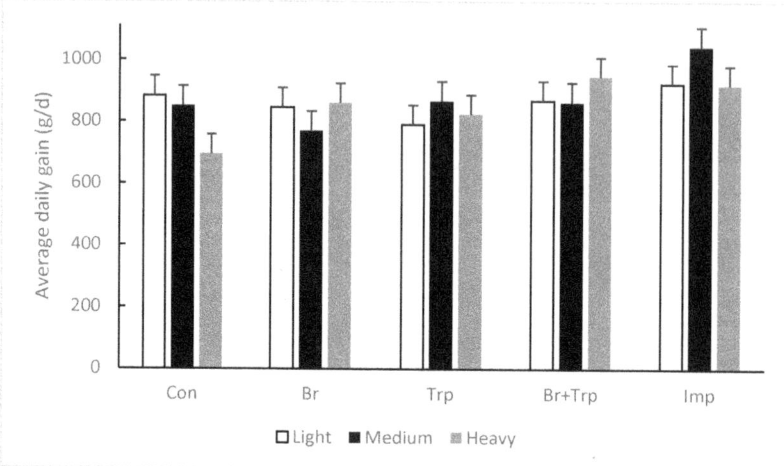

Figure 1. Effect of dietary neuroleptic or Improvac and weight class on average daily gain between 17 and 22 weeks of age. The error bars are the least significant difference (LSD) for weight × treatment = 134 g/d.

Over the period from 17 to 22 weeks, the Improvac-treated boars ate more than all other classes of pigs, particularly over the latter three weeks of the study. Thus, over the entire 5 week treatment period, the Improvac-treated pigs ate 17% more feed than the control boars, whereas over the last 3 weeks of the study, the Improvac-treated boars ate 22% more feed than the control boars. While there were no significant individual effects of either bromide (+3.7%) or tryptophan (+2.9%) treatments on feed intake, pigs treated with both compounds tended to consume more feed (6.7%, $p = 0.19$) than controls. Indeed, over the first 4 weeks of the study, the pigs fed the combined Br+Trp diet consumed significantly more (+8.3%, $p = 0.05$) feed than the control boars. Light boars ate less feed than either the heavy or medium boars. There was no effect of any of the dietary or vaccine treatments on feed conversion ratio. There was a significant effect of pig size on FCR, with the lightest pigs using feed most efficiently and the heavy pigs being the least efficient.

Despite the differences in growth rate, there were no significant treatment effects on carcass weight or dressing percentage. There was no significant effect of dietary sedatives on any measures of backfat, although pigs fed the diet containing both Trp and Br tended to have a greater P2 backfat than the control boars (+1.0 mm, $p = 0.06$). However, this was principally due to the greater backfat depth in the heavy pigs compared to the other classes of pigs (+2.7, +1.2 and −1.0 mm for the heavy, medium and light pigs, respectively; LSD = 1.87 mm) as indicated by the significant interaction between treatment and weight. Improvac significantly ($p < 0.05$) increased ultrasonic backfat (+1.7 mm) and leg fat (+2.1 mm) and tended to increase slaughter P2 (+0.9 mm, $p = 0.10$). However, there was again an interaction, with the medium pigs treated with Improvac being fatter than the control boars, but not the heavy or light pigs (+1.0, +2.3 and −0.5 mm for the heavy, medium and light pigs, respectively; LSD = 1.87 mm).

Behavioural observations are presented in Table 5. The amount of time spent fighting, mounting or engaged in aggressive acts was not different between the treatment groups or between the weight categories. However, there was an increase in aggressive activity and mounting activity between weeks 18 and 22 of age. Feeder occupancy was significantly higher in the Trp boars and Improvac-treated boars than in the control boars or the boars fed diets containing both Trp and Br. Also, feeder occupancy was greater in pens of medium-sized boars than in pens of control boars. There were no treatments or weight group effects on the number of pigs queued for feeder space, but the number decreased with age.

Table 4. Effect of sex, dietary additives and liveweight on growth performance over the finisher phase between 17 and 22 weeks of age and carcass characteristics at slaughter [1].

	Treatment							Weight Class				
	Control	Br	Trp	Br+Trp	Improvac	LSD [2]	*p*-Value	Heavy	Medium	Light	LSD [3]	*p*-Value
Liveweight (kg)												
17 weeks	64.0	64.3	64.3	63.7	63.8	3.46	0.99	71.9	64.2	56.0	2.68	<0.001
18 weeks	69.0	69.7	69.8	69.8	68.7	4.04	0.96	77.3	68.9	61.9	3.13	<0.001
19 weeks	74.8	75.7	74.8	75.9	76.1	3.94	0.93	83.6	75.0	67.7	3.05	<0.001
20 weeks	80.5	82.0	81.0	82.4	84.1	3.96	0.37	89.9	82.3	73.8	3.07	<0.001
21 weeks	85.9	87.8	87.2	88.6	90.4	3.83	0.16	95.4	88.5	79.5	2.97	<0.001
22 weeks	92.3	93.1	93.2	94.9	97.5	3.98	0.093	101.6	94.9	86.1	3.08	<0.001
Rate of gain (g/d)												
17–19 weeks	771	812	751	871	873	189.9	0.55	832	777	838	147.1	0.63
17–21 weeks [D]	781	803	820	889	949	70.7	<0.001	839	868	838	54.8	0.45
17–22 weeks [D]	808	823	826	891	961	77.5	0.004	848	877	861	60.0	0.60
19–22 weeks	834	831	875	905	1019	107.4	0.012	859	944	876	83.2	0.10
Feed intake (g/d)												
17–19 weeks	2103	2162	2151	2205	2257	235.4	0.70	2354	2117	2056	182.4	0.008
17–21 weeks	2285	2354	2363	2475	2650	190.9	0.009	2550	2458	2268	147.9	0.003
17–22 weeks	2354	2441	2423	2511	2747	240.4	0.030	2579	2537	2370	186.3	0.069
19–22 weeks	2521	2627	2604	2714	3074	360.4	0.041	2728	2817	2579	279.2	0.22
Feed conversion ratio (g/g)												
17–19 weeks	2.89	2.71	2.91	2.54	2.59	0.562	0.54	2.87	2.81	2.51	0.435	0.20
17–21 weeks	2.96	2.93	2.88	2.80	2.80	0.259	0.59	3.06	2.85	2.71	0.200	0.007
17–22 weeks	2.95	2.96	2.94	2.81	2.86	0.227	0.58	3.06	2.90	2.76	0.176	0.009
19–22 weeks	3.09	3.16	2.98	2.99	3.02	0.436	0.89	3.21	3.00	2.95	0.338	0.24
Carcass characteristics												
Carcass weight (kg)	71.5	71.5	72.0	72.7	74.6	3.40	0.31	78.3	72.8	66.2	2.63	<0.001
Dressing (g/kg)	769.2	765.7	768.2	765.2	764.9	4.24	0.79	770.2	766.9	762.8	3.28	0.12
Ultrasound P2 (mm)	8.3	8.8	8.6	9.2	10.0	1.04	0.024	9.5	8.4	9.0	0.80	0.039
Slaughter P2 (mm) [D]	9.3	10.2	9.5	10.3	10.2	1.08	0.18	10.4	10.2	9.2	0.83	0.016
Slaughter leg fat (mm)	11.6	12.0	11.3	12.3	13.7	1.14	0.004	12.3	12.8	11.4	0.88	0.015

[1] Improvac injections were given at 13 and 17 weeks of age. [2] Least significant difference ($p = 0.05$) between treatment groups. [3] Least significant difference ($p = 0.05$) between weight groups. [D] Treatment × weight interaction ($p \leq 0.05$).

Table 5. Effect of sex, dietary additives, liveweight and age on some direct behavioural observations in Experiment 2 [1].

	Week	Treatment (T)					Weight (W)			LSD [2]	LSD [3]	LSD [4]	Significance [5]			
		Control	Br	Trp	Br+Trp	Improvac	Heavy	Medium	Light				Treat	Weight	Week	T × W
Aggressive acts	18	9.5	10.7	14.2	9.2	11.7	11.0	8.8	13.4	4.98	3.85	3.15	0.23	0.35	0.025	0.18
(sec/10 min)	22	16.3	12.5	18.8	13.2	12.5	14.7	15.2	14.1							
Fights	18	6.3	4.6	8.4	12.5	8.6	6.5	10.5	7.3	4.31	3.34	2.73	0.54	0.16	0.39	0.90
(sec/10 min)	22	7.1	6.7	9.9	4.1	6.7	5.1	7.6	7.9							
Mounts	18	6.7	15.2	6.6	10.0	9.2	10.6	13.4	4.6	9.04	7.00	5.72	0.92	0.39	0.007	0.26
(sec/10 min)	22	17.1	12.4	18.6	21.8	16.8	21.4	13.2	17.5							
Feeder occupancy	18	0.76	0.74	0.86	0.65	0.82	0.72	0.78	0.79	0.103	0.080	0.065	<0.001	0.057	0.47	0.045
(pigs/feeder)	22	0.65	0.79	0.85	0.65	0.78	0.69	0.82	0.72							
Queued for feeder	18	0.40	0.35	0.38	0.34	0.43	0.32	0.43	0.39	0.128	0.099	0.081	0.57	0.59	0.040	0.098
(pigs/feeder)	22	0.23	0.23	0.40	0.34	0.28	0.29	0.28	0.31							

[1] Improvac injections were given at 13 and 17 weeks of age. [2] Least significant difference (p = 0.05) between treatment groups. [3] Least significant difference (p = 0.05) between weight groups. [4] Least significant difference (p = 0.05) between week groups. [5] There were no other significant (p > 0.05) interactions.

4. Discussion

The important new findings from these studies are that dietary neuroleptics may also ameliorate the reduction in growth performance of commercially housed boars. While there were no significant main effects of Mg, Br or Trp on daily gain, in Experiment 1, all group means were numerically greater (+4 to +9%) than that of the control boars. In addition, dietary Trp significantly decreased FCR by 10%. A smaller overall effect was seen with bromide, but this became more pronounced throughout treatment and FCR declined to 2.67 in the last two weeks of growth. The Br pigs also tended to be heavier than entire boars at slaughter, with a heavier carcass and higher dressing. While there were no significant individual effects of either Br (+1.8%) or Trp (+2.2%) treatments on daily gain in Experiment 2, pigs treated with both compounds grew significantly faster (10.3%) than controls. Importantly, there was an interaction between treatment and weight class such that this effect was most pronounced in the heavy pigs, where pigs from all sedative groups grew faster than the control boars. A similar response was seen for feed intake.

Eidrigevich et al. [12] reported on several experiments involving growing pigs and fattening cattle administered daily doses of a Br/salt mixture. Their findings showed that a daily intake of 5 mg/kg of body weight, consisting of sodium, potassium and ammonium Br, enhanced the growth rate of the pigs. They also observed a temporary inhibition of sexual function during Br administration, but once the treatment ceased, both male and female animals could breed successfully. On the other hand, Barber et al. [13] found no significant effect of a mix of Br salts (ammonium, potassium and sodium), either alone or in combination with copper sulphate, on the growth performance or carcass characteristics of finisher pigs. The only other literature on the effects of Br on livestock was the work of Genicot et al. [11], who found that, while potassium Br supplementation did not affect ADG of Belgian Blue cattle, there was an improvement in feed efficiency (+9%) over the latter stages of treatment. Also, there were some behavioural alterations such that rear engagements and side and direct attacks were reduced during Br supplementation. Therefore, it appears that, under some circumstances, there may be some positive effects of Br in reducing sexual and aggressive activities in livestock with resultant improvements in growth performance.

The effects of Trp on pigs' behaviour and growth have been much more studied than Br, particularly short-term studies focussed on pork eating quality. Some of these studies have also included Mg [14–17]. In general, there has been little or no effect on meat quality, although muscle pH has been increased in some cases, particularly in stress-susceptible pigs. Also, Peters et al. [15] found that dietary Trp-supplemented pigs were better able to handle simulated transport stress than their control counterparts. There have also been some long-term studies, and in one such comprehensive series of studies, Li et al. [18] found that supplemental Trp decreased the duration and intensity—but not the frequency—of aggression in unfamiliar finisher pigs. The pigs' responses to handling stressors, including electric shock, were unaffected by Trp treatment. High dietary Trp did not affect growth performance or objective meat quality measures [18]. Polletto et al. [8] found that supplemental Trp increased blood Trp and serotonin concentrations and reduced aggressive behavioural activity and time spent standing while increasing lying. Supplemental Trp also reduced the number of agonistic interactions and aggressiveness in 3-month-old gilts. Dietary supplementation of Trp tended to increase ADG in 3-month-old gilts but not in 6-month-old gilts. More recently, Henry et al. [9] found that supplemental Trp increased plasma Trp and serotonin concentrations but did not affect ADG, feed intake or behaviour in weaner pigs. Therefore, as with Br, it appears that under some circumstances, there may be some positive effects of Trp in reducing sexual and aggressive activities in livestock, although it only occasionally results in improvements in growth performance. However, it should be borne in mind that none of these studies have used entire males, where aggressive and sexual behaviours are most pronounced.

Dietary Mg treatment between 2 and 5 days before slaughter has been demonstrated to reduce plasma catecholamine concentrations and the incidence of meat quality defects in pigs [10,19]. Subsequently, several short-term studies have shown some improvements in meat quality, particularly in stress susceptible pigs [14–16]. A recent systematic review of the effect of more long-term dietary Mg supplementation in pigs indicated that in most, but not all, studies, there were beneficial effects of dietary magnesium [20]. In the present study, the effects of long-term supplemental Mg were not as pronounced as the effects of Br and Trp, although ADG was increased over that of the control boars during the first 2 weeks of administration. These data are consistent with a transient increase in plasma Mg before declining to basal rates after 10 days of Mg feeding [19], meaning that the effects of dietary Mg supplementation may be short-lived. While there appear to be some positive effects of pre-slaughter dietary Mg supplementation on transport and lairage meat quality, the efficacy of longer-term Mg supplementation in reducing negative aggressive and mounting behaviour in entire male pigs is less compelling.

This study confirmed that surgically castrated pigs consume more and grow less efficiently over both the grower and finisher phases and are fatter at slaughter than entire and immunised males [21]. In turn, while there was no difference in growth performance between control and immunised entire male pigs over the grower phase, there was an increase in feed intake and ADG after the secondary immunisation particularly beyond 2 weeks after secondary immunisation. These findings are consistent with the literature as summarised in the meta-analysis of Dunshea et al. [22]. Although there seems to be clear evidence of immunocastration improving performance in group-housed boars, there is still reluctance in some quarters to accept the practice [23–25]. However, education of consumers may overcome some of these issues [24]. In Australia, for example, at least a 60% of male pigs are immunocastrated [26].

There were very few significant effects noted during the behavioural observations, perhaps because of the variation in behaviours or because the presence of observers may have impacted behaviour. When assessed using video analysis, there was a profound reduction in sexual and aggressive activities with immunocastration [4], but this was not observed here. Therefore, it is perhaps not surprising that there were also very few dietary effects on behaviour in the present study. Despite this, feeder occupancy was significantly higher in the Trp boars and Improvac-treated boars than in the other groups, suggesting that these animals were more inclined to enter the feeder. This is related to a greater fed intake, at least in the case of the immunocastrated male pigs.

5. Conclusions

Thus, it appears that the benefits of the combined sedative treatment were particularly pronounced in heavy entire male pigs, which are most likely to suffer a reduction in growth performance due to overcrowding and/or aggressive and behavioural activities (although this was not apparent from the limited behavioural observations). Further studies are required to determine the dose response and duration of treatment of these neuroleptic compounds and whether they can further enhance the beneficial effects of Improvac, the effects of which do not become pronounced until approximately 1–2 weeks after the secondary vaccination. In particular, NaBr is a relatively inexpensive compound, whereas Trp is relatively expensive. It is important to determine the most efficacious and cost-effective combination of these two dietary additives to improve the growth performance of finisher boars. It will also be important to understand the pharmacokinetics of NaBr to ensure that there are no issues with residues.

Author Contributions: Conceptualization, F.R.D.; methodology, F.R.D. and I.M.; formal analysis, F.R.D.; investigation, F.R.D. and R.J.S.; resources, F.R.D. and R.J.S.; data curation, F.R.D., I.M. and R.J.S.; writing—original draft preparation, F.R.D.; writing—review and editing, F.R.D., I.M. and R.J.S.; project administration, F.R.D. and R.J.S.; funding acquisition, F.R.D. All authors have read and agreed to the published version of the manuscript.

Funding: This research was funded by Australian Pork Limited (formerly Pig Research and Development Corporation) project number DV160/1344.

Institutional Review Board Statement: The animal study protocol was approved by the Ethics Committee of Rivalea Australia (formerly Bunge Meat Industries) (protocol # 99N99C approved October 1999).

Informed Consent Statement: No human subjects were used in this study.

Data Availability Statement: Data are available on request.

Acknowledgments: The authors acknowledge the assistance of the technical staff of the Research and Development Unit at Rivalea Australia.

Conflicts of Interest: The co-author Robert J. Smits is an employee of Tri Advice Pty Ltd. The other authors declare no conflicts of interest.

References

1. De Haer, L.C.M.; De Vries, A.G. Feed intake patterns and digestibility in growing pigs housed individually or in groups. *Livest. Prod. Sci.* **1993**, *33*, 277–292. [CrossRef]
2. Suster, D.; Leury, B.J.; Kerton, D.J.; Borg, M.R.; Butler, K.L.; Dunshea, F.R. Longitudinal DXA measurements demonstrate lifetime differences in lean and fat tissue deposition in individually-penned and group-penned boars and barrows. *Aust. J. Agric. Res.* **2006**, *57*, 1009–1015. [CrossRef]
3. Dunshea, F.R.; Cronin, G.M.; Barnett, J.L.; Hemsworth, P.H.; Hennessy, D.P.; Campbell, R.G.; Luxford, B.; Smits, R.J.; Tilbrook, A.J.; King, R.H.; et al. Immunisation against gonadotrophin-releasing hormone (GnRH) increases growth and reduces variability in group-housed boars. *Anim. Prod. Sci.* **2011**, *51*, 695–701. [CrossRef]
4. Cronin, G.M.; Dunshea, F.R.; Butler, K.L.; McCauley, I.; Barnett, J.L.; Hemsworth, P.H. The effects of immuno- and surgical-castration on the behaviour and consequently growth of group-housed, male finisher pigs. *Appl. Anim. Behav. Sci.* **2003**, *81*, 111–126. [CrossRef]
5. Fredriksen, B.; Nafstad, O. The Norwegian Research Programme for Entire Male Pig Production. *Acta Vet. Scand.* **2006**, *48* (Suppl. S1), S16. [CrossRef]
6. Lin-Schilstra, L.; Ingenbleek, P.T.M. Examining Alternatives to Painful Piglet Castration Within the Contexts of Markets and Stakeholders: A Comparison of Four EU Countries. *Animals* **2021**, *11*, 486. [CrossRef]
7. Primary Industries Ministerial Council. *Model Code of Practice for the Welfare of Animals—Pigs (Revised)*; CSIRO Publishing: Melbourne, Australia, 2007.
8. Poletto, R.; Meisel, R.L.; Richert, B.T.; Cheng, H.W.; Marchant, J.N. Aggression in replacement grower and finisher gilts fed a short-term high-tryptophan diet and the effect of long-term human-animal interaction. *Appl. Anim. Behav. Sci.* **2010**, *122*, 98–110. [CrossRef]
9. Henry, M.; Shoveller, A.K.; O'Sullivan, T.L.; Niel, L.; Friendship, R. Effect of varying levels of dietary tryptophan on aggression and abnormal behavior in growing pigs. *Front. Vet. Sci.* **2022**, *9*, 849970. [CrossRef]
10. D'Souza, D.N.; Warner, R.D.; Leury, B.J.; Dunshea, F.R. The effect of dietary magnesium aspartate supplementation on pork quality. *J. Anim. Sci.* **1998**, *76*, 104–109. [CrossRef]
11. Genicot, B.; Mouligneau, F.; Lekeux, P. Efficiency of a sedative on the behaviour and the performances of Belgian white and blue double-muscled cattle during fattening. *J. Vet. Medicine. Ser. A* **1991**, *38*, 668–675. [CrossRef]
12. Eidrigevich, E.V.; Osipov, K.V.; Zaplatnikov, A.K.; Bazarnyi, V.M.; D'yakova, A.A.; Rumyantsev, B.N. Bromide fed to young bulls and swine stimulates weight gain. *Vet. Meditsina* **1967**, *12*, 82–87. (In Russian)
13. Barber, R.S.; Braude, R.; Mitchell, K.G. Arsanilic acid, sodium salicylate and bromide salts as potential growth stimulants for pigs receiving diets without copper sulphate. *Br. J. Nutr.* **1971**, *25*, 381–389. [CrossRef] [PubMed]
14. Peeters, E.; Neyt, A.; Beckerst, F.; De Smet, S.; Aubert, A.E.; Geers, R. Influence of supplemental magnesium, tryptophan, vitamin C, and vitamin E on stress responses of pigs to vibration. *J. Anim. Sci.* **2005**, *83*, 1568–1580. [CrossRef] [PubMed]
15. Peeters, E.; Driessen, B.; Geers, R. Influence of supplemental magnesium, tryptophan, vitamin C, vitamin E, and herbs on stress responses and pork quality. *J. Anim. Sci.* **2006**, *84*, 1827–1838. [CrossRef]
16. Panella-Riera, N.; Dalmau, A.; Fàbrega, E.; Furnols, M.F.I.; Gispert, M.; Tibau, J.; Soler, J.; Velarde, A.; Oliver, M.A.; Gil, M. Effect of supplementation with $MgCO_3$ and L-Tryptophan on the welfare and on the carcass and meat quality of two halothane pig genotypes (NN and nn). *Livest. Sci.* **2008**, *115*, 107–117. [CrossRef]
17. Panella-Riera, N.; Velarde, A.; Dalmau, A.; Fàbrega, E.; Furnols, M.F.I.; Gispert, M.; Soler, J.; Tibau, J.; Oliver, M.A.; Gil, M. Effect of magnesium sulphate and L-tryptophan and genotype on the feed intake, behaviour and meat quality of pigs. *Livest. Sci.* **2009**, *124*, 277–287. [CrossRef]
18. Li, Y.Z.; Kerr, B.J.; Kidd, K.T.; Gonyou, H.W. Use of supplementary tryptophan to modify the behavior of pigs. *J. Anim. Sci.* **2006**, *84*, 212–220. [CrossRef]
19. D'Souza, D.N. The Influence of Pre-and Post-Slaughter Factors on Pork Meat Quality. Ph.D. Thesis, University of Melbourne, Parkville, Australia, 1998.
20. Bushby, E.V.; Dye, L.; Collins, L.M. Is magnesium supplementation an effective nutritional method to reduce stress in domestic pigs? A systematic review. *Front. Vet. Sci.* **2021**, *7*, 596205. [CrossRef]

21. Dunshea, F.R.; Colantoni, C.; Howard, K.; McCauley, I.; Jackson, P.; Long, K.A.; Lopaticki, S.; Nugent, E.A.; Simons, J.A.; Walker, J.; et al. Vaccination of boars with a GnRH vaccine (Improvac) eliminates boar taint and increases growth performance. *J. Anim. Sci.* **2001**, *79*, 2524–2535. [CrossRef]
22. Dunshea, F.R.; Allison, J.R.; Bertram, M.; Boler, D.D.; Brossard, L.; Campbell, R.; Crane, J.P.; Hennessy, D.P.; Huber, L.; de Lange, C.; et al. The effect of immunization against GnRF on nutrient requirements of male pigs: A review. *Animal* **2013**, *7*, 1769–1778. [CrossRef]
23. Vanhonacker, F.; Verbeke, W.; Tuyttens, F.A.M. Belgian consumers' attitude towards surgical castration and immunocastration of piglets. *Anim. Welf.* **2009**, *18*, 371–380. [CrossRef]
24. Pasquale, J.; Nannoni, E.; Sardi, L.; Rubini, G.; Salvatore, R.; Bartoli, L.; Adinolfi, F.; Martelli, G. Towards the Abandonment of Surgical Castration in Pigs: How is Immunocastration Perceived by Italian Consumers? *Animals* **2019**, *9*, 198. [CrossRef] [PubMed]
25. European Commission: Directorate-General for Health and Food Safety. *Pig Castration—Methods of Anaesthesia and Analgesia for All Pigs and Other Alternatives for Pigs Used in Traditional Products*; Publications Office: Luxembourg, 2017. Available online: https://data.europa.eu/doi/10.2875/057159 (accessed on 2 November 2024).
26. Sinnett, A.; Dunshea, F.; D'Souza, D.; Malcolm, B.; Griffith, G.A. Benefit cost analysis of entire male pig immuno-castration. *Aust. Farm Bus. Manag. J.* **2022**, *19*, 79–93. [CrossRef]

animals

MDPI

Article

The Effects of Octapeptin Supplementation on Growth Performance, Serum Biochemistry, Serum Immunity, and Gut Microbiota in Weaned Piglets

Sheng Huang [1,2], Li Yang [2], Li Wang [2], Yu Chen [2], Xiuliang Ding [1,2], Feiyun Yang [1,2], Shiyan Qiao [3] and Jinxiu Huang [1,2,*]

1 Chongqing Academy of Animal Sciences, Chongqing 402460, China; huangs@cqaa.cn (S.H.); sddingxl@aliyun.com (X.D.); yfeiyun@yeah.net (F.Y.)

2 National Center of Technology Innovation for Pigs, Chongqing 402460, China; lekalter@163.com (L.Y.); q15928806587@163.com (L.W.); 15222269277@126.com (Y.C.)

3 State Key Laboratory of Animal Nutrition and Feeding, Ministry of Agriculture and Rural Affairs Feed Industry Centre, China Agricultural University, Beijing 100193, China; qiaoshiyan@cau.edu.cn

* Correspondence: huangjx@cqaa.cn; Tel.: +86-023-15823497719

Simple Summary: Weaning is a critical and challenging period for piglets, often leading to stress, poor growth, and increased disease incidence. This study investigated using octapeptin as a feed additive for weaned piglets. The results indicated that octapeptin significantly improved diarrhea and enhanced feed conversion ratio by modulating immunity and reducing inflammation compared to standard diets. Octapeptin also decreased TNF-α levels, boosted the immune system, and increased beneficial bacteria, such as *Collinsella* and *Olsenella*, positively impacted gut health. These findings suggested that octapeptin is a promising, safe, natural antibiotic alternative that promotes health and growth in weaned piglets.

Abstract: With the prohibition of antibiotics in animal feed, the livestock industry faces significant challenges, including increased morbidity and mortality rates and reduced farming efficiency. Developing green, natural, and safe antibiotic alternatives has become a research hotspot. This study evaluated the effects of octapeptin as a feed additive on growth performance, diarrhea incidence, serum biochemistry, serum immune factors, and gut microbiota of weaned piglets. Seventy-two weaned piglets were randomly assigned to three groups based on body weight and sex, with each group receiving different dietary treatments: a negative control group (CON, basal diet), a positive control group (MC, basal diet + 5 mg/kg Microcin C7), and an octapeptin supplement group (OP, basal diet + 40 mg/kg octapeptin). After 28 days of feeding experimental diets, the results demonstrated that supplementing the diet of weaned piglets with octapeptin significantly improved the feed conversion ratio compared to the control group ($p < 0.05$) over the entire experimental period. Furthermore, a reduction in diarrhea incidence was observed during the late nursery period (14–28 d), resulting in an overall improvement in diarrhea compared to the other two groups ($p < 0.01$). Serum biochemical analysis results revealed a trend towards decreased alanine aminotransferase level in the octapeptin group, with no significant differences in other indicators, suggesting potential improvements in liver function without causing liver damage. In addition, compared to the control group, octapeptin enhanced mucosal immunity by decreasing TNF-α level ($p < 0.05$). Fecal microbiota analysis results showed a significant increase in beneficial bacteria such as *Collinsella* and *Olsenella* in the octapeptin group compared to the other two groups ($p < 0.05$), indicating a positive impact on gut health. These findings supported the potential of octapeptin as an alternative to antibiotic growth promoters in weaned piglets' diets.

Keywords: weaned piglets; octapeptin; growth performance; fecal microbiota; feed additives

1. Introduction

Since the 1950s, following Moore's report on the benefits of antibiotics in promoting growth and enhancing disease resistance in broiler chickens [1], the use of antibiotics in animal husbandry has surged [2,3]. This increase has led to antibiotic residues in livestock [4], food safety concerns [5], and environmental pollution [6,7], posing potential threats to both ecosystems and human health [8]. Controlling antibiotic resistance and reducing antibiotic use have become national priorities and research hotspots in many countries. Since 1986, several countries, including Sweden, the European Union, the United States, Japan, and South Korea, have banned antibiotics in animal feed for growth promotion [9,10]. China implemented a comprehensive ban on growth-promoting antibiotics in July 2020 [11]. Consequently, identifying and developing alternative agents that can promote growth, alleviate diarrhea, and maintain intestinal health has become urgent.

Antimicrobial peptides (AMPs) exhibit antibacterial [12–15], antiviral, and immunomodulatory activities, with low toxicity and minimal side effects [16] and no bacterial resistance [17–19]. These peptides are derived from diverse sources, including animals, plants (such as buferin, beta-defensin 2, AMP-A3, cecropin), bacteria (such as colicin and octapeptin), and fungi (such as plectasin and peptaibols), positioning them as promising candidates for alternative antimicrobial therapies [20]. They have broad potential applications in disease treatment [21], food preservation [22,23], and as feed additives [24–26]. In livestock production, studies on piglets have shown that AMPs like colicin, cecropin, and defensin have significant growth-promoting effects, increasing final weight, average daily gain, and feed conversion rate while reducing diarrhea incidence [27]. AMPs also elevate glutathione levels, total antioxidant capacity, and peroxidase activity in serum and intestines, significantly increasing IgA and IgG levels while decreasing TNF-α, IL-1β, and IL-6 [28,29]. Further research has shown that AMPs increase beneficial bacteria like *lactobacilli* and *bifidobacteria* and reduce *E. coli* in the gut [26,30,31]. AMPs positively impact intestinal villus height, crypt depth, and gut barrier function [32,33]. As a dietary supplement, AMPs can enhance the humoral immunity of piglets [34], increasing the levels of immunoproteins such as IgG, IgM, and IgA. Additionally, AMPs can improve cellular immunity by increasing the number and proliferative function of peripheral T-cell subsets in piglets [35].

Due to the accumulation of antibiotic-resistant genes, there is an urgent need for alternatives to antibiotics that can effectively combat resistant bacterial strains without promoting further resistance. Octapeptin, a broad-spectrum lipopeptide produced by *Bacillus circulans*, has been reported to meet these requirements [36]. Research has demonstrated that octapeptin exhibits activity against polymyxin-resistant bacteria [37]. Meanwhile, octapeptin is less likely to induce resistance and does not exhibit cross-resistance with polymyxins [37], positioning it as a promising candidate for combating resistant bacterial strains. Currently, the in vitro and in vivo activity of octapeptin has been evaluated through antibacterial assays and mouse models [37]. In vitro studies have found that octapeptin has a MIC of about 2~8 μg/mL against some Gram-negative bacteria, which is 8–32 times lower than polymyxin B [38]. In vivo studies found octapeptin to have lower nephrotoxicity and a longer half-life than polymyxin B [39].

Weaning is one of the most challenging stages in pig production, significantly impacting growth performance [40,41]. It induces stress responses in pigs, disrupts intestinal digestion and absorption capabilities, compromises gut health [42], markedly reduces growth and immunity in weaned piglets [43,44], and increases the incidence of diarrhea [45]. Therefore, mitigating weaning stress may result in improved performance, and octapeptin may have beneficial effects when used at appropriate levels. Although numerous studies have demonstrated that antimicrobial peptides possess both antimicrobial properties and immunomodulatory effects [46–48], research specifically focusing on the immunomodulatory effects of octapeptin is almost non-existent. Therefore, this study aims to evaluate the effects of octapeptin as a feed additive on growth performance, diarrhea incidence, serum biochemical indices, serum immune factors, and intestinal microbiota in weaned piglets, with the expectation of preliminarily exploring the immunomodulatory activity of octapeptin.

2. Materials and Methods

All procedures of this experiment were approved by the Institutional Animal Care and Use Committee of Chongqing Academy of Animal Sciences (Chongqing, China). The experiment was conducted at the Shuanghe Research Base of the Chongqing Academy of Animal Sciences.

2.1. Preparation of Antimicrobial Octapeptin

Our laboratory provided the antimicrobial octapeptin sample (Chongqing, China), which had a purity of 95%, and used corn flour as the carrier. The octapeptin was derived from *Bacillus circulans*.

2.2. Animals, Diets, and Experimental Design

Seventy-two 28-day-old weaned piglets (Duroc × Landrace × Yorkshire), with an initial body weight of 6.57 ± 0.08 kg, were used in this study, which lasted for 28 days. Upon arrival at the experimental facility, littermates were randomly assigned to different groups to avoid confounding effects. The piglets underwent a 7-day feed adaptation period, after which they were assigned to three groups using a randomized complete block design, with sex and body weight at weaning as blocking factors. Each treatment group consisted of 6 replicates, with each replicate containing 4 piglets (2 males and 2 females). The dietary treatments were as follows: a negative control group (CON, basal diet), a positive control group (MC, basal diet + 5 mg/kg Microcin C7), and an OP treatment group (OP, basal diet + 40 mg/kg octapeptin). Microcin C7, used in the positive control group at a dosage of 5 mg/kg (purity 95%), is a narrow-spectrum antimicrobial peptide composed of seven amino acids, known for its low cross-resistance [49], making it suitable as a feed additive. Research has demonstrated that an addition of 500 mg/kg (with a purity of 1.19%) of Microcin C7 has beneficial effects on growth performance, diarrhea incidence, apparent total tract digestibility (ATTD) of ether extract and Ca, immune performance, intestinal morphology, and microbiota structure in piglets [28]. All diets were formulated to meet the nutrient requirements recommended by NRC (2012), and the ingredient composition and nutrient content of the basal diet are presented in Table 1.

2.3. Animal Management

Seventy-two piglets were kept in the experimental pens with plastic slatted floors, with 4 piglets per pen (1.4 m × 2 m). Each pen was fitted with an adjustable stainless-steel feeder and a duckbill drinker to allow piglets to eat and drink ad libitum. The pig housing environment was meticulously controlled, including regulation of CO_2 and ammonium levels in the air, ventilation rates, humidity, and temperature. The average indoor temperature was maintained between 24 and 26 °C, while relative humidity was kept at 60–70%. To mitigate disease risks, the experimental facilities underwent daily cleaning, supplemented by weekly health assessments of the piglets conducted by a veterinarian. In this study, piglets would undergo a 12 h fasting period before weight measurements on days 0, 14, and 28, alongside weighing the remaining feed in each pen. These data points were used to calculate the average daily gain (ADG), average daily feed intake (ADFI), and feed-to-gain ratio (F/G). The diarrhea index (DI) and diarrhea rate (DR) were assessed using the formula diarrhea index = sum of diarrhea scores/(total number of piglets × number of days tested), diarrhea rate (%) = the total number of diarrhea piglets/(total number of piglets × number of days tested) × 100%. A higher diarrhea rate indicated more severe diarrhea among the piglets. Fecal consistency was assessed and scored based on the following criteria [50]: normal feces, characterized by formed or granular stools, were assigned a score of 0; mild diarrhea, indicated by soft but formed stools, was given a score of 1; moderate diarrhea, identified by thick, unformed stools without separation of fecal water, received a score of 2;

and severe diarrhea, marked by watery, unformed stools with separation of fecal water, was scored as 3. Any score above 0 indicated the presence of diarrhea and was included in the calculation of the diarrhea rate.

Table 1. Ingredient composition and nutrient content of the basal diet (as-fed basis, %).

Ingredient, %	Content	Nutrient Level	Measured
Corn	56.18	Metabolizable Energy, Kcal/kg [1]	3444.44
Extruded Corn	8.00	Crude Protein	20.6
Soybean Meal, 46%	17.50	Calcium	0.59
Extruded Soybean Flour	4.00	Available Phosphorus	0.34
Fermented Soybean Meal	5.50	Total Lysine	1.34
Yeast Culture	2.00	Total Methionine	0.43
Glucose	1.00	Total Methionine + Cysteine	0.77
Fish Meal	1.00	Total Threonine	0.82
High-Fat Powder	1.00	Total Tryptophan	0.22
Calcium Formate	0.80		
Dicalcium Phosphate	0.60		
Salt	0.35		
Acidifier [3]	0.80		
Lysine, 98.5%	0.40		
Choline Chloride, 50%	0.10		
DL-Methionine, 99%	0.10		
Threonine, 99%	0.08		
Complex Enzymes	0.07		
Premix [2]	0.52		
Total	100		

[1] The metabolizable energy value of the diet was calculated according to the proportion of each ingredient in the formulation and the data in the NRC (2012). [2] The premix provided the following nutrients per kilogram of feed: Copper (Cu) 5.12 mg, Iodine (I) 0.15 mg, Iron (Fe) 87.59 mg, Manganese (Mn) 3.69 mg, Selenium (Se) 0.30 mg, Zinc (Zn) 84.43 mg, Vitamin A (VA) 9000 IU, Vitamin D (VD) 3000 IU, Vitamin E (VE) 24 IU, Vitamin K (VK) 3 mg, Thiamine 3 mg, Riboflavin 7.5 mg, Pantothenic Acid 15 mg, Niacin 30 mg, Pyridoxine 3.60 mg, Biotin 0.15 mg, Folic Acid 1.50 mg, and Vitamin B12 (VB12) 0.036 mg. [3] Acidifier: The composition includes 35% phosphoric acid, 5% citric acid, 10% fumaric acid, 5% benzoic acid, 5% calcium formate, and 40% SiO_2.

2.4. Sample Collection

On the morning of day 28, two piglets (one male and one female) with the average body weight from each replicate group were selected for blood sampling. Blood was collected from the jugular vein of fasted piglets using a 10 mL syringe and placed into pro-coagulant blood collection tubes. The samples were left to stand for 2–3 h and then centrifuged at $1000\times g$ for 10 min at 4 °C. The supernatant was aliquoted into four 1.5 mL centrifuge tubes and stored at −20 °C for future analysis.

On day 28th, fresh fecal samples were collected from pigs on a per-pen basis. Using a sterile spoon, the central portion of the feces, which had not come into contact with the ground, was extracted and placed into sterile zip-lock bags. After collecting half a bag of feces, the samples were thoroughly mixed, resulting in 30 fecal samples. These samples were then transported to the laboratory and stored at −80 °C for subsequent 16S rRNA high-throughput sequencing to analyze the fecal microbiota.

2.5. Serum Biochemical and Immune Parameters

An automated biochemical analyzer was used to measure the levels of total bile acid (TBA), total protein (TP), albumin (ALB), globulin (GLB), total bilirubin (TBil), blood urea nitrogen (BUN), cholesterol (CHO), and triglycerides (TG) in serum. The activities of alanine aminotransferase (ALT), aspartate aminotransferase (AST), and gamma-glutamyl transferase (GGT) were also assessed. All procedures followed the manufacturer's instructions for the automatic biochemical analyzer (Beckman Coulter AU5800; Brea, CA, USA).

Serum concentrations of immunological factors, including IgG, IgM, IgA, IL-2, IL-6, IL-10, and TNF-α, were determined using ELISA kits according to the manufacturer's instructions. All cytokine ELISA kits were obtained from Thermo Fisher Scientific (Waltham, MA, USA).

2.6. Microbiota Composition by 16S rRNA Sequencing Analysis

The DNA extracted from fecal genomic samples was checked using 1% agarose gel electrophoresis. Sequencing adapters were added to the ends of primers 338F and 806R, and the V3-V4 gene fragments were PCR-amplified. The PCR products were purified, quantified, and normalized to construct a paired-end (PE) library. After quality control, the library was sequenced using an Illumina MiSeq platform.

The paired-end reads obtained from MiSeq sequencing were split according to samples and underwent quality control and filtering based on sequencing quality. The paired-end reads merged based on their overlap to generate optimized sequences. These optimized sequences were then processed using sequence denoising methods (such as DADA2 or Deblur) to obtain amplicon sequence variant (ASV) representative sequences and abundance information. Based on the ASV representative sequences and abundance data, alpha diversity analyses were conducted on the Majorbio Cloud Platform (https://cloud.majorbio.com/, accessed on 29 June 2024) to assess species diversity within individual samples. Indices such as Chao, Shannon, Simpson, and Coverage were calculated at a 97% similarity level, and rarefaction curves were plotted. Beta diversity analyses were performed to compare community composition and structural differences among different groups, with principal coordinate analysis (PCoA) plots generated based on distance matrices.

2.7. Statistical Analyses

Data on piglet growth performance, blood biochemistry, and serum immunity were analyzed using the GLM procedure in SAS software (version 9.2, SAS Institute Inc., Cary, NC, USA) through analysis of variance (ANOVA) followed by Tukey multiple range tests, with $n = 24$ replicates per treatment group. The experimental unit for growth performance measurements was the pen, while for serum immunity and blood biochemistry, the experimental unit was the serum sample collected from each piglet. The chi-square test was used to analyze the diarrhea index and rate, with each piglet serving as the experimental unit. The Wilcoxon rank-sum test compared the relative abundance of microbial communities, considering each piglet as an independent experimental unit. Differences were considered statistically significant at $p < 0.05$ and were regarded as trends when $0.05 < p \leq 0.1$. Superscripts were added in each table to indicate tendencies where applicable.

3. Results

3.1. Effect of Octapeptin on the Growth Performance of Piglets

The growth performance of piglets' results, shown in Table 2, proved that adding octapeptin to the diet of weaned piglets significantly reduced the F/G throughout the entire experimental period ($p < 0.05$). There was no effect on ADG and body weight. During the entire experimental period, the ADFI of the MC group was significantly reduced by 10.82% compared to the CON group ($p < 0.05$), with a tendency of reduction observed in the 14–28 d ($p < 0.1$). The ADFI of the OP group was intermediate between the MC and CON groups, with no significant difference observed among the groups. According to Table 3, the OP group exhibited a remarkable reduction in diarrhea index and diarrhea rate compared to the CON and MC groups for days 14 to 28 ($p < 0.01$), resulting in a decrease in diarrhea index and diarrhea rate over the entire experimental period ($p < 0.05$).

3.2. Effect of Octapeptin on Serum Biochemical Indexes of Piglets

The effects of octapeptin supplementation on the serum biochemical parameters of weaned piglets are detailed in Table 4. Compared to the CON group, ALT levels in the MC group were markedly reduced by 41.52% ($p < 0.05$). The OP group showed ALT levels that were intermediate in value between the CON and MC groups. No significant changes were observed in other biochemical parameters. Additionally, a decrease of 13.06% in CHO level was observed in the MC group ($p < 0.1$), while the CHO level in the OP group decreased by 7.90%.

Table 2. The effect of dietary octapeptin supplementation on growth performance in piglets.

Items	CON	MC	OP	SEM	*p*-Value
BW, kg					
0 d	6.62	6.56	6.54	0.08	0.72
14 d	8.97	9.34	9.00	0.18	0.33
28 d	14.79	15.04	15.31	0.46	0.73
ADG, g/d					
0–14 d	173.77	197.96	171.61	9.50	0.13
14–28 d	415.71	407.06	412.48	20.37	0.96
Whole period	291.93	302.51	309.42	16.36	0.75
ADFI, g/d					
0–14 d	269.40	261.37	245.91	11.30	0.35
14–28 d	682.46	590.34	630.26	26.54	0.08
Whole period	475.93 ^{a}	424.45 ^{b}	433.71 ^{ab}	13.64	0.04
F/G					
0–14 d	1.56 ^{a}	1.36 ^{b}	1.41 ^{ab}	0.04	<0.01
14–28 d	1.66	1.46	1.53	0.08	0.19
Whole period	1.65 ^{a}	1.40 ^{b}	1.43 ^{b}	0.07	0.04

CON = basal diet without additive; MC = basal diet + 5 mg/kg Microcin C7; OP = basal diet + 40 mg/kg octapeptin. BW: body weight; ADG: average daily body gain; ADFI: average daily feed intake; F/G: ratio of feed to gain. SEM, standard error of the mean. The following tables were the same as this table. ^{a,b} Means listed in the same row with different superscripts are significantly different ($p \leq 0.05$).

Table 3. The effect of dietary octapeptin supplementation on diarrhea index and diarrhea rate in piglets.

Items	CON	MC	OP	SEM	*p*-Value
DI					
0–14 d	0.52	0.47	0.46	0.10	0.12
14–28 d	0.57 ^{a}	0.60 ^{a}	0.43 ^{b}	0.13	<0.01
Whole period	0.54 ^{a}	0.53 ^{a}	0.45 ^{b}	0.10	0.02
DR, %					
0–14 d	10.71	10.42	9.40	2.99	0.52
14–28 d	11.16 ^{a}	12.26 ^{a}	6.75 ^{b}	3.10	<0.01
Whole period	10.94 ^{a}	11.26 ^{a}	8.06 ^{b}	2.41	<0.01

CON = basal diet without additive; MC = basal diet + 5 mg/kg Microcin C7; OP = basal diet + 40 mg/kg octapeptin. DI: diarrhea index; DR: diarrhea rate. ^{a,b} Means listed in the same row with different superscripts are significantly different ($p \leq 0.05$).

3.3. *Effect of Octapeptin on Serum Immunity Indexes of Piglets*

The effects of adding octapeptin to the diet on the immune factors of weaned piglets are shown in Table 5. The IgG level in the OP group was markedly reduced by 61.77% compared to the MC group ($p < 0.01$), with no significant difference observed relative to the CON group. Meanwhile, the TNF-α level in the OP group was comparable to the MC group but notably decreased by 27.85% compared to the CON group ($p < 0.05$). It was also noted that the IgA level in the MC group increased significantly by 33.33% compared to the CON group ($p < 0.05$), while the IgA level in the OP group showed no difference compared to either the MC or CON group.

Table 4. The effect of dietary octapeptin supplementation on blood biochemistry in piglets.

Items	CON	MC	OP	SEM	*p*-Value
ALT, U/L	72.83 [a]	42.58 [b]	66.67 [ab]	7.06	0.02
AST, U/L	57.00	73.58	66.67	8.35	0.39
GGT, U/L	53.58	55.50	50.83	3.80	0.69
TBA, μmol/L	42.02	43.99	59.06	9.30	0.39
TP, g/L	49.15	47.84	48.20	0.90	0.58
ALB, g/L	30.77	29.33	29.03	1.22	0.53
GLB, g/L	18.38	18.51	19.11	0.71	0.74
TBil, μmol/L	3.89	3.80	3.53	1.31	0.98
BUN, mmol/L	2.64	2.91	3.09	0.37	0.69
CHO, mmol/L	2.91	2.53	2.68	0.12	0.09
TG, mmol/L	0.74	0.68	0.69	0.06	0.76

ALT: alanine aminotransferase; AST: aspartate aminotransferase; GGT: gamma-glutamyl transferase; TBA: total bile acids; TP: total protein; ALB: albumin; GLB: globulin; TBil: total bilirubin; BUN: blood urea nitrogen; CHO: cholesterol; TG: triglycerides. [a,b] Means listed in the same row with different superscripts are significantly different ($p \leq 0.05$).

Table 5. The effect of dietary octapeptin supplementation on serum immunity in piglets.

Items	CON	MC	OP	SEM	*p*-Value
IgG, mg/mL	3.62 [b]	5.99 [a]	2.29 [b]	0.61	<0.01
IgA, mg/mL	0.51 [b]	0.68 [a]	0.62 [ab]	0.05	0.04
IgM, mg/mL	17.74	18.40	15.01	2.02	0.41
IL-2, ng/L	576.16	484.72	305.5	92.17	0.13
IL-6, ng/L	32.60	43.31	24.66	6.70	0.13
IL-10, ng/L	22.27	25.01	15.57	5.62	0.42
TNF-α, ng/L	36.37 [a]	25.07 [b]	26.24 [b]	2.63	0.02

IgG: Immunoglobulin G; IgA: Immunoglobulin A; IgM: Immunoglobulin M; IL-2: Interleukin 2; IL-6: Interleukin 6; IL-10: Interleukin 10; TNF-α: Tumor Necrosis Factor Alpha. [a,b] Means listed in the same row with different superscripts are significantly different ($p \leq 0.05$).

3.4. Fecal Microbiome Analysis

The rarefaction curves from the 16S rRNA high-throughput sequencing, shown in Figure 1A, indicated that all treatment groups (CON, MC, OP) reached a plateau, demonstrating that the sample sizes were sufficient to cover most microbial diversity. The PCA plot illustrated the distribution of microbial community structures across different treatment groups, with PC1 and PC2 explaining 48.96% and 16.82% of the variance, respectively (Figure 1B). No significant differences were observed in the Ace index, Simpson index, Shannon index, or Goods coverage index among the CON, MC, and OP groups (Figure 1).

The relative abundances at the class level for all treatment groups were displayed in Figure 2A, highlighting that the top four dominant classes were *Clostridia*, *Bacilli*, *Bacteroidia*, and *Coriobacteriia*. *Clostridia* constituted the most significant portion of the fecal microbiota, with relative abundances exceeding 80%. At the genus level, 45 genera were identified across all samples after data normalization, with the top 13 genera shown in Figure 2B. Statistical analysis using the Wilcoxon rank-sum test compared the relative abundances of microbial communities at the genus level among the three experimental groups, as depicted in Figure 3.

Compared to the CON group, the MC and OP groups substantially reduced the abundance of UCG-005 ($p < 0.05$), while the MC group had a notable increase in the genus norank_f_*Erysipelotrichaceae* ($p < 0.05$), and the OP group exhibited a marked increase in the genus *Collinsella* ($p < 0.05$). Furthermore, the OP group had significantly higher abundances of *Collinsella* and *Olsenella* than the MC group ($p < 0.05$).

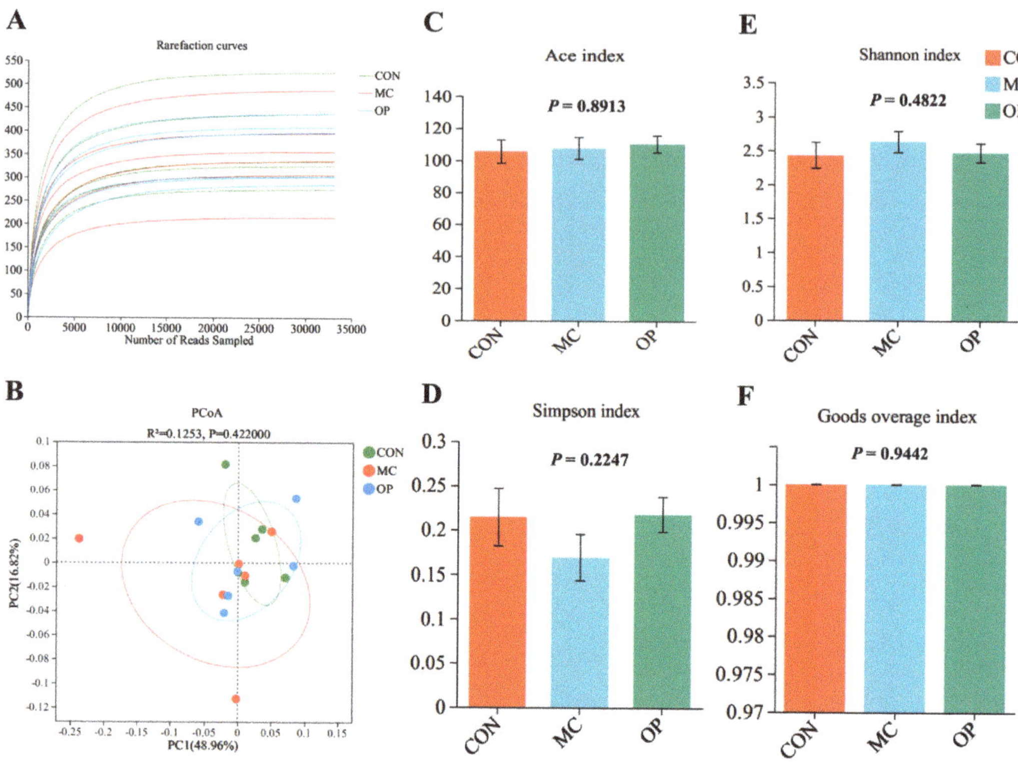

Figure 1. Analysis of fecal microbiome diversity and composition in piglets. Rarefaction curves (**A**). Principal coordinate analysis (PCoA) plot based on Bray–Curtis distances, illustrating the clustering of microbial communities (**B**). Alpha diversity of cecal flora (**C–F**). Error bars represent the standard error of the mean. Statistical significance was determined using ANOVA.

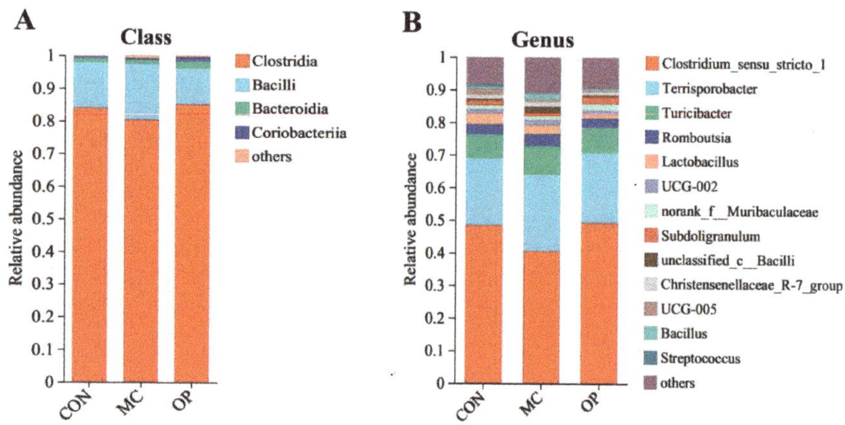

Figure 2. Relative abundance of piglet fecal microbiota at class and genus levels across different treatment groups. *Clostridia*, *Bacilli*, *Bacteroidia*, and *Coriobacteriia* are the dominant classes (**A**). Relative abundance of fecal microbiota at the genus level. The dominant genera include *Clostridium* sensu stricto 1, *Terrisporobacter*, *Turicibacter*, and others (**B**).

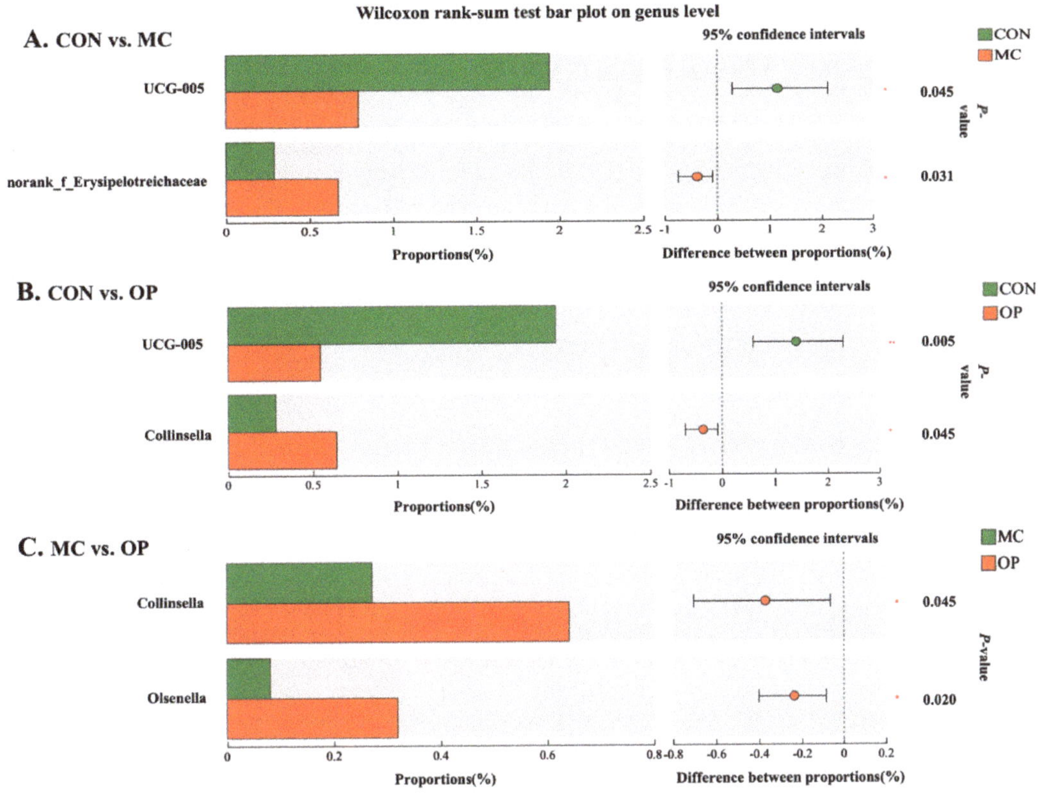

Figure 3. Wilcoxon rank-sum test bar plots of fecal microbiota at the genus level across different treatment groups. Differential genera UCG-005 ($p = 0.045$) and norank_f_*Erysipelotrichaceae* ($p = 0.031$) between CON and MC groups (**A**). Differential genera UCG-005 ($p = 0.005$) and *Collinsella* ($p = 0.045$) between CON and OP groups (**B**). Differential genera *Collinsella* ($p = 0.045$) and *Olsenella* ($p = 0.020$) between MC and OP groups (**C**).

4. Discussion

Numerous studies have demonstrated that feed additives can enhance intestinal immunity and regulate the gut microbiota in piglets [51–53], thereby mitigating the adverse effects of weaning and other environmental challenges [54,55]. Antimicrobial peptides are considered one of the best antibiotic alternatives [56]. In this study, octapeptin, an antimicrobial peptide with bactericidal and immunomodulatory activities, was evaluated for its efficacy and safety as a potential feed additive for weaned piglets.

This study suggests that octapeptin enhances the feed conversion ratio by mitigating diarrhea. It is worth noting that both the control and treatment groups experienced varying degrees of diarrhea, which is speculated to be caused by environmental changes and dietary shifts following weaning. The first two weeks post-weaning are the most stressful for piglets, as environmental and dietary changes can significantly affect their feed intake and body weight [41]. Research indicates that antimicrobial peptides can alleviate inflammation and reduce the energy expenditure associated with immune responses, which allows more nutrients to be utilized for growth, thus improving the feed conversion ratio [57]. Moreover, the observed reduction in feed intake may be linked to a satiety effect induced by changes in the gut microbiome, which can modulate appetite and feeding behavior [58]. These changes could decrease feed intake without compromising the energy available for growth. Research has shown that the supplementation of Microcin C7 can improve ATTD by increasing the availability of nutrients for intestinal absorption and inducing changes in intestinal morphology, epithelial thickness, and epithelial cell turnover, thereby enhancing feed conversion [59]. This mechanism differs from that of octapeptin, which improves feed conversion primarily by alleviating diarrhea.

Blood biochemical parameters are critical indicators of nutritional levels, endocrine status, and overall health, reflecting changes in tissue cell permeability and metabolism and indicating organ function and condition [60,61]. ALT is an enzyme primarily found in liver cells involved in protein metabolism and accelerates the conversion of amino acids in the body. Even a 1% destruction of liver cells can double serum enzyme levels, making ALT a marker enzyme for acute liver cell damage [62]. The study showed a notable reduction in the ALT level in the Microcin C7 group compared to the control group. The octapeptin group showed ALT levels that were intermediate values between the control and Microcin C7 groups, with no considerable changes in other biochemical parameters, suggesting a possible improvement in liver function without causing hepatic injury.

Cytokines play crucial roles in immune responses and inflammation, mediating susceptibility to infections and gastrointestinal disorders [63]. AMPs have been found to regulate cytokine levels. For example, Yi et al. [64] reported that feed supplemented with Cathelicidin-WA, an AMP derived from the snake *Bungarus fasciatus*, reduced serum levels of the pro-inflammatory cytokine IL-6. In the same vein, the present study observed that octapeptin markedly reduced TNF-α levels. It was also noted that IgA level significantly increased in the Microcin C7 group compared to the control group, with the octapeptin group exhibiting intermediate levels between the two. These findings suggest that octapeptin supplementation may have alleviated inflammation to some extent and helped maintain immune homeostasis in piglets. The immature intestinal immune function of newly weaned piglets at 2–4 weeks increases their disease susceptibility [44,65]. Small fluctuations in immune factors can trigger an imbalance in immune homeostasis, leading to inflammation or diarrhea [66]. Several in vitro and in vivo studies have shown that the concentration of pro-inflammatory cytokines in the small intestine of piglets usually increases after weaning [44]. High levels of IL-6 can damage tissues, while TNF-α disrupts epithelial barrier function [67]. Additionally, weaned piglets' IgA concentrations remain low until 50 days of age [68]. Intestinal inflammation is often associated with impaired intestinal barrier function, making it easier for pathogens to invade and cause diarrhea [35]. Therefore, this study preliminarily suggests that octapeptin may regulate gut immune homeostasis in weaned piglets by modulating immune factor levels, such as TNF-α. This regulation could enhance the intestinal defense mechanisms, reduce inflammatory responses, and consequently lower the incidence of diarrhea.

Weaning impairs piglets' intestinal epithelial barrier function, disrupting gut microbial balance [69]. Once this balance is disturbed, potential pathogenic bacteria can invade and colonize the intestine [70]. Weaning stress also causes a sharp reduction in feed intake, limiting nutrients for bacterial survival and proliferation [71]. *Salmonella* and enterotoxigenic *E. coli* can use ethanolamine as a carbon or nitrogen source, gaining a nutritional advantage over other microbiota [72]. Enterotoxigenic *E. coli* also uses fucose to activate the type III secretion system, promoting pathogen adhesion to host intestinal cells [73]. Consequently, weaned piglets are more prone to intestinal inflammation and post-weaning diarrhea due to rapid pathogen proliferation and loss of microbial diversity [74]. Recent studies have reported that dietary supplementation with AMPs, such as lactoferrin and lactoferrin fusion peptides, small peptide-chelated iron, AMP A5 (A3), and cecropin AD, benefits host animals by reducing pathogenic bacteria and increasing beneficial lactic acid bacteria, thereby improving the gut environment in weaned pigs by enhancing gut barrier function [27]. Our fecal microbiota alpha and beta diversity analysis showed no notable differences between groups. However, this study provides preliminary evidence that the addition of octapeptin leads to significant microbial changes, particularly the notable increase in beneficial genera *Olsenella* spp. and *Collinsella*. Based on the findings from previous analyses of growth performance, diarrhea incidence, and immune factors, we can reasonably speculate that the initial mechanism of octapeptin as a feed additive in weaning piglets lies in maintaining immune homeostasis through the regulation of immune factors, which in turn enhances gut defense, promotes the colonization of beneficial bacteria, and collectively contributes to the reinforcement and maintenance of gut health. The increase in the number of *Olsenella* spp. and *Collinsella* was positively correlated with the increase in IL-10, which may help to maintain the diversity and stability of gut microbes and reduce the colonization of harmful bacteria [75,76]. These bacteria are involved in various metabolic activities, including short-chain fatty acid production, which is crucial for gut health and barrier function [77,78]. Given that the mechanisms by which octapeptin regulates the microbiome are not fully understood, this study provides experimental evidence for octapeptin as a potential treatment for post-weaning diarrhea and offers insights for future microbiome research.

5. Conclusions

In this study, adding octapeptin to the diets of weaned piglets significantly improved feed conversion ratio and alleviated diarrhea compared with the control group. Octapeptin supplementation has the potential to improve liver function. Additionally, it significantly reduced TNF-α levels, enhancing mucosal immunity. Octapeptin positively influenced the fecal microbiota, considerably increasing the abundance of beneficial bacteria such as *Olsenella* and *Collinsella*. These findings suggest that octapeptin is a promising alternative to traditional antibiotic growth promoters in piglets, offering a natural and safe option for enhancing piglet health and growth performance.

Author Contributions: F.Y., S.Q. and J.H. conceived the study and provided resources. X.D. refined the experimental design. S.H. conducted the experiments and organized the data. L.W. and Y.C. performed the data analysis. L.Y. finalized the manuscript. All authors have read and agreed to the published version of the manuscript.

Funding: This research was supported by the National Key Research and Development Program of China (Grant No. 2023YFD1301105) and the Strategic Priority Research Program of the National Center of Technology Innovation for Pigs (NCTIP XD/B05).

Institutional Review Board Statement: The animal study protocol was approved by the Institutional Review Board of Chongqing Academy of Animal Sciences (protocol code caas-20210725 and 1 August 2021).

Informed Consent Statement: Informed consent to publish this paper was obtained from the animal's owner.

Data Availability Statement: The data presented in this study are available upon request from the corresponding author.

Acknowledgments: We would like to thank the members of the Antimicrobial Peptide Laboratory for their guidance and support during the experiments.

Conflicts of Interest: The authors declare no conflicts of interest.

References

1. Gibbons, N.E.; Moore, R.L. A Note on Artificially Infected Fowl as Carriers of Salmonella. *Poult. Sci.* **1946**, *25*, 115–118. [CrossRef]
2. Dibner, J.J.; Richards, J.D. Antibiotic growth promoters in agriculture: History and mode of action. *Poult. Sci.* **2005**, *84*, 634–643. [CrossRef]
3. Van Boeckel, T.P.; Brower, C.; Gilbert, M.; Grenfell, B.T.; Levin, S.A.; Robinson, T.P.; Teillant, A.; Laxminarayan, R. Global trends in antimicrobial use in food animals. *Proc. Natl. Acad. Sci. USA* **2015**, *112*, 5649–5654. [CrossRef]
4. Lyu, J.; Yang, L.; Zhang, L.; Ye, B.; Wang, L. Antibiotics in soil and water in China—A systematic review and source analysis. *Environ. Pollut.* **2020**, *266*, 115147. [CrossRef] [PubMed]
5. Okocha, R.C.; Olatoye, I.O.; Adedeji, O.B. Food safety impacts of antimicrobial use and their residues in aquaculture. *Public Health Rev.* **2018**, *39*, 21. [CrossRef]
6. Ashworth, D.J.; Ibekwe, A.M.; Men, Y.; Ferreira, J.F.S. Dissemination of antibiotics through the wastewater–soil–plant–earthworm continuum. *Sci. Total Environ.* **2023**, *858*, 159841. [CrossRef] [PubMed]

7. Zhang, J.; Zhang, X.; Zhou, Y.; Han, Q.; Wang, X.; Song, C.; Wang, S.; Zhao, S. Occurrence, distribution and risk assessment of antibiotics at various aquaculture stages in typical aquaculture areas surrounding the Yellow Sea. *J. Environ. Sci.* **2023**, *126*, 621–632. [CrossRef]
8. Perez-Bou, L.; Gonzalez-Martinez, A.; Gonzalez-Lopez, J.; Correa-Galeote, D. Promising bioprocesses for the efficient removal of antibiotics and antibiotic-resistance genes from urban and hospital wastewaters: Potentialities of aerobic granular systems. *Environ. Pollut.* **2024**, *342*, 123115. [CrossRef] [PubMed]
9. Casewell, M.; Friis, C.; Marco, E.; McMullin, P.; Phillips, I. The European ban on growth-promoting antibiotics and emerging consequences for human and animal health. *J. Antimicrob. Chemother.* **2003**, *52*, 159–161. [CrossRef]
10. Dewulf, J.; Sternberg-Lewerin, S.; Ryan, M. Tackling antimicrobial resistance in the food and livestock sector. In *Challenges to Tackling Antimicrobial Resistance*; Cambridge University Press: Cambridge, UK, 2020; pp. 99–124.
11. Wen, R.; Li, C.; Zhao, M.; Wang, H.; Tang, Y. Withdrawal of antibiotic growth promoters in China and its impact on the foodborne pathogen Campylobacter coli of swine origin. *Front. Microbiol.* **2022**, *13*, 1004725. [CrossRef]
12. Park, P.; Matsubara, D.K.; Barzotto, D.R.; Lima, F.S.; Chaimovich, H.; Marrink, S.J.; Cuccovia, I.M. Vesicle protrusion induced by antimicrobial peptides suggests common carpet mechanism for short antimicrobial peptides. *Sci. Rep.* **2024**, *14*, 9701. [CrossRef] [PubMed]
13. Wang, Z.; Chen, X.; Yan, L.; Wang, W.; Zheng, P.; Mohammadreza, A.; Liu, Q. Antimicrobial peptides in bone regeneration: Mechanism and potential. *Expert Opin. Biol. Ther.* **2024**, *24*, 285–304. [CrossRef] [PubMed]
14. Talapko, J.; Meštrović, T.; Juzbašić, M.; Tomas, M.; Erić, S.; Horvat Aleksijević, L.; Bekić, S.; Schwarz, D.; Matić, S.; Neuberg, M.; et al. Antimicrobial Peptides—Mechanisms of Action, Antimicrobial Effects and Clinical Applications. *Antibiotics* **2022**, *11*, 1417. [CrossRef] [PubMed]
15. Li, X.; Zuo, S.; Wang, B.; Zhang, K.; Wang, Y. Antimicrobial Mechanisms and Clinical Application Prospects of Antimicrobial Peptides. *Molecules* **2022**, *27*, 2675. [CrossRef]
16. Gan, B.H.; Gaynord, J.; Rowe, S.M.; Deingruber, T.; Spring, D.R. The multifaceted nature of antimicrobial peptides: Current synthetic chemistry approaches and future directions. *Chem. Soc. Rev.* **2021**, *50*, 7820–7880. [CrossRef] [PubMed]
17. Kovács, R.; Nagy, F.; Tóth, Z.; Bozó, A.; Balázs, B.; Majoros, L. Synergistic effect of nikkomycin Z with caspofungin and micafungin against *Candida albicans* and *Candida parapsilosis* biofilms. *Lett. Appl. Microbiol.* **2019**, *69*, 271–278. [CrossRef]
18. Buda De Cesare, G.; Cristy, S.A.; Garsin, D.A.; Lorenz, M.C. Antimicrobial Peptides: A New Frontier in Antifungal Therapy. *mBio* **2020**, *11*, e02123-20. [CrossRef]
19. Li, T.; Li, L.; Du, F.; Sun, L.; Shi, J.; Long, M.; Chen, Z. Activity and Mechanism of Action of Antifungal Peptides from Microorganisms: A Review. *Molecules* **2021**, *26*, 3438. [CrossRef]
20. Kim, J.; Cho, B.-H.; Jang, Y.-S. Understanding the Roles of Host Defense Peptides in Immune Modulation: From Antimicrobial Action to Potential as Adjuvants. *J. Microbiol. Biotechnol.* **2023**, *33*, 288–298. [CrossRef]
21. Hernández-Aristizábal, I.; Ocampo-Ibáñez, I.D. Antimicrobial Peptides with Antibacterial Activity against Vancomycin-Resistant *Staphylococcus aureus* Strains: Classification, Structures, and Mechanisms of Action. *Int. J. Mol. Sci.* **2021**, *22*, 7927. [CrossRef]
22. Rai, M.; Pandit, R.; Gaikwad, S.; Kövics, G. Antimicrobial peptides as natural bio-preservative to enhance the shelf-life of food. *J. Food Sci. Technol.* **2016**, *53*, 3381–3394. [CrossRef] [PubMed]
23. Vishweshwaraiah, Y.L.; Acharya, A.; Hegde, V.; Prakash, B. Rational design of hyperstable antibacterial peptides for food preservation. *npj Sci. Food* **2021**, *5*, 26. [CrossRef] [PubMed]
24. Liu, Q.; Yao, S.; Chen, Y.; Gao, S.; Yang, Y.; Deng, J.; Ren, Z.; Shen, L.; Cui, H.; Hu, Y.; et al. Use of antimicrobial peptides as a feed additive for juvenile goats. *Sci. Rep.* **2017**, *7*, 12254. [CrossRef] [PubMed]
25. Zhang, Y.; Meng, Z.; Li, S.; Liu, T.; Song, J.; Li, J.; Zhang, X. Two Antimicrobial Peptides Derived from Bacillus and Their Properties. *Molecules* **2023**, *28*, 7899. [CrossRef]
26. Silveira, R.F.; Roque-Borda, C.A.; Vicente, E.F. Antimicrobial peptides as a feed additive alternative to animal production, food safety and public health implications: An overview. *Anim. Nutr.* **2021**, *7*, 896–904. [CrossRef]
27. Vanrolleghem, W.; Tanghe, S.; Verstringe, S.; Bruggeman, G.; Papadopoulos, D.; Trevisi, P.; Zentek, J.; Sarrazin, S.; Dewulf, J. Potential dietary feed additives with antibacterial effects and their impact on performance of weaned piglets: A meta-analysis. *Vet. J.* **2019**, *249*, 24–32. [CrossRef]
28. Shang, L.; Zhou, J.; Tu, J.; Zeng, X.; Qiao, S. Evaluation of Effectiveness and Safety of Microcin C7 in Weaned Piglets. *Animals* **2022**, *12*, 3267. [CrossRef] [PubMed]
29. Yu, H.T.; Ding, X.L.; Li, N.; Zhang, X.Y.; Zeng, X.F.; Wang, S.; Liu, H.B.; Wang, Y.M.; Jia, H.M.; Qiao, S.Y. Dietary supplemented antimicrobial peptide microcin J25 improves the growth performance, apparent total tract digestibility, fecal microbiota, and intestinal barrier function of weaned pigs. *J. Anim. Sci.* **2017**, *95*, 5064–5076. [CrossRef]
30. Shao, Y.; Wang, X.; Li, F.; Ma, S.; Li, J.; He, J.; Jiang, Y.; Cui, W.; Wang, L.; Qiao, X.; et al. Recombinant Enterococcus faecium expressing porcine lactoferricin exerts bactericidal effects and protects against enterotoxigenic *Escherichia coli* in mice. *Process Biochem.* **2022**, *116*, 94–107. [CrossRef]
31. Zhang, L.; Guo, D.; Liu, Y.; Shao, Y.; Wang, Y.; Xu, Y.; Jiang, Y.; Cui, W.; Li, Y.; Tang, L. Probiotic Lactobacillus casei expressing porcine antimicrobial peptide PR39 elevates antibacterial activity in the gastrointestinal tract. *Can. J. Microbiol.* **2016**, *62*, 961–969. [CrossRef]

32. Lin, Q.; Su, G.; Wu, A.; Chen, D.; Yu, B.; Huang, Z.; Luo, Y.; Mao, X.; Zheng, P.; Yu, J.; et al. *Bombyx mori* gloverin A2 alleviates enterotoxigenic *Escherichia coli*-induced inflammation and intestinal mucosa disruption. *Antimicrob. Resist. Infect. Control* **2019**, *8*, 189. [CrossRef] [PubMed]
33. Li, W.; Zeng, Z.; Zhou, D.; Wang, G.; Wang, Z.; Li, Y.; Han, Y.; Qin, M.; Luo, C.; Feng, S.; et al. Effect of oral administration of microcin Y on growth performance, intestinal barrier function and gut microbiota of chicks challenged with Salmonella Pullorum. *Vet. Res.* **2024**, *55*, 66. [CrossRef]
34. Yuan, W.; Jin, H.T.; Ren, Z.H.; Deng, J.L.; Zuo, Z.C.; Wang, Y.; Deng, H.D.; Deng, Y.T. Effects of antibacterial peptide on humoral immunity in weaned piglets. *Food Agric. Immunol.* **2015**, *26*, 682–689. [CrossRef]
35. Ren, Z.H.; Yuan, W.; Deng, H.D.; Deng, J.L.; Dan, Q.X.; Jin, H.T.; Tian, C.L.; Peng, X.; Liang, Z.; Gao, S.; et al. Effects of antibacterial peptide on cellular immunity in weaned piglets. *J. Anim. Sci.* **2015**, *93*, 127–134. [CrossRef] [PubMed]
36. Velkov, T.; Roberts, K.D.; Li, J. Rediscovering the octapeptins. *Nat. Prod. Rep.* **2017**, *34*, 295–309. [CrossRef]
37. Blaskovich, M.A.T.; Pitt, M.E.; Elliott, A.G.; Cooper, M.A. Can octapeptin antibiotics combat extensively drug-resistant (XDR) bacteria? *Expert Rev. Anti Infect. Ther.* **2018**, *16*, 485–499. [CrossRef]
38. Velkov, T.; Gallardo-Godoy, A.; Swarbrick, J.D.; Blaskovich, M.A.T.; Elliott, A.G.; Han, M.; Thompson, P.E.; Roberts, K.D.; Huang, J.X.; Becker, B.; et al. Structure, Function, and Biosynthetic Origin of Octapeptin Antibiotics Active against Extensively Drug-Resistant Gram-Negative Bacteria. *Cell Chem. Biol.* **2018**, *25*, 380–391.e5. [CrossRef]
39. Chitty, J.L.; Butler, M.S.; Suboh, A.; Edwards, D.J.; Cooper, M.A.; Fraser, J.A.; Robertson, A.A.B. Antimicrobial Octapeptin C4 Analogues Active against Cryptococcus Species. *Antimicrob. Agents Chemother.* **2018**, *62*, e00986-17. [CrossRef]
40. Yu, C.H.; Chen, C.Y.; Chang, C.C. The immediate effects of weaning stress on the hypothalamus-pituitary-adrenal alteration of newly weaned piglets. *J. Anim. Physiol. Anim. Nutr.* **2019**, *103*, 1218–1223. [CrossRef]
41. Tang, X.; Xiong, K.; Fang, R.; Li, M. Weaning stress and intestinal health of piglets: A review. *Front. Immunol.* **2022**, *13*, 1042778. [CrossRef]
42. Jiang, X.; Lu, N.; Zhao, H.; Yuan, H.; Xia, D.; Lei, H. The Microbiome–Metabolome Response in the Colon of Piglets under the Status of Weaning Stress. *Front. Microbiol.* **2020**, *11*, 2055. [CrossRef] [PubMed]
43. Hu, J.; Nie, Y.; Chen, J.; Zhang, Y.; Wang, Z.; Fan, Q.; Yan, X. Gradual Changes of Gut Microbiota in Weaned Miniature Piglets. *Front. Microbiol.* **2016**, *7*, 1727. [CrossRef] [PubMed]
44. Saladrigas-García, M.; D'Angelo, M.; Ko, H.L.; Nolis, P.; Ramayo-Caldas, Y.; Folch, J.M.; Llonch, P.; Solà-Oriol, D.; Pérez, J.F.; Martín-Orúe, S.M. Understanding host-microbiota interactions in the commercial piglet around weaning. *Sci. Rep.* **2021**, *11*, 23488. [CrossRef]
45. Xiong, X.; Tan, B.; Song, M.; Ji, P.; Kim, K.; Yin, Y.; Liu, Y. Nutritional Intervention for the Intestinal Development and Health of Weaned Pigs. *Front Vet. Sci.* **2019**, *6*, 46. [CrossRef] [PubMed]
46. Otvos, L. Immunomodulatory effects of anti-microbial peptides. *Acta Microbiol. Et Immunol. Hung.* **2016**, *63*, 257–277. [CrossRef]
47. Wu, D.; Fu, L.; Wen, W.; Dong, N. The dual antimicrobial and immunomodulatory roles of host defense peptides and their applications in animal production. *J. Anim. Sci. Biotechnol.* **2022**, *13*, 141. [CrossRef]
48. Márton, R.A.; Sebők, C.; Mackei, M.; Tráj, P.; Vörösházi, J.; Kemény, Á.; Neogrády, Z.; Mátis, G. Pap12-6: A host defense peptide with potent immunomodulatory activity in a chicken hepatic cell culture. *PLoS ONE* **2024**, *19*, e0302913. [CrossRef]
49. Ran, R.; Zeng, H.; Zhao, D.; Liu, R.; Xu, X. The Novel Property of Heptapeptide of Microcin C7 in Affecting the Cell Growth of *Escherichia coli*. *Molecules* **2017**, *22*, 432. [CrossRef]
50. Ma, L.; Zhou, B.; Liu, H.; Chen, S.; Zhang, J.; Wang, T.; Wang, C. Dietary rutin improves the antidiarrheal capacity of weaned piglets by improving intestinal barrier function, antioxidant capacity and cecal microbiota composition. *J. Sci. Food Agric.* **2024**, *104*, 6262–6275. [CrossRef]
51. Bampidis, V.; Azimonti, G.; Bastos, M.L.; Christensen, H.; Durjava, M.; Dusemund, B.; Kouba, M.; López-Alonso, M.; López Puente, S.; Marcon, F.; et al. Efficacy of a feed additive consisting of ferrous lysinate sulfate for all animal species (Phytobiotis Futterzusatzstoffe GmbH). *EFSA J.* **2024**, *22*, e8789. [CrossRef]
52. Bampidis, V.; Azimonti, G.; Bastos, M.L.; Christensen, H.; Durjava, M.; Dusemund, B.; Kouba, M.; López-Alonso, M.; López Puente, S.; Marcon, F.; et al. Safety and efficacy of a feed additive consisting of *Macleaya cordata* (Willd.) R. Br. extract and leaves (Sangrovit® Extra) for suckling and weaned piglets and other growing Suidae (Phytobiotics Futterzusatzstoffe GmbH). *EFSA J.* **2024**, *22*, e8793. [CrossRef] [PubMed]
53. Bampidis, V.; Azimonti, G.; Bastos, M.L.; Christensen, H.; Durjava, M.; Dusemund, B.; Kouba, M.; López-Alonso, M.; López Puente, S.; Marcon, F.; et al. Efficacy of a feed additive consisting of endo-1,4-beta-xylanase and endo-1,3(4)-beta-glucanase produced with *Talaromyces versatilis* IMI 378536 and DSM 26702 (ROVABIO® ADVANCE) for weaned piglets (Adisseo France SAS). *EFSA J.* **2024**, *22*, e8795. [CrossRef]
54. Bampidis, V.; Azimonti, G.; Bastos, M.L.; Christensen, H.; Dusemund, B.; Durjava, M.; Kouba, M.; López-Alonso, M.; López Puente, S.; Marcon, F.; et al. Safety and efficacy of a feed additive consisting of muramidase produced with *Trichoderma reesei* DSM 32338 (Balancius™) for laying hens (DSM nutritional products). *EFSA J.* **2024**, *22*, e8788. [CrossRef]
55. Cai, L.; Zhao, Y.; Chen, W.; Li, Y.; Han, Y.; Zhang, B.; Pineda, L.; Li, X.; Jiang, X. Effect of an organic acid blend as an antibiotic alternative on growth performance, antioxidant capacity, intestinal barrier function, and fecal microbiota in weaned piglets. *J. Anim. Sci.* **2024**, *102*, skae149. [CrossRef] [PubMed]

56. Rahman, M.R.T.; Fliss, I.; Biron, E. Insights in the Development and Uses of Alternatives to Antibiotic Growth Promoters in Poultry and Swine Production. *Antibiotics* **2022**, *11*, 766. [CrossRef] [PubMed]
57. Poudel, P.; Levesque, C.L.; Samuel, R.; St-Pierre, B. Dietary inclusion of Peptiva, a peptide-based feed additive, can accelerate the maturation of the fecal bacterial microbiome in weaned pigs. *BMC Vet. Res.* **2020**, *16*, 60. [CrossRef] [PubMed]
58. Huting, A.M.S.; Middelkoop, A.; Guan, X.; Molist, F. Using Nutritional Strategies to Shape the Gastro-Intestinal Tracts of Suckling and Weaned Piglets. *Animals* **2021**, *11*, 402. [CrossRef]
59. Choi, J.Y.; Kim, J.S.; Ingale, S.L.; Kim, K.H.; Shinde, P.L.; Kwon, I.K.; Chae, B.J. Effect of potential multimicrobe probiotic product processed by high drying temperature and antibiotic on performance of weanling pigs. *J. Anim. Sci.* **2011**, *89*, 1795–1804. [CrossRef]
60. Kaneko, J.J.; Harvey, J.W.; Bruss, M.L. *Clinical Biochemistry of Domestic Animals*; Elsevier: Amsterdam, The Netherlands, 2013.
61. Zaitsev, S.Y.; Belous, A.A.; Voronina, O.A.; Rykov, R.A.; Bogolyubova, N.V. Correlations between Antioxidant and Biochemical Parameters of Blood Serum of Duroc Breed Pigs. *Animals* **2021**, *11*, 2400. [CrossRef]
62. Zong, E.; Yan, S.; Wang, M.; Yin, L.; Wang, Q.; Yin, J.; Li, J.; Li, Y.; Ding, X.; Huang, P.; et al. The effects of dietary supplementation with hyodeoxycholic acid on the differentiation and function of enteroendocrine cells and the serum biochemical indices in weaned piglets. *J. Anim. Sci.* **2019**, *97*, 1796–1805. [CrossRef]
63. Nordgreen, J.; Edwards, S.A.; Boyle, L.A.; Bolhuis, J.E.; Veit, C.; Sayyari, A.; Marin, D.E.; Dimitrov, I.; Janczak, A.M.; Valros, A. A Proposed Role for Pro-Inflammatory Cytokines in Damaging Behavior in Pigs. *Front. Vet. Sci.* **2020**, *7*, 646. [CrossRef]
64. Yi, H.; Zhang, L.; Gan, Z.; Xiong, H.; Yu, C.; Du, H.; Wang, Y. High therapeutic efficacy of Cathelicidin-WA against postweaning diarrhea via inhibiting inflammation and enhancing epithelial barrier in the intestine. *Sci. Rep.* **2016**, *6*, 25679. [CrossRef] [PubMed]
65. Zheng, L.; Duarte, M.E.; Sevarolli Loftus, A.; Kim, S.W. Intestinal Health of Pigs Upon Weaning: Challenges and Nutritional Intervention. *Front. Vet. Sci.* **2021**, *8*, 628258. [CrossRef]
66. Zhang, L.; Guo, T.; Zhan, N.; Sun, T.; Shan, A. The Effects of the antimicrobial peptide WK3 on diarrhea, growth performance and intestinal health of weaned piglets challenged with enterotoxigenic *Escherichia coli* K88. *Food Nutr. Res.* **2021**, *65*. [CrossRef] [PubMed]
67. Liu, Y. Fatty acids, inflammation and intestinal health in pigs. *J. Anim. Sci. Biotechnol.* **2015**, *6*, 41. [CrossRef] [PubMed]
68. Chang, S.Y.; Song, M.H.; Lee, J.H.; Oh, H.J.; Kim, Y.J.; An, J.W.; Go, Y.B.; Song, D.C.; Cho, H.A.; Cho, S.Y.; et al. Phytogenic feed additives alleviate pathogenic *Escherichia coli*-induced intestinal damage through improving barrier integrity and inhibiting inflammation in weaned pigs. *J. Anim. Sci. Biotechnol.* **2022**, *13*, 107. [CrossRef] [PubMed]
69. Zhao, X.; Fu, H.Y.; Qiu, S.N.; Teng, T.; Bai, G.D.; Ju, D.X.; Sun, Y.C.; Shi, B.M. Effects of early protein restriction on the growth performance and gut development of pigs fed diets with or without antibiotic. *Animal* **2019**, *14*, 1392–1401. [CrossRef]
70. Chen, L.; Collij, V.; Jaeger, M.; van den Munckhof, I.C.L.; Vich Vila, A.; Kurilshikov, A.; Gacesa, R.; Sinha, T.; Oosting, M.; Joosten, L.A.B.; et al. Gut microbial co-abundance networks show specificity in inflammatory bowel disease and obesity. *Nat. Commun.* **2020**, *11*, 4018. [CrossRef]
71. Fassarella, M.; Blaak, E.E.; Penders, J.; Nauta, A.; Smidt, H.; Zoetendal, E.G. Gut microbiome stability and resilience: Elucidating the response to perturbations in order to modulate gut health. *Gut* **2021**, *70*, 595–605. [CrossRef]
72. Thiennimitr, P.; Winter, S.E.; Winter, M.G.; Xavier, M.N.; Tolstikov, V.; Huseby, D.L.; Sterzenbach, T.; Tsolis, R.M.; Roth, J.R.; Bäumler, A.J. Intestinal inflammation allows Salmonella to use ethanolamine to compete with the microbiota. *Proc. Natl. Acad. Sci. USA* **2011**, *108*, 17480–17485. [CrossRef]
73. Garsin, D.A. Ethanolamine utilization in bacterial pathogens: Roles and regulation. *Nat. Rev. Microbiol.* **2010**, *8*, 290–295. [CrossRef] [PubMed]
74. Meng, Q.; Luo, Z.; Cao, C.; Sun, S.; Ma, Q.; Li, Z.; Shi, B.; Shan, A. Weaning Alters Intestinal Gene Expression Involved in Nutrient Metabolism by Shaping Gut Microbiota in Pigs. *Front. Microbiol.* **2020**, *11*, 694. [CrossRef] [PubMed]
75. Wang, K.; Jiang, M.; Chen, Y.; Huang, Y.; Cheng, Z.; Datsomor, O.; Jama, S.M.; Zhu, L.; Li, Y.; Zhao, G.; et al. Changes in the rumen development, rumen fermentation, and rumen microbiota community in weaned calves during steviol glycosides treatment. *Front. Microbiol.* **2024**, *15*, 1395665. [CrossRef]
76. Chen, Q.; Yin, Q.; Xie, Q.; Jiang, C.; Zhou, L.; Liu, J.; Li, B.; Jiang, S. 2′-Fucosyllactose Promotes the Production of Short-Chain Fatty Acids and Improves Immune Function in Human-Microbiota-Associated Mice by Regulating Gut Microbiota. *J. Agric. Food Chem.* **2022**, *70*, 13615–13625. [CrossRef] [PubMed]
77. Guevarra, R.B.; Hong, S.H.; Cho, J.H.; Kim, B.R.; Shin, J.; Lee, J.H.; Kang, B.N.; Kim, Y.H.; Wattanaphansak, S.; Isaacson, R.E.; et al. The dynamics of the piglet gut microbiome during the weaning transition in association with health and nutrition. *J. Anim. Sci. Biotechnol.* **2018**, *9*, 54. [CrossRef]
78. Yu, S.J.; Morris, A.; Kayal, A.; Milošević, I.; Van, T.T.H.; Bajagai, Y.S.; Stanley, D. Pioneering gut health improvements in piglets with phytogenic feed additives. *Appl. Microbiol. Biotechnol.* **2024**, *108*, 142. [CrossRef]

animals

MDPI

Article

Study of the Effects of Condensed Tannin Additives on the Health and Growth Performance of Early-Weaned Piglets

Min Ma [1,2], Yuriko Enomoto [1], Tomotsugu Takahashi [1], Kazuyuki Uchida [3], James K. Chambers [3], Yuki Goda [4], Daisuke Yamanaka [5], Shin-Ichiro Takahashi [4], Masayoshi Kuwahara [2] and Junyou Li [1,*]

[1] Animal Resource Science Center, Graduate School of Agricultural and Life Sciences, The University of Tokyo, Kasama 319-0206, Japan; aminma@g.ecc.u-tokyo.ac.jp (M.M.); aenoyan@g.ecc.u-tokyo.ac.jp (Y.E.); tomtaka@g.ecc.u-tokyo.ac.jp (T.T.)

[2] Veterinary Pathophysiology and Animal Health, Graduate School of Agriculture and Life Science, The University of Tokyo, Tokyo 113-8654, Japan; akuwam@g.ecc.u-tokyo.ac.jp

[3] Laboratory of Veterinary Pathology, Graduate School of Agriculture and Life Science, The University of Tokyo, Tokyo 113-8654, Japan; auchidak@mail.ecc.u-tokyo.ac.jp (K.U.); achamber@mail.ecc.u-tokyo.ac.jp (J.K.C.)

[4] Laboratory of Cell Regulation, Graduate School of Agriculture and Life Science, The University of Tokyo, Tokyo 113-8654, Japan; leonardo.fibonacci11235813@docomo.ne.jp (Y.G.); atkshin@mail.ecc.u-tokyo.ac.jp (S.-I.T.)

[5] Laboratory of Food and Physiological Models, Graduate School of Agriculture and Life Science, The tUniversity of Tokyo, Kasama 113-8654, Japan; adyama@mail.ecc.u-tokyo.ac.jp

* Correspondence: ajunyou@g.ecc.u-tokyo.ac.jp; Tel.: +81-299-45-2606; Fax: +81-299-45-5950

Simple Summary: Tannins, astringent polyphenols found in plants, show potential as a natural antibiotic to act as a substitute for antibiotics in animal feed additives. For ruminants, it has been widely documented that the addition of tannins to feed improves feed efficiency by increasing the amount of bypass protein. However, for monogastric animals, tannins are widely recognized as antinutritional factors, and, in some regions, it is still common practice to remove tannins from feed, e.g., through silage. The results of our previous study showed that the addition of 0.2% and 0.3% MGM-P (MGM is a commercial brand of tannin), especially the 0.3% addition, provided preventative effects regarding the incidence of diarrhea in early-weaned piglets. It has also been shown to have the ability to improve villus morphology and alleviate piglet diarrhea. Therefore, this study evaluated the effectiveness of higher doses of tannin extract MGM-P (0.5% and 1.0%) in preventing diarrhea and improving the growth performance of weaned piglets. Comparisons were also made with antibiotic additives. The results suggest that an addition level of 0.5% shows potential as an alternative to the use of antibiotics in monogastric animal feed.

Abstract: Using 0.5% and 1.0% MGM-P, the objective of the present study was to determine a more appropriate additive level for early-weaned piglets as an alternative to the use of antibiotics. Thirty-six weaned piglets were allotted to one of four groups and given a basal diet (NC), with the basal diet containing either 0.5% (LT) or 1.0% (HT) MGM-P or antibiotics (PC). Diarrhea incidence, growth performance, hematology, blood biochemistry, and blood amino acid concentrations were monitored during the experimental period. Three piglets per group with a body weight nearest to the average level were slaughtered after the experiment to assess their organ index. The results showed that no diarrhea was observed either in the treatment groups or in the control group. The 0.5% group showed an upward trend in body weight and average daily gain at all stages. The WBC counts at 21 days of age were higher ($p > 0.05$) both in the MGM-P addition groups and the LT and HT groups. For some of the plasma amino acids, such as arginine, phenylalanine concentrations were significantly lower ($p < 0.05$) in the HT group at the end of the trial. The pathological examination of all organs confirmed no differences. Consequently, the 0.5% MGM-P addition level may be suggested as a potential alternative to the use of antibiotic additives. Even with additives as high as 1%, there is no negative effect on ADG and FCR.

Keywords: quebracho tannin; early-weaned piglets; growth performance; diarrhea; free amino acids

1. Introduction

The early weaning of piglets is used to increase the reproductive efficiency of sows by maximizing the number of sows that deliver and increasing their slaughter weight each year. However, this process can result in post-weaning diarrhea (PWD), growth retardation, and even the death of early-weaned piglets, which brings huge economic losses to the pig industry [1]. Considering efficacy and cost, since the early 1950s, the use of antibiotics in the swine industry has been the most common practice worldwide [2]. Flavomycin is considered a relatively safe antibiotic because of its non-absorbability in the gastro-intestinal tract of animals and is widely used as a growth promoter in pig feed in most countries [3]. Wahlstrom et al. supplemented weaned piglets with 2 mg/kg of flavomycin in the diet and found an upward trend in average daily weight gain during both the growing and fattening periods and that high doses of the antibiotic resulted in faster growth during the fattening period [4]. However, drug-resistant microorganisms resulting from the excessive use of antibiotic additives over time have debilitated the curative effectiveness of clinically important antibiotics in human and animal medicine, threatening human health [5].

Over the past several decades, various alternatives to antibiotics and additional measures have been tried to reduce the use of antibiotics [6]. Despite the wide variety of projects under investigation, few alternatives can completely replace antibiotics in practice without posing any risk. Tannins are naturally occurring astringent polyphenols in plants with antimicrobial properties. The properties of tannins, as a natural antimicrobial, are attributed to their ability to combine with extracellular microbial enzymes to inhibit their activity [7]. This process has neither a specific target, nor do tannins have access to the inside of the cell; therefore, it is relatively difficult for this process to cause drug resistance. Nevertheless, this characteristic also renders tannins susceptible to exhibiting anti-nutritional properties through their binding to feed proteins and digestive enzymes [8]. Furthermore, they also have anti-oxidative [9] and anti-inflammatory [10] properties, which may help improve intestinal barrier function in piglets.

MGM-P is one of the condensed tannin (CT) products extracted from the heartwood of the quebracho tree (*Schinopsis lorentzii*). Compared to hydrolyzed tannins, condensed tannins are more structurally stable, which permits them to perform functions more permanently in the complex environment of the gastrointestinal tract. Su et al. [11] studied the effects of adding quebracho tannin to the diet of nursing pigs and found that the addition of tannin at the 0.1% level had no positive effect on the diarrhea incidence and growth performance of pigs. The results of our previous study showed that the addition of 0.2% and 0.3% MGM-P, especially the high addition level of 0.3% MGM-P, improved villus morphology and alleviated piglet diarrhea incidence [1]. Therefore, higher doses of MGM-P supplementation may have the potential to replace antibiotic additives. However, it should be noted that tannins can dose-dependently inhibit the utilization of feed amino acids in monogastric animals to produce antinutritional effects [12].

The aim of the present study was to evaluate higher doses of MGM-P (0.5% and 1.0%) in preventing the effects of diarrhea and improving the growth performance of weaned piglets. In our study, diarrhea incidence, growth performance, hematology, blood biochemistry, blood amino acid concentrations, and organ weights were measured.

2. Materials and Methods

The experiment described herein was conducted at the Animal Resource Science Center of the University of Tokyo (Kasama, Japan) and approved (P20-097) by the Animal Care and Use Committee of the Graduate School of Agricultural and Life Sciences at the University of Tokyo.

2.1. Materials

2.1.1. Tannin

The condensed tannins in the quebracho tannin extract, MGM-P, were more than 50% (Table 1). The product was purchased from Kawamura Co., Ltd., Tokyo, Japan.

Table 1. Technical specification of MGM-P.

Characteristic	Criterion
Polyphenols (mg catechin/g)	>500
Humidity (maximum)	15

2.1.2. Diet

The basal diet was purchased from KIMURA NOSAN SHOJI CO., LTD, Tokyo, Japan. The feed complied with National Research Council standards [13]. The ingredients (Table 2) and chemical composition (Table 3) were the same as those used in our previous publication [1].

Table 2. Ingredients of basal diet (as-fed basis) [1].

Ingredient	Content (%)
Corn	34.45
Defatted milk powder	18.00
Fatty powder	6.20
Sugar	10.00
Soybean meal	25.00
Fish meal	4.50
Calcium diphosphate	0.20
Calcium carbonate	0.65
Salt	0.20
B vitamins	0.15
Vitamins A, D, and E	0.10
Trace minerals	0.15
L-lysine hydrochloride	0.06
DL-methionine	0.09
L-threonine	0.03
Copper sulphate	0.21
Vitamin K3	0.01
Total	100

[1] The other diets were based on this diet.

Table 3. Chemical composition of basal diet.

Chemical Composition	Content (%)	Amino Acid	Content (%)
DM	90.50	Contained	
CP	22.60	Arginine	1.32
EE	6.60	Histidine	0.63
CF	1.10	Isoleucine	0.99
Ash	5.60	Leucine	2.03
NFE	54.60	Lysin	1.56
DE (Mcal/kg)	3.70	Methionine + cysteine	0.83
Ca	0.81	Phenylalanine + tyrosine	1.92
NpP	0.45	Threonine	0.96
Na	0.26	Tryptophan	0.28
Cl	0.36	Valine	1.15
K	0.99	Digestible	
Mg	0.14	Arginine	1.22
Fe (mg/kg)	182.18	Histidine	0.58
Zn (mg/kg)	105.32	Isoleucine	0.88

Table 3. *Cont.*

Chemical Composition	Content (%)	Amino Acid	Content (%)
Mn (mg/kg)	87.51	Leucine	1.83
Cu (mg/kg)	125.29	Lysin	1.42
I (mg/kg)	1.95	Methionine + cysteine	0.74
Se (mg/kg)	0.30	Phenylalanine + tyrosine	1.47
Vitamin A (IU/kg)	100,051.62	Threonine	0.85
Vitamin D (IU/kg)	2000	Tryptophan	0.25
Vitamin E (IU/kg)	20.04	Valine	1.01
Vitamin K (IU/kg)	0.57		
Thiamine (mg/kg)	5.15		
Riboflavin (mg/kg)	15.38		
Pantothenic acid (mg/kg)	27.83		
Nicotinic acid (mg/kg)	25.63		
Vitamin B6 (mg/kg)	5.93		
Choline (mg/kg)	1204.8		
Vitamin B12 (μg/kg)	21.88		
Biotin (mg/kg)	0.16		
Folic acid (mg/kg)	0.36		

Abbreviations: DM—dry matter; CP—crude protein; EE—ether extract; CF—crude fiber; NFE—nitrogen-free extract; DE—digestible energy.

2.1.3. Animals and Experimental Design

Four gestating specific-pathogen-free sows were purchased from Nakamura Chikusan (Ibaraki, Japan) to obtain piglets (Duroc × Landrace × Yorkshire) for this study. All piglets were measured by birth weight and numbered. The male piglets were castrated at 14 days of age and all piglets were introduced to the basal diet to acclimatize. At 21 days of age, the 36 piglets were weaned and divided into four groups according to body weight and sex using the Experimental Animal Allotment Program in accordance with the method established by Kim and Lindemann [14] (Table 4). The negative control (NC) group received only the basal diet; the low-dose treatment (LT) and high-dose treatment (HT) groups received the basal diet with 5 g/kg and 10 g/kg MGM-P, respectively. The positive control (PC) group received the basal diet with 0.1 g/kg flavomycin^{80} (Huvepharma Japan Inc., Kyoto, Japan).

Table 4. Experimental animal allotment.

Group	NC	LT	HT	PC
Body weight	5.86	5.84	5.85	5.82
Number of piglets	9/(3pens)	9/(3pens)	9/(3pens)	9/(3pens)
SEM	0.42	0.44	0.49	0.42
p-value		1.00		
CV (%)		0.31		

Abbreviations: n—number of piglets; SEM—standard error of the mean; CV—coefficient of variation.

One pen fed three piglets, and each treatment comprised three pens. The piglets in each pen were close in body weight. Each pen was equipped with a feed trough and a drinking bowl with a valve for ad libitum access to food and water. The experimental period was 21 days.

2.2. Methods

2.2.1. Diarrhea Manifestations

Feces were observed twice daily at 9:00 a.m. and 3:00 p.m. The occurrence of diarrhea was determined when sloppy feces were found on two or more consecutive days.

2.2.2. Growth Performance

The piglets were weighed and feed consumption was recorded at the same time before the experiment (d0) and on days 7, 14, and 21 of the experiment. Average daily feed intake (ADFI) and feed conversion rate (FCR) were calculated.

2.2.3. Blood Sampling

Blood was collected from all 36 piglets from the jugular vein during weighing on days 0, 7, 14, and 21. A 21-gauge needle (VENOJECT II; Terumo, Tokyo, Japan) was used to harvest blood for storage in 5 mL collection tubes containing EDTA-Na.

2.2.4. Blood Hematology Analysis

Hematology analyses, including white blood cell (WBC), lymphocyte, neutrophil, red blood cell, and platelet counts, were performed using a pocH-100iV Diff hematology analyzer (Sysmex Corp., Kobe, Japan).

2.2.5. Plasma Collection and Biochemical Examination

After hematological analysis, the collected blood was centrifuged for 20 min (3000 rpm) at 4 °C to obtain plasma and then analyzed to determine biochemical parameters including glutamic pyruvic transaminase (GPT), glutamic oxaloacetic transaminase (GOT), gamma-glutamyl transferase (GGT), ammonia (NH_3), blood urea nitrogen (BUN), amylase (AMYL), glucose (GLU), total protein (TP), and triglyceride (TG) using an automatic dry-chemistry analyzer (DRI-CHEM 3500s; Fujifilm, Tokyo, Japan).

2.2.6. Plasma-Free Amino Acids

To prevent changes in the concentration of free amino acids in the collected blood, deproteinization was performed immediately after measuring blood biochemistry [15]. The blood was then stored at −80 °C until further analysis. In total, 20 amino acids were tested in the experiment, including 10 essential amino acids (arginine, histidine, isoleucine, leucine, lysine, methionine, phenylalanine, threonine, tryptophan, and valine), 3 semi-essential amino acids (cysteine, tyrosine, and glutamine), and 7 non-essential amino acids (aspartic acid, serine, alanine, glycine, glutamic acid, proline, and asparagine). The analysis was carried out using the LC/MS/MS Method Package for Primary Metabolites version 2.0 (Shimadzu, Kyoto, Japan) with a Shimadzu LCMS-8030 system.

2.2.7. Actual and Relative Weights/Lengths of Organs and Intestines

One medium-weight piglet per pen, with a total of three piglets in each treatment group, was selected after the feeding trial and sacrificed following the induction of deep anesthesia via thiopental sodium (Ravonal 0.5 g; Mitsubishi Tanabe Pharma, Osaka, Japan) injection into the jugular vein. Necropsies were performed, and the piglets' organs (liver, pancreas, spleen, kidney, stomach, small intestine, and large intestine) were carefully removed. The weights of all organs and the lengths of the intestinal sections were measured. Relative organ weight was calculated as organ weight divided by BW (%), and the relative length of the intestinal section to piglet BW was also calculated (cm/kg).

2.2.8. Statistical Analysis

Data analysis was performed using JMP Pro software (version 15.2.0, SAS Institute Inc., Cary, NC, USA). One-way analysis of variance was used to compare differences among the experimental groups. When the p-value from the analysis of variance was <0.05, pairwise differences were assessed using the Tukey–Kramer honestly significant difference test. The results of the experiment are presented as the means ± standard errors of the mean.

3. Results

3.1. Diarrhea

During the experimental period, no PWD was confirmed in piglets in all of the treatment groups, including the LT, HT, and PC groups and even the control group. Mushy feces were occasionally observed in the HT group, which was easily rectified in a short period of time, and no contagion was observed in the same pen.

3.2. Growth Performance

The final body weight changes are shown in Table 5. The piglets' final average body weight in the control group was 17.68 kg; in comparison, in the LT group, this figure was 18.25; in the HT group, it was 17.35; and in the PC group, it was 17.77. The upward trend in body weight in the 0.5% addition group, however, was not significant. ADG also showed an upward trend in the LT group at 0.59; in comparison, in the NC group, it was 0.56; in the HT group, it was 0.55; and in the PC group, it was 0.57. The ADFI was 0.74 in the NC group, 0.74 in the LT group, 0.73 in the HT group, and 0.74 in the PC group.

Table 5. Effects of MGM-P supplementation on the growth performance of weaned piglets.

Measurement	NC	LT	HT	PC	p-Value
Initial weight (kg)	5.86 ± 0.42	5.84 ± 0.44	5.85 ± 0.49	5.82 ± 0.42	1.00
Final weight (kg)	17.68 ± 1.24	18.25 ± 1.1	17.35 ± 1.11	17.77 ± 0.72	0.97
ADG (kg/d)	0.56 ± 0.04	0.59 ± 0.03	0.55 ± 0.03	0.57 ± 0.02	0.80
ADFI (kg/d)	0.74 ± 0.05	0.74 ± 0.04	0.73 ± 0.06	0.74 ± 0.05	1.00
FCR (kg/kg)	1.31 ± 0.01	1.26 ± 0.02	1.33 ± 0.04	1.29 ± 0.03	0.34

Abbreviations: ADG—average daily gain; ADFI—average daily feed intake; FCR—feed conversion ratio. Data on the piglets' weight and ADG are expressed as the mean ± SEM (n = 9); data on ADFI and FCR are expressed as the mean ± SEM (n = 3). There were no statistically significant differences among the four groups based on the results of the one-way analysis of variance.

The weekly body weight changes are shown in Figure 1. During the experimental period, all piglets remained healthy, and no fatalities occurred. Although the changes in the LT group's body weight showed an upward trend, this trend was not significant.

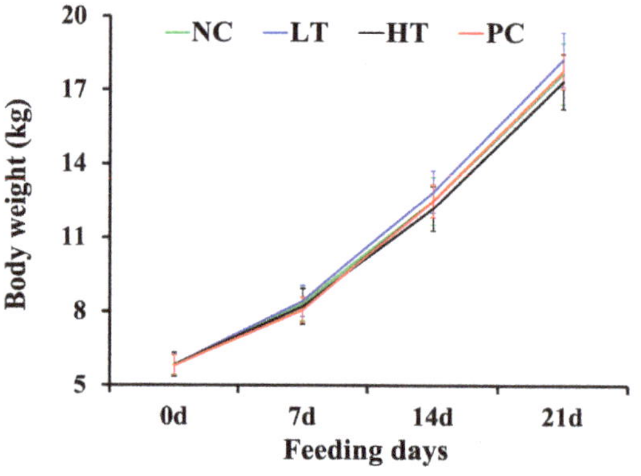

Figure 1. Effects of MGM-P supplementation on the body weight of the weaned piglets. Values are expressed as the mean ± SEM; n = 9. There were no statistically significant differences among the four groups based on the results of the one-way analysis of variance.

3.3. Blood Hematology Analysis

Information on the changes in blood hematology in the piglets after weaning can be observed in Table 6. There were no significant differences in blood hematology counts;

however, a higher tendency of WBC and lymphocyte and neutrophil counts was observed in the CT additive groups and the LT and HT groups at the end of the experiment ($p > 0.05$).

Table 6. Effects of MGM-P supplementation on blood hematology parameters in the weaned piglets.

	Measurement	NC	LT	HT	PC	p-Value
0 d [1]						
	WBC ($\times 10^2/\mu L$)	81.8 ± 8.0	80.4 ± 7.3	75.6 ± 8.7	80.8 ± 6.9	0.95
	Lymphocyte ($\times 10^2/\mu L$)	73.6 ± 7.7	70.6 ± 6.4	66.1 ± 7.4	73.0 ± 6.5	0.88
	Neutrophil ($\times 10^2/\mu L$)	4.1 ± 0.4	4.6 ± 0.9	4.9 ± 1.1	3.6 ± 0.3	0.65
	RBC ($\times 10^4/\mu L$)	482.0 ± 38.7	500.9 ± 22.5	549.8 ± 32.1	493.7 ± 18.3	0.38
	PLT ($\times 10^4/\mu L$)	136.7 ± 20.8	148.6 ± 16.3	135.2 ± 20.0	158.7 ± 15.7	0.78
21 d						
	WBC ($\times 10^2/\mu L$)	101.1 ± 8.9	127.9 ± 15.4	124.4 ± 16.2	107.9 ± 9.2	0.40
	Lymphocyte ($\times 10^2/\mu L$)	69.7 ± 4.3	91.3 ± 9.2	87.1 ± 9.2	80.3 ± 7.8	0.22
	Neutrophil ($\times 10^2/\mu L$)	19.0 ± 4.1	21.3 ± 4.3	22.6 ± 4.5	16.6 ± 0.9	0.71
	RBC ($\times 10^4/\mu L$)	678.2 ± 16.1	704.4 ± 13.2	728.0 ± 9.8	709.8 ± 11.0	0.07
	PLT ($\times 10^4/\mu L$)	76.0 ± 6.8	81.9 ± 2.7	71.2 ± 5.7	76.3 ± 4.8	0.56

Abbreviations: WBC—white blood cells; RBC—red blood cells; PLT—platelets. All data are expressed as the mean ± SEM; $n = 9$. [1] Blood was collected before the provision of feed with MGM-P on the day of weaning. There were no statistically significant differences among the four groups based on the results of the one-way analysis of variance.

3.4. Blood Biochemical Analysis

As shown in Table 7, no significant differences were observed in indicators related to liver metabolism, including GPT, GOT, GGT, and NH_3 ($p > 0.05$). The BUN, AMYL, GLU, and TP concentrations were also not affected by the different treatments ($p > 0.05$). However, antibiotic supplementation significantly increased ($p < 0.05$) the TG concentration in the piglets' plasma at 21 d in comparison with that of the NC and LT groups.

Table 7. Effects of MGM-P supplementation on the blood biochemical parameters of the weaned piglets.

	Measurement	NC	LT	HT	PC	p-Value
0 d [1]						
	GPT (U/L)	35.6 ± 1.1	31.4 ± 1.6	34.6 ± 1.1	33.4 ± 1.3	0.14
	GOT (U/L)	31.9 ± 2.4	32.3 ± 2.1	39.8 ± 7.4	32.6 ± 2.0	0.49
	GGT (U/L)	24.6 ± 4.6	22.7 ± 2.2	20.7 ± 1.6	21.7 ± 2.1	0.80
	NH_3 ($\mu g/dL$)	121.0 ± 14.7	113.8 ± 4.7	150.2 ± 26.8	168.6 ± 40.3	0.40
	BUN (mg/dL)	6.8 ± 0.8	5.6 ± 0.3	6.0 ± 0.7	6.7 ± 0.6	0.47
	AMYL (U/L)	868.2 ± 102.2	971.2 ± 132.8	1060.2 ± 159.6	1040.7 ± 66.6	0.67
	GLU (mg/dL)	123.6 ± 2.6	123.2 ± 3.8	121.0 ± 4.0	132.7 ± 5.0	0.18
	TP (g/dL)	4.5 ± 0.2	4.8 ± 0.1	4.5 ± 0.1	4.9 ± 0.1	0.07
	TG (mg/dL)	53.4 ± 6.0	63.1 ± 13.7	47.8 ± 7.8	77.7 ± 15.4	0.29
21 d						
	GPT (U/L)	39.6 ± 1.3	36.0 ± 1.3	38.4 ± 2.2	38.3 ± 1.9	0.53
	GOT (U/L)	39.4 ± 3.7	38.6 ± 4.3	50.8 ± 13.6	51.1 ± 4.4	0.50
	GGT (U/L)	30.6 ± 3.0	32.3 ± 1.7	30.2 ± 1.5	30.4 ± 2.6	0.91
	NH_3 ($\mu g/dL$)	110.4 ± 10.4	131.6 ± 11.8	151.2 ± 32.6	153.4 ± 14.7	0.38
	BUN (mg/dL)	16.4 ± 1.4	16.8 ± 1.2	14.7 ± 1.1	17.1 ± 0.8	0.47
	AMYL (U/L)	744.2 ± 82.7	868.0 ± 116.0	1013.9 ± 124.3	824.7 ± 42.8	0.27
	GLU (mg/dL)	141.3 ± 3.2	153.7 ± 8.8	148.2 ± 11.2	143.8 ± 4.1	0.68
	TP (g/dL)	5.0 ± 0.1	5.1 ± 0.1	5.0 ± 0.1	5.1 ± 0.1	0.52
	TG (mg/dL)	16.0 ± 1.3 ^{b}	16.2 ± 1.7 ^{b}	21.8 ± 3.6 ^{ab}	35.2 ± 8.8 ^{a}	0.03

Abbreviations: GPT—glutamic pyruvic transaminase; GOT—glutamic oxaloacetic transaminase; GGT—gamma-glutamyl transferase; NH_3—ammonia; BUN—blood urea nitrogen; AMYL—amylase; GLU—glucose; TP—total protein; TG—triglyceride. All data are expressed as the mean ± SEM; $n = 9$. [1] Blood was collected before the provision of feed with MGM-P on the day of weaning. $^{a, b}$ Mean values within a row with dissimilar superscript letters are significantly different ($p < 0.05$).

3.5. Plasma-Free Amino Acids

Information on the changes in the plasma-free amino acid concentrations of the weaned pigs is shown in Table 8. Dietary 1.0% MGM-P supplementation significantly reduced ($p < 0.05$) the concentration of arginine in the piglets' blood at age 21 d compared with that of the PC group. In addition, piglets in the HT group had a significantly reduced ($p < 0.05$) phenylalanine concentration at age 21 d compared with piglets provided with a basal diet and a diet containing antibiotics.

Table 8. Effects of MGM-P supplementation on plasma-free amino acid concentrations in the weaned piglets (uM).

Measurement	NC	LT	HT	PC	*p*-Value
0 d [1]					
Asparagine	99.1 ± 7.3	85.0 ± 7.1	93.2 ± 5.3	99.9 ± 6.3	0.36
Aspartic acid	82.8 ± 2.5	80.9 ± 3.2	89.7 ± 2.6	83.4 ± 2.5	0.13
Serine	198.1 ± 7.7	176.7 ± 11.7	195.4 ± 10.6	215.2 ± 12.6	0.12
Alanine	542.4 ± 24.4	478.9 ± 39.1	502.4 ± 19.5	550.5 ± 32.1	0.29
Glycine	854.2 ± 47.3	804.0 ± 47.9	787.6 ± 35.3	854.4 ± 47.0	0.62
Glutamine	382.9 ± 20.7	336.2 ± 26.3	378.3 ± 13.0	363.8 ± 18.5	0.37
Threonine	241.8 ± 17.6	212.7 ± 18.3	234.3 ± 14.7	252.8 ± 18.0	0.42
Cysteine	3.5 ± 0.4	2.5 ± 0.5	3.6 ± 0.5	3.0 ± 0.6	0.39
Glutamic acid	102.3 ± 6.8	99.0 ± 9.4	106.2 ± 7.6	94.9 ± 6.2	0.75
Proline	264.4 ± 17.7	225.7 ± 21.2	234.0 ± 11.3	259.2 ± 21.8	0.39
Lysine	146.2 ± 11.0	117.3 ± 13.9	130.7 ± 8.6	144.9 ± 10.0	0.23
Histidine	35.3 ± 1.4 [ab]	29.0 ± 2.9 [b]	32.7 ± 1.5 [ab]	37.5 ± 1.9 [a]	0.04
Arginine	117.0 ± 8.5	91.6 ± 7.4	106.7 ± 7.0	111.2 ± 10.2	0.19
Valine	205.9 ± 13.3	171.3 ± 17.3	185.6 ± 13.1	208.7 ± 17.3	0.28
Methionine	67.5 ± 7.2	58.7 ± 6.8	66.0 ± 6.3	64.2 ± 5.1	0.78
Tyrosine	199.3 ± 16.6	151.7 ± 17.4	167.1 ± 7.8	180.2 ± 16.3	0.17
Isoleucine	108.7 ± 7.3	93.1 ± 11.2	103.0 ± 6.6	105.7 ± 9.7	0.63
Leucine	142.5 ± 9.8	119.5 ± 15.1	128.8 ± 6.2	144.1 ± 14.9	0.44
Phenylalanine	93.9 ± 7.2	77.6 ± 9.4	75.3 ± 5.7	94.7 ± 10.1	0.22
Tryptophan	58.8 ± 4.4	51.2 ± 4.9	54.2 ± 2.3	58.4 ± 4.8	0.54
7 d					
Asparagine	77.4 ± 4.1	72.0 ± 3.5	75.6 ± 5.3	75.3 ± 3.7	0.83
Aspartic acid	66.2 ± 2.7	68.1 ± 1.4	69.2 ± 2.6	70.7 ± 1.4	0.5
Serine	138.4 ± 11.8	142.2 ± 7.6	147.2 ± 16.4	145.2 ± 6.7	0.95
Alanine	362.5 ± 26.6	329.7 ± 21.1	382.8 ± 23.1	366.3 ± 21.2	0.44
Glycine	687.6 ± 43.7	778.8 ± 40.0	795.1 ± 34.4	742.9 ± 25.9	0.19
Glutamine	370.7 ± 12.0	355.1 ± 17.7	389.0 ± 22.2	374.6 ± 12.3	0.56
Threonine	268.4 ± 12.9	280.9 ± 17.6	285.6 ± 25.6	297.5 ± 10.2	0.71
Cysteine	2.5 ± 0.3	3.0 ± 0.4	2.6 ± 0.5	3.8 ± 0.5	0.18
Glutamic acid	100.6 ± 10.0	101.0 ± 8.7	104.8 ± 6.9	111.8 ± 5.6	0.74
Proline	177.9 ± 10.3	163.0 ± 6.5	173.1 ± 11.8	173.3 ± 8.4	0.72
Lysine	90.0 ± 12.1	56.3 ± 7.3	69.2 ± 17.1	62.3 ± 8.8	0.23
Histidine	25.8 ± 1.8	21.6 ± 1.3	25.7 ± 3.7	23.5 ± 1.4	0.51
Arginine	91.5 ± 9.3	83.2 ± 7.5	92.5 ± 15.2	83.5 ± 7.1	0.87
Valine	223.3 ± 9.9	205.3 ± 11.2	221.5 ± 17.5	231.9 ± 13.3	0.56
Methionine	180.5 ± 29.7	169.0 ± 23.7	156.3 ± 22.6	161.5 ± 18.2	0.9
Tyrosine	136.6 ± 12.9	126.9 ± 8.7	135.0 ± 13.5	150.2 ± 6.6	0.5
Isoleucine	123.2 ± 10.7	105.8 ± 6.8	115.8 ± 8.0	126.9 ± 6.8	0.3
Leucine	123.4 ± 8.5	104.6 ± 6.8	121.1 ± 14.0	118.0 ± 9.3	0.56
Phenylalanine	77.8 ± 5.8	61.1 ± 4.8	70.9 ± 8.2	68.2 ± 4.6	0.29
Tryptophan	45.4 ± 4.1	41.5 ± 3.3	46.9 ± 5.0	49.0 ± 2.5	0.57

Table 8. *Cont.*

Measurement	NC	LT	HT	PC	*p*-Value
14 d					
Asparagine	80.2 ± 5.4	89.1 ± 6.0	83.8 ± 3.4	84.1 ± 8.0	0.77
Aspartic acid	66.5 ± 2.3	73.4 ± 3.3	72.0 ± 1.8	65.0 ± 2.0	0.05
Serine	161.2 ± 13.3	188.3 ± 12.6	173.1 ± 10.8	169.7 ± 15.2	0.53
Alanine	322.3 ± 26.3	342.8 ± 28.7	369.2 ± 20.1	304.0 ± 28.8	0.35
Glycine	1036.6 ± 46.1	1115.9 ± 62.1	1166.7 ± 37.4	1001.7 ± 34.4	0.07
Glutamine	403.1 ± 24.9	431.4 ± 20.5	430.9 ± 18.4	370.9 ± 23.1	0.18
Threonine	315.9 ± 19.1	353.6 ± 18.3	322.2 ± 12.0	328.7 ± 25.9	0.55
Cysteine	3.6 ± 0.5	3.7 ± 0.7	2.5 ± 0.4	4.1 ± 0.5	0.2
Glutamic acid	103.7 ± 9.8	122.3 ± 16.3	116.2 ± 10.1	106.0 ± 7.1	0.62
Proline	196.3 ± 13.4	207.6 ± 12.2	201.6 ± 7.9	185.7 ± 17.6	0.69
Lysine	103.4 ± 12.2	119.6 ± 10.0	105.1 ± 9.6	113.7 ± 16.6	0.77
Histidine	32.7 ± 3.4	34.3 ± 2.4	31.3 ± 2.9	33.0 ± 3.4	0.92
Arginine	88.3 ± 9.4	105.0 ± 8.3	90.5 ± 6.7	92.6 ± 11.4	0.57
Valine	244.9 ± 14.2	263.8 ± 11.3	244.5 ± 10.8	252.1 ± 19.6	0.76
Methionine	157.7 ± 19.6	189.9 ± 27.4	144.8 ± 17.7	156.6 ± 18.9	0.49
Tyrosine	150.6 ± 12.5	162.2 ± 11.5	144.2 ± 7.3	149.0 ± 11.2	0.68
Isoleucine	119.8 ± 9.0	125.3 ± 7.5	122.4 ± 6.6	126.6 ± 11.1	0.95
Leucine	138.9 ± 15.0	146.5 ± 9.4	144.9 ± 10.8	145.3 ± 18.5	0.98
Phenylalanine	71.9 ± 5.4	77.9 ± 6.1	69.1 ± 4.8	71.5 ± 5.7	0.71
Tryptophan	59.0 ± 4.1	64.0 ± 6.4	60.7 ± 3.1	59.1 ± 3.8	0.84
21 d					
Asparagine	121.3 ± 9.6	107.7 ± 7.6	107.6 ± 9.8	122.2 ± 7.1	0.45
Aspartic acid	90.2 ± 2.6	87.1 ± 2.3	89.4 ± 2.9	88.0 ± 2.3	0.83
Serine	246.1 ± 11.8	247.2 ± 20.0	228.4 ± 25.1	278.5 ± 18.4	0.34
Alanine	412.7 ± 36.7	382.8 ± 29.9	421.3 ± 39.3	447.5 ± 17.2	0.56
Glycine	1332.0 ± 97.8	1197.4 ± 34.3	1157.9 ± 34.0	1253.1 ± 53.2	0.22
Glutamine	584.5 ± 31.7	559.9 ± 39.3	571.1 ± 26.2	570.7 ± 16.8	0.95
Threonine	534.6 ± 38.0	506.3 ± 27.2	460.3 ± 38.0	565.8 ± 33.8	0.19
Cysteine	4.2 ± 0.5 [b]	5.1 ± 0.4 [ab]	4.8 ± 0.7 [b]	7.1 ± 0.7 [a]	0.01
Glutamic acid	117.4 ± 9.3	143.4 ± 10.7	147.5 ± 12.6	135.8 ± 10.6	0.23
Proline	312.4 ± 21.4	296.3 ± 26.0	286.2 ± 29.2	300.3 ± 14.7	0.89
Lysine	181.6 ± 16.4	162.5 ± 13.2	151.3 ± 12.5	196.2 ± 12.9	0.12
Histidine	56.2 ± 3.4	52.3 ± 3.5	51.5 ± 5.9	59.7 ± 4.6	0.55
Arginine	140.2 ± 9.9 [ab]	141.3 ± 9.9 [ab]	118.6 ± 8.6 [b]	164.4 ± 13.8 [a]	0.04
Valine	428.9 ± 23.2	411.2 ± 30.9	384.4 ± 24.2	436.7 ± 14.3	0.43
Methionine	191.7 ± 27.3	192.5 ± 18.4	187.6 ± 31.4	164.6 ± 12.2	0.81
Tyrosine	208.4 ± 16.5	201.6 ± 14.6	183.8 ± 13.8	206.7 ± 10.1	0.59
Isoleucine	209.5 ± 15.5	191.1 ± 15.8	174.9 ± 11.5	218.5 ± 9.6	0.12
Leucine	253.5 ± 21.9	228.6 ± 21.9	213.7 ± 16.3	271.6 ± 12.2	0.14
Phenylalanine	123.8 ± 9.2 [a]	102.3 ± 10.1 [ab]	88.4 ± 7.4 [b]	121.5 ± 7.4 [a]	0.02
Tryptophan	99.9 ± 6.6	89.6 ± 7.4	80.7 ± 5.6	96.8 ± 5.4	0.16

All data are expressed as the mean ± SEM; $n = 9$. [1] Blood was collected before the provision of feed with MGM-P on the day of weaning. [a, b] Mean values within a row with dissimilar superscript letters are significantly different ($p < 0.05$).

3.6. *Actual and Relative Weights/Lengths of Organs and Intestines*

No abnormalities were found in the piglets' organs during necropsies performed at the end of the experiment. Information on the effect of dietary MGM-P supplementation on the relative weight or length of the organs and intestines of the piglets is presented in Table 9. The different dietary treatments had no influence on these relevant parameters of the piglet organs under examination ($p > 0.05$).

Table 9. Effects of MGM-P supplementation on the organ weight/length of the weaned piglets.

Measurement	NC	LT	HT	PC	p-Value
Organ weight/length					
Liver (g)	597.27 ± 86.60	531.93 ± 72.17	514.50 ± 87.61	536.10 ± 45.63	0.87
Pancreas (g)	38.47 ± 5.43	35.67 ± 6.24	35.53 ± 6.16	39.07 ± 5.32	0.96
Spleen (g)	41.47 ± 2.25	47.87 ± 7.60	37.63 ± 2.19	40.67 ± 3.41	0.47
Kidney (g)	135.73 ± 25.50	121.87 ± 22.44	119.10 ± 25.52	128.30 ± 6.33	0.95
Stomach (g)	104.10 ± 18.65	86.87 ± 12.66	87.90 ± 9.40	96.27 ± 10.26	0.78
Small intestine weight (g)	595.97 ± 73.98	603.53 ± 99.37	582.07 ± 36.16	545.03 ± 11.39	0.92
Small intestine length (cm)	1188.00 ± 27.26	1140.17 ± 62.98	1123.00 ± 31.47	1105.00 ± 57.36	0.65
Large intestine weight (g)	202.07 ± 20.78	185.23 ± 28.52	182.50 ± 16.20	177.07 ± 12.16	0.84
Large intestine length (cm)	219.17 ± 10.14	220.00 ± 17.24	215.00 ± 12.58	200.33 ± 10.27	0.69
Relative organ weight/length					
Liver (%)	2.88 ± 0.15	2.65 ± 0.12	2.49 ± 0.06	2.59 ± 0.10	0.18
Pancreas (%)	0.19 ± 0.01	0.18 ± 0.01	0.17 ± 0.02	0.19 ± 0.01	0.78
Spleen (%)	0.21 ± 0.02	0.24 ± 0.00	0.19 ± 0.02	0.20 ± 0.03	0.43
Kidney (%)	0.65 ± 0.03	0.60 ± 0.03	0.57 ± 0.03	0.62 ± 0.03	0.44
Stomach (%)	0.49 ± 0.03	0.43 ± 0.02	0.43 ± 0.03	0.46 ± 0.02	0.33
Small intestine weight (%)	3.10 ± 0.84	3.07 ± 0.48	2.90 ± 0.24	2.66 ± 0.17	0.92
Small intestine length (cm/kg)	59.32 ± 7.93	58.61 ± 7.72	56.61 ± 7.05	53.58 ± 1.60	0.93
Large intestine weight (%)	0.99 ± 0.08	0.92 ± 0.02	0.91 ± 0.08	0.86 ± 0.01	0.50
Large intestine length (cm/kg)	10.86 ± 1.19	11.19 ± 1.10	10.72 ± 0.94	9.86 ± 1.22	0.86

All data are expressed as the mean ± SEM; $n = 3$. There were no statistically significant differences among the four groups based on the results of the one-way analysis of variance.

4. Discussion

In the present study, no PWD was confirmed in piglets subjected to all of the treatments, including the LT, HT, and PC groups and even the control group. Yi et al. [16] reported that dietary 0.1% CT from kenwood supplementation decreased the diarrhea rate after the day of weaning to 28 days ($p < 0.05$), with no significant effect on growth performance. Su et al. [11] studied the effects of adding quebracho tannin to the feed of nursing pigs and found that the addition of tannin at the 0.1% level had no positive effect on the diarrhea incidence and growth performance of pigs. The results of the author's previous study also proved that the addition of quebracho CT reduced the incidence of diarrhea, and the results also indicated that the diarrhea reduction effect was dose-dependent and 0.3% more efficacious than that reported for the 0.2% supplement [1]. This is one reason why an experiment involving a greater addition of tannin was conducted in the present study.

One explanation for the lack of PWD being confirmed in piglets from all of the treatments could be that, in the present study, piglets were exposed and acclimatized to the solid form of the basal diet starting 7 days before weaning. This process may mitigate the stress induced by the piglets' conversion of nutrients from breast milk to solid feed. Indeed, this process mitigated piglet feed intake and digestibility caused by the nutritional effect on the intestinal mucosa villus. Secondly, the basal diet used in the present study was commercial feed, containing several anti-PWD ingredients, such as probiotics, several other types of herbal extract, and a small proportion of high-level zinc sulfate, but without any antibiotics. The final zinc content was 119.6 mg/kg. Thus, to verify the real tannin effect on post-weaning diarrhea incidence and intestinal microorganisms in early-weaned piglets, we started an additional experiment involving a basal diet without probiotics and other herbal extract content in which the zinc level was consistent with the NRC level. Another reason for this finding is the fact that the present study was conducted in a university facility, where better conditions came from the lower rearing density and clean sanitary conditions, and these conditions helped to reduce the weaning stress to which the piglets were exposed. Prescott J.F. et al. [17] indicated in their study that antibiotics can only exert their greatest effect when the animals to which they are administered are in poor health and their living conditions are unhygienic. During the experiment, some piglets occasionally showed sticky and mushy feces in the HT group, which was easily rectified within a short period of time.

Yi et al. [16] suggested in their study that a 0.1% CT addition could reduce the incidence of diarrhea to 16.7%. In the present study, sticky and mushy feces were observed in the HT (1% = 10 g CT/kg) group, although not frequently. In addition, no contagion was found in the same pen. This above phenomenon could be considered the result of indigestion. Thus, it is not known whether the occurrence of soft stools is associated with high levels of MGM-P, and the exact cause requires further confirmation.

In the present study, we found that LT treatment showed an upward trend in ADG and ADFI with CT supplementation, but HT treatment showed a downward trend in ADG and ADFI, even if this figure was not significant. Tannin has been considered an anti-nutritional factor for a long period of time, especially CT. In their study, Ortizd et al. [18] fed chicks with feed containing 8 g/kg and 16 g/kg of faba bean tannin extract, which comprises condensed tannin, and found that it significantly affected the chicks' growth performance, with 24-day weight gains of only 68% and 58% that of the normal diet group, respectively. Yi et al. [16] showed in their study that when administrating an additive of 0.1% condensed tannin, there was no significant effect on BW, ADG, ADFI, and F/G ($p > 0.05$). E. Seoni et al. [19] reported in their study that sainfoin, which contains a non-negligible amount of condensed tannin, is a suitable homegrown protein source for grower–finisher pigs and can be included at a rate of up to 15% to replace 7% soybean in a diet, without having any noteworthy effects on growth. Therefore, the CT additive level is critical in determining whether it produces antinutritional effects.

CT in several forage plants (e.g., L. *corniculatus* and *sulla*) has been shown to offer advantages for ruminants and result in increased milk production, wool growth, ovulation rate, and lambing percentage, as well as reduced bloat risk and reduced internal parasite burdens. When CT-bonded protein as a bypass protein enters the abomasum, the protein will be released and digested. Jones and Mangan [20] reported in their study that CT can bind with protein at near-neutral pH (pH 3.5–7.5) to form CT–protein complexes, which dissociate and release protein at a pH less than 3.5. Thus, in monogastric animals, whose stomach pH is usually less than 3.5, the appropriate CT additive could not increase the number of CT-bonded protein complexes high enough to affect protein digestibility.

In the present study, 0.5% MGM-P addition resulted in an upward trend in ADG and ADFI compared to the antibiotic additive group. As mentioned above, the present study was conducted in a university facility where better conditions helped to reduce the incidence of several stresses to which the piglets were exposed during weaning, meaning that antibiotics could not exert their greatest effect.

The RBC of the piglet blood obtained in our study, as with our previous study, was at a similar level to that reported by Czech et al. [21]. These results indicate that RBC is stable when CT is present. WBCs play a primary role in both fighting inflammation and clearing extracellular pathogens [22,23]. The onset of PWD is often accompanied by an increase in the number of WBCs in the blood of piglets [1,24]. In the present study, the WBC count in all of the treatments showed low levels on the weaning day, the age of 21 days, compared to those reported by Czech et al. [21]. These results prove that because all of the piglets used in the present study were in good health at the weaning stage, the white blood cell's primary role in the body's defense could not be shown. The values of the CT addition groups, both the LT and HT groups, gradually increased to a range of normal values consistent with our previous research results. However, the WBC count in the NC and PC groups showed a downward trend after weaning. The above results somehow suggest that tannin treatment affects WBC levels, and, thus, this effect requires further examination.

Due to the relative lack of studies on the administration of CT additives to monogastric animals, in the present study, a relatively high additive level of 1% was used in order to determine the liver cell injury parameters of GOT, GPT, and GGT. The results in this regard did not show any anomalies, thus indicating that a supplement level as high as 1.0% is acceptable for animals.

No differences in the piglets' blood ammonia and urea nitrogen concentrations were confirmed. This result indicates that protein digestion and the functions of the liver and kidneys were not affected by CT addition. Ye et al. [25] found in their study that the addition of 50 mg/kg of flavomycin to the feed of Hy-Line Brown chickens resulted in elevated plasma triglyceride levels. Similar results were observed in the PC group at 21 d; however, levels in the chickens included in this particular group were not affected by CT addition.

Regarding the plasma amino acid concentration at the end of the experiment, at the age of 21 days after weaning, the HT group showed significantly lower arginine and phenylalanine levels. Mariscal-Landín et al. [12] evaluated the effect of different tannin levels on the coefficient of apparent ileal digestibility of sorghum amino acids in growing pigs and found that tannin levels of up to 1.05% did not affect the digestion of arginine, whereas tannin levels of 4% or more inhibited its digestion. Arginine is one of the factors linked to growth hormone release in young children through the somatotropic axis and, if deficient, may affect early-stage growth [26]. In another study, when broiler diets were supplemented with 0.5%, 1.5%, 2%, and 2.5% mimosa tannins, 2.5% supplementation caused a significant decrease in the ileal digestibility of phenylalanine compared to the basal diet [27]. Phenylalanine is also necessary for the sufficient growth of weaned piglets [28,29]. Consequently, the weaker body weight gain observed in the HT group may have been related to the lower concentration of these two amino acids in the piglets.

In their study, Wang et al. [30] added 0, 0.05, 0.1, and 0.15% tannic acid and antibiotics to the diets of 21-day-old weaned piglets and found that neither tannic acid nor antibiotics had any effect on the piglets' relative organ weight. The above results are consistent with those of our previous study [1] and the present work. Thus far, few studies have been conducted on the effect of high-level condensed tannin addition on the relative organ weight of piglets. The present research results show that the addition of 1.0% MGM-P still has no effect on the development of organs, and the pathological features of organs were not observed during dissection.

To summarize our study, the effectiveness of antibiotic additives is diminished under current feeding conditions. Supplementation of 0.5% MGM-P in piglet feed is expected to replace antibiotics. In the central role of tannins as antimicrobials as an alternative to antibiotics, it is necessary to investigate the effect of tannins on the intestinal microflora of piglets in future research.

5. Conclusions

In conclusion, according to the results of present research on MGM-P supplementation, there is a tendency to increase the nADG and FCR of piglets when the additive level is 0.5%, especially without antinutritional effects and anemia. Even with additives as high as 1%, there is no negative effect on ADG and FCR. The results on growth and health imply that the use of 0.5% MGM-P in early-weaned piglet diets has the potential to replace antibiotic additives.

Author Contributions: M.M., Y.E., T.T. and J.L. performed the animal and laboratory experiments and collected the data. M.K., J.K.C., K.U., Y.G., D.Y. and S.-I.T. provided methodical and technical support. M.M. analyzed the data and drafted the manuscript. J.L. supervised and guided the experiment and revised the manuscript. All authors have read and agreed to the published version of the manuscript.

Funding: This research was financially supported by Kawamura Ltd. Tokyo, Japan (nou31-77), and the Japan Racing and Livestock Promotion Foundation (JRL5, 242).

Institutional Review Board Statement: All animal care and handling procedures conformed to the Guidelines for Animal Experiments of the University of Tokyo and were approved by the Animal Care and Use Committee of Life Science, Faculty of Agriculture, University of Tokyo (approval number: P20-097).

Informed Consent Statement: Not applicable.

Data Availability Statement: The data that support the findings of the present study are available from the corresponding author upon reasonable request.

Acknowledgments: The authors wish to thank Huvepharma Japan Inc. for the gratuitous donation of flavomycin.

Conflicts of Interest: The authors have no conflicts of interest directly relevant to the content of the above article.

References

1. Ma, M.; Chambers, J.K.; Uchida, K.; Ikeda, M.; Watanabe, M.; Goda, Y.; Yamanaka, D.; Takahashi, S.-I.; Kuwahara, M.; Li, J. Effects of Supplementation with a Quebracho Tannin Product as an Alternative to Antibiotics on Growth Performance, Diarrhea, and Overall Health in Early-Weaned Piglets. *Animals* **2021**, *11*, 3316. [CrossRef] [PubMed]
2. Dibner, J.J.; Richards, J.D. Antibiotic Growth Promoters in Agriculture: History and Mode of Action. *Poult. Sci.* **2005**, *84*, 634–643. [CrossRef] [PubMed]
3. Ochetim, S.; Odur, I.D. Response of Young Pigs to Flavomycin. *East Afr. Agric. For. J.* **1979**, *44*, 31–33. [CrossRef]
4. Wahlstrom, R.C.; Libal, G.W. *Effect of Flavomycin on Performance of Growing-Finishing Pigs*; South Dakota Swine Field Day Proceedings and Research Reports; 1972. Available online: https://openprairie.sdstate.edu/sd_swine_1972/4 (accessed on 1 August 2024).
5. Gong, J.; Yin, F.; Hou, Y.; Yin, Y. Review: Chinese Herbs as Alternatives to Antibiotics in Feed for Swine and Poultry Production: Potential and Challenges in Application. *Can. J. Anim. Sci.* **2013**, *94*, 223–241. [CrossRef]
6. Callaway, T.R.; Lillehoj, H.; Chuanchuen, R.; Gay, C.G. Alternatives to Antibiotics: A Symposium on the Challenges and Solutions for Animal Health and Production. *Antibiotics* **2021**, *10*, 471. [CrossRef] [PubMed]
7. Hassan, Z.M.; Manyelo, T.G.; Selaledi, L.; Mabelebele, M. The Effects of Tannins in Monogastric Animals with Special Reference to Alternative Feed Ingredients. *Molecules* **2020**, *25*, 4680. [CrossRef]
8. Cirkovic Velickovic, T.D.; Stanic-Vucinic, D.J. The Role of Dietary Phenolic Compounds in Protein Digestion and Processing Technologies to Improve Their Antinutritive Properties. *Compr. Rev. Food Sci. Food Saf.* **2018**, *17*, 82–103. [CrossRef] [PubMed]
9. Rice-Evans, C.A.; Miller, N.J.; Paganga, G. Structure-Antioxidant Activity Relationships of Flavonoids and Phenolic Acids. *Free Radic. Biol. Med.* **1996**, *20*, 933–956. [CrossRef] [PubMed]
10. Mota, L.R.; Thomas, G.; Barbosa Filho, J.M. Anti-Inflammatory Actions of Tannins Isolated from the Bark of *Anacardwm occidentale* L. *J. Ethnopharmacol.* **1985**, *13*, 289–300. [CrossRef] [PubMed]
11. Chengwen, S.U.; Xiangting, X.U.; Song, Z. Study on Quebracho Tannin on Production Performance and Immunity of Nursing Pigs. *China Feed* **2016**, 23–32. [CrossRef]
12. Mariscal-Landín, G.; Avellaneda, J.H.; Reis de Souza, T.C.; Aguilera, A.; Borbolla, G.A.; Mar, B. Effect of Tannins in Sorghum on Amino Acid Ileal Digestibility and on Trypsin (E.C.2.4.21.4) and Chymotrypsin (E.C.2.4.21.1) Activity of Growing Pigs. *Anim. Feed Sci. Technol.* **2004**, *117*, 245–264. [CrossRef]
13. NRC. *Nutrient Requirements of Swine*, 11th rev. ed.; National Academies Press: Washington, DC, USA, 2012; ISBN 978-0-309-22423-9.
14. Kim, B.G.; Lindemann, M.D. A spreadsheet method for experimental animal allotment. *J. Anim. Sci.* **2007**, *85* (Suppl. S2), 112.
15. Li, J.; Piao, C.; Jin, H.; Wongpanit, K.; Manabe, N. Delayed Deproteinization Causes Methodological Errors in Amino Acid Levels in Plasma Stored at Room Temperature or −20 °C. *Asian-Australas. J. Anim. Sci.* **2009**, *22*, 1703–1708. [CrossRef]
16. Yi, H.; Wang, Z.; Yang, B.; Yang, X.; Gao, K.; Xiong, Y.; Wu, Q.; Qiu, Y.; Hu, S.; Wang, L.; et al. Effects of Zinc Oxide and Condensed Tannins on the Growth Performance and Intestinal Health of Weaned Piglets in ETEC-Challenged Environment. *Front. Microbiol.* **2023**, *14*, 1181519. [CrossRef] [PubMed]
17. Prescott, J.F.; Baggot, J.D. *Antimicrobial Therapy in Veterinary Medicine*, 2nd ed.; Iowa State University Press: Amsterdam, The Netherlands, 1993; ISBN 0-8138-0889-8.
18. Ortiz, L.T.; Alzueta, C.; Trevino, J.; Castano, M. Effects of Faba Bean Tannins on the Growth and Histological Structure of the Intestinal Tract and Liver of Chicks and Rats. *Br. Poult. Sci.* **1994**, *35*, 743–754. [CrossRef] [PubMed]
19. Seoni, E.; Battacone, G.; Ampuero Kragten, S.; Dohme-Meier, F.; Bee, G. Impact of Increasing Levels of Condensed Tannins from Sainfoin in the Grower–Finisher Diets of Entire Male Pigs on Growth Performance, Carcass Characteristics, and Meat Quality. *Animal* **2021**, *15*, 100110. [CrossRef] [PubMed]
20. Jones, W.T.; Mangan, J.L. Complexes of the Condensed Tannins of Sainfoin (Onobrychis Viciifolia Scop.) with Fraction 1 Leaf Protein and with Submaxillary Mucoprotein, and Their Reversal by Polyethylene Glycol and pH. *J. Sci. Food Agric.* **1977**, *28*, 126–136. [CrossRef]
21. Czech, A.; Grela, E.; Klebaniuk, R.; Ognik, K.; Samolińska, W. Polish Crossbred Pigs' Blood Haematological Parameters Depending on Their Age and Physiological State. *Ann. Wars. Univ. Life Sci.-SGGW-Anim. Sci.* **2018**, *56*, 185–195. [CrossRef]
22. Fournier, B.M.; Parkos, C.A. The Role of Neutrophils during Intestinal Inflammation. *Mucosal Immunol.* **2012**, *5*, 354–366. [CrossRef]
23. Teng, T.-S.; Ji, A.; Ji, X.-Y.; Li, Y.-Z. Neutrophils and Immunity: From Bactericidal Action to Being Conquered. *J. Immunol. Res.* **2017**, *2017*, e9671604. [CrossRef]

24. Wu, M.; Yi, D.; Zhang, Q.; Wu, T.; Yu, K.; Peng, M.; Wang, L.; Zhao, D.; Hou, Y.; Wu, G. Puerarin Enhances Intestinal Function in Piglets Infected with Porcine Epidemic Diarrhea Virus. *Sci. Rep.* **2021**, *11*, 6552. [CrossRef] [PubMed]
25. Ye, M.; Wei, C.; Khalid, A.; Hu, Q.; Yang, R.; Dai, B.; Cheng, H.; Wang, Z. Effect of Bacillus Velezensis to Substitute In-Feed Antibiotics on the Production, Blood Biochemistry and Egg Quality Indices of Laying Hens. *BMC Vet. Res.* **2020**, *16*, 400. [CrossRef] [PubMed]
26. Soliman, A.; De Sanctis, V.; Alaaraj, N.; Ahmed, S.; Alyafei, F.; Hamed, N.; Soliman, N. Early and Long-Term Consequences of Nutritional Stunting: From Childhood to Adulthood. *Acta Bio Med. Atenei Parm.* **2021**, *92*, e2021168. [CrossRef]
27. Iji, P.A.; Khumalo, K.; Slippers, S.; Gous, R.M. Intestinal Function and Body Growth of Broiler Chickens on Maize-Based Diets Supplemented with Mimosa Tannins and a Microbial Enzyme. *J. Sci. Food Agric.* **2004**, *84*, 1451–1458. [CrossRef]
28. Mertz, E.T.; Beeson, W.M.; Jackson, H.D. Classification of Essential Amino Acids for the Weanling Pig. *Arch. Biochem. Biophys.* **1952**, *38*, 121–128. [CrossRef] [PubMed]
29. Matthews, D.E. An Overview of Phenylalanine and Tyrosine Kinetics in Humans. *J. Nutr.* **2007**, *137*, 1549S–1575S. [CrossRef] [PubMed]
30. Wang, M.; Huang, H.; Hu, Y.; Huang, J.; Yang, H.; Wang, L.; Chen, S.; Chen, C.; He, S. Effects of Dietary Microencapsulated Tannic Acid Supplementation on the Growth Performance, Intestinal Morphology, and Intestinal Microbiota in Weaning Piglets. *J. Anim. Sci.* **2020**, *98*, skaa112. [CrossRef]

animals

MDPI

Article

Clostridium butyricum Probiotic Feed Additive: Modulation of Sow Milk Metabolomics and Mitigation of Pre-Weaning Piglet Diarrhea

Jakavat Ruampatana [1], Junpen Suwimonteerabutr [1,2], Kunaporn Homyog [3], Wanwimon Mekboonsonglarp [4], Korntip Kanjanavaikoon [5], Wouter Van der Veken [6], Sutthasinee Poonyachoti [7], Takele Feyera [8], Sarn Settachaimongkon [9,10,*] and Morakot Nuntapaitoon [1,2,*]

1 Department of Obstetrics, Gynaecology and Reproduction, Faculty of Veterinary Science, Chulalongkorn University, Bangkok 10330, Thailand; jakavat.r@gmail.com (J.R.)
2 Center of Excellence in Swine Reproduction, Department of Obstetrics, Gynaecology and Reproduction, Faculty of Veterinary Science, Chulalongkorn University, Bangkok 10330, Thailand
3 Center of Veterinary Diagnosis, Faculty of Veterinary Science, Mahidol University, Nakhon Pathom 73170, Thailand
4 Scientific and Technological Research Equipment Center (STREC), Chulalongkorn University, Bangkok 10330, Thailand
5 Huvepharma (Thailand) Co., Ltd., Bangkok 10900, Thailand
6 Huvepharma N.V., 2600 Antwerp, Belgium
7 Department of Physiology, Faculty of Veterinary Science, Chulalongkorn University, Bangkok 10330, Thailand
8 Department of Animal Science and Veterinary Sciences, Aarhus University, AU-Viborg, DK-8830 Tjele, Denmark
9 Department of Food Technology, Faculty of Science, Chulalongkorn University, Bangkok 10330, Thailand
10 Omics Sciences and Bioinformatics Center, Faculty of Science, Chulalongkorn University, Bangkok 10330, Thailand
* Correspondence: sarn.s@chula.ac.th (S.S.); morakot.n@chula.ac.th (M.N.)

Simple Summary: Pre-weaning piglet diarrhea is one of the major concerns in the swine industry. Recently, various probiotics have been applied to improve animal gut health and performance. This study investigated the impact of *C. butyricum* probiotic feed additive on sow and piglet performance, together with alterations in lipidomic and metabolomic profiles of sow milk. Results showed that sows given the probiotics had lower backfat loss and their piglets experienced less diarrhea incidence, although there were no other significant benefits for piglet growth. Changes in certain fatty acids and metabolites present in sow colostrum and milk, which could impact their nutritional profiles and the health of the piglets, were significantly observed. This study provided new insights regarding the impacts of probiotics application that could potentially lead to better outcomes in swine farms.

Abstract: The present study aimed to investigate the impact of *Clostridium butyricum* probiotic feed additive on sow and piglet performances, together with alterations in the lipidomic and metabolomic profiles of sow milk. Sixty-four Landrace × Yorkshire crossbred sows and 794 piglets were included. Sows were divided into two groups; i.e., (i) conventional gestation diet (control; $n = 35$) and (ii) conventional diet added with 10 g/sow/day of probiotic *C. butyricum* spores (treatment; $n = 29$) from one week before the estimated farrowing day until weaning (29.6 ± 4.8 days). The sow and piglet performances and incidence of piglet diarrhea were recorded. Changes in gross chemical composition, fatty acid and non-volatile polar metabolite profiles of sow colostrum, transient milk and mature milk were evaluated. The results showed that relative backfat loss in the treatment group (−2.3%) was significantly lower than in control group (11.6%), especially in primiparous sows ($p = 0.019$). The application of *C. butyricum* probiotics in sows significantly reduced the incidence of diarrhea in piglets ($p < 0.001$) but no other effect on piglet performance was found. Lipidomic and metabolomic analyses revealed variations in sow colostrum and milk biomolecular profiles, with indicative compounds significantly altered by feeding with the *C. butyricum* probiotics. In conclusion, the use of *C. butyricum* probiotics in sows may improve sow body condition and reduce diarrhea incidence in piglets, with underlying changes in milk composition that warrant further investigation. These findings support the potential of *C. butyricum* as a beneficial feed additive in swine production.

Keywords: probiotics; feed additives; metabolomics; lipidomics; sow milk; piglet performance

1. Introduction

Pre-weaning piglet diarrhea is well recognized as a major concern in the swine industry. The diarrhea causes malabsorption and excessive secretion of water and electrolytes into the intestine resulting in watery feces, nausea, abdominal cramps, shivering, decreased feed intake, growth retardation and increased piglet mortality [1]. Several pathogens—as causative agents for pre-weaning piglet diarrhea—include *Escherichia coli*, *Enterococcus hirae*, *Clostridium difficile*, *Clostridium perfringens*, *Salmonella* spp., *Campylobacter* spp., rotavirus, coronavirus and *Cryptosporidium* spp. [2]. In addition, the use of antibiotics, various preventive measures and applications of feed additives have been alternatively introduced to improve intestinal health and reduce the incidence of diarrhea in newborn piglets.

Probiotics are defined as live microorganisms, which confer a health benefit on the host, when administered in adequate amounts [3]. Probiotics have been introduced as alternative feed additives, aiming to reduce antimicrobial resistance and drug residues in the swine production chain [4]. *Clostridium butyricum* (*C. butyricum*)—a Gram-positive obligate anaerobic bacillus—has been acknowledged for its probiotic capacity; i.e., modulation of gut microbiota, improvement of intestinal barrier functions or protection against pathogenic bacteria, which enhance growth and reduce diarrhea incidence in piglets [5]. The principal effects of *C. butyricum* probiotics are associated with its ability to produce short-chain fatty acids (especially butyric acid), amino acids, enzymes, and vitamins which play a crucial role in energy metabolism and the development of healthy intestinal epithelial cells [5]. In addition, *C. butyricum* feed additives are reported to be associated with enhanced digestibility and nutrient absorption of pigs [4]. Therefore, many studies focused on the influence of the presence of *C. butyricum* probiotics in the feed of a sow herd—especially during late-gestation and lactation period—on the production and quality of colostrum and milk of the sows [6–8]. It is well recognized that sow colostrum and milk are essential sources of energy, passive immunity and nutrients that support the growth and survival of newborn piglets during the lactation period [9]. Various strategic dietary supplements aiming to ameliorate yields, biochemical and immunological composition in sow colostrum and milk have been attempted [10–12]. Regarding the effects of dietary probiotics, alterations in gross chemical composition—i.e., fat, protein, lactose, milk-solids-not-fat, IgA and IgG—in sow colostrum and milk have been indicated in many studies [6,7,13,14]. Nevertheless, changes in minor milk components affected by dietary probiotic intake have not been well investigated.

Metabolomics—a comprehensive characterization of small molecular weight metabolites (<1.5 kDa) present in biological matrices—has recently been acknowledged in lactation biology, milk and dairy research [15]. This high-throughput approach allows a better understanding of dynamic changes in milk metabolome, influenced by various inherent and environmental factors in dairy production [15]. Although metabolomics has been applied in swine milk research [16–18], publications using this approach to investigate the impact of probiotic feed additives on the alteration in sow milk metabolome are rather limited [19]. This information could provide a better understanding and novel insights into the relationships between probiotic-induced changes in sow milk composition and piglet performance.

Therefore, the aims of this study were to investigate the impact of dietary *C. butyricum* probiotic administration during the late-gestation and lactation period on: (i) changes in the performances of lactating sows and pre-weaning piglets; along with (ii) the alterations in fatty acid and non-volatile polar metabolite profiles of sow colostrum and milk.

2. Materials and Methods

2.1. Animal Care

All experimental protocols in this study were approved by the Institutional Animal Care and Use Committee (IACUC) at the Faculty of Veterinary Science, Chulalongkorn University (Approval number 2031056) and followed the guidelines documented in "The ethical Principles and Guidelines for the Use of Animals for Scientific Purposes" edited by the National Research Council of Thailand.

2.2. Experimental Design

The study was carried out in a commercial swine farm in the western part of Thailand. Sixty-four crossbred Landrace × Yorkshire F1 sows, with parities between 1 and 7; and 794 piglets were included in this study (control group; n = 426 and treatment group; n = 368). All sows were raised in a conventional open-housing system in the same week. Sows were transferred to the farrowing house seven days before the estimated farrowing date and were housed individually in crates until weaning. Additionally, backfat thickness was measured when sows were introduced to the farrowing house at day 109 of gestation and weaning, using A-mode ultrasonography (Renco Lean-Maeter®, Minneapolis, MN, USA). Animals were divided into two groups according to feeding regimen; i.e., (i) conventional gestation diet (control; n = 35) and (ii) conventional diet incorporated with 10 g/sow/day (5×10^9 CFU) of probiotic *C. butyricum* spores (Top Gut®, Huvepharma Ltd., Bangkok, Thailand) (treatment; n = 29) by top dressing from a week before the estimated farrowing day until weaning (29.6 ± 4.8 days). The *C. butyricum* was authorized as a zootechnical feed additive for pigs (Regulation (EU) No. 2021/1411) [20]. Sow performance parameters recorded during the experiment included: parity number; number of total born piglets per litter; live-born piglets per litter; percentage of stillborn piglets; percentage of mummified fetuses; number of weaned piglets per litter; litter weight at 3, 10, 17, 21 days and weaning; backfat thickness at day 109 of gestation, and weaning; and backfat loss. Piglets were identified by number. The incidence of diarrhea in piglets was determined by observation of their fecal characteristics throughout the entire lactation period. The diarrhea score was assigned as normal feces (score = 0), soft (score = 1) and runny and/or watery feces (score = 2). Piglet body weight was measured at birth and subsequently on days 1, 3 and 21 of lactation.

2.3. Animals, Housing and Management

The experiment was conducted in a commercial farm in Thailand. The number of productive sows was 3000. Sows were kept in a conventional evaporative cooling system. Sows were housed in individual pens (1.50 m^2) during gestation and were fed a commercial gestation diet according to requirements. Feed was provided twice a day following a standardized feeding pattern, resulting in an average of 3 kg of feed per sow, daily. The water was supplied ad libitum from individual nipple drinkers. On day 109 of gestation, sows were moved to the farrowing house.

The farrowing facilities were an evaporation housing system. Each sow was housed in an individual farrowing pen (2.95 m^2) distributed in 3 rows, with a central alley for sows and 2 side alleys for piglets. The farrowing pens were slatted, with concrete at their center for the sows; and were equipped with steel slats on both sides, for the piglet. Each farrowing pen included a creep area for the piglets (0.60 m^2) on one side, covered by a plastic plate and equipped with a heating lamp, a rubber mattress, and a feeding bowl. Lactating sows were fed twice a day with a dry corn-soybean meal diet that met or exceeded nutritional requirements [21]. The nutritional content of the experimental diet in the present study is presented in Table 1. The amount of feed offered was increased daily until libitum feed was reached after one week of lactation. Sows and piglets had ad libitum access to water via separated nipple drinkers.

Table 1. Nutritional content of the experimental diets during the gestation and lactation period.

Nutritional Content	Gestation Diet	Lactation Diet
Metabolizable energy, kcal/kg	2783	3363
Crude protein, %	11.76	16.21
Crude fat, %	5.75	8.39
Crude fiber, %	4.63	4.70
Ash, %	11.26	7.53
Moisture, %	9.72	9.97
Lysine, %	1.50	2.60

The farrowing process was carefully supervised. During farrowing, the sows and piglets were interfered with as little as possible. Routine interventions were limited to visual supervision of farrowing and the removal of placenta, mummified piglets or dead piglets. Farrowing assistance was provided by skilled personnel when the birth interval exceeded 45 min, and/or there was no progress of uterine contraction. Newborn piglets were dried with a towel before being numbered. No extra management was performed on the newborn piglets. Routine procedures for piglets involved tail docking, tooth clipping, and administering a 1 mL iron supplement intramuscularly (ABI-DEX® 100, T.P. Drug laboratories (1996) Co., Ltd., Bangkok, Thailand) on their first day of life. On the third day, piglets received an oral dose of coccidiocide (Baycox®, OLIC Co., Ltd., Ayutthaya, Thailand). Throughout the entire study, the animals were checked daily for health or eating problems. No pathological symptoms were observed on the farm during the study.

2.4. Determination of Colostrum Consumption and Colostrum Yield

Colostrum consumption of individual piglets was estimated using the equation previously proposed by Theil et al. [22]. The colostrum yield was defined as the sum of individual colostrum consumption by all piglets in the litter. Milk yield was calculated using the Bayesian hierarchical model previously reported by Hansen et al. [23].

2.5. Collection of Colostrum and Milk Samples

Colostrum was manually collected by hand from all functional teats within one hour after the onset of farrowing. Transient and mature milk was collected from all functional teats on days 3 and 10 of lactation, respectively. For the transient and mature milk collections, sows received an intravenous injection of 0.2 mL oxytocin (10 IU/mL, VetOne®, Boise, ID, USA) to facilitate milk let-down. Before colostrum and milk collection, all udders were cleaned with sterile water and dried with a towel to reduce contamination. Approximately 30 mL of colostrum, transient and mature milk were collected from all functional mammary glands of the sows, into plastic cups. The samples were filtered through gauze, transferred into a clean bottle (30 mL), and stored in a cool Styrofoam box (4 °C) during the collection. Once the samples arrived at the laboratory, milk samples were centrifuged at 2700× g for 20 min at 4 °C (Centrifuge 5810R, Eppendorf SE, Hamburg, Germany) and stored at −20 °C until further analysis.

2.6. Determination of Major Chemical Composition in Sow Colostrum and Milk

The major chemical composition—i.e., fat, protein, lactose, dry matter and casein concentration (%wt/wt)—of colostrum and milk samples were analyzed using infrared spectroscopy (MilkoScan FT2 instrument, Foss MilkoScan, Hillerød, Denmark). The concentrations of IgG and IgA in colostrum samples were determined according to the methods described in our previous work [24].

2.7. Analysis of Fatty Acid Profiles in Sow Colostrum and Milk

Fatty acid (FA) composition in sow colostrum and milk samples were characterized using gas chromatography coupled with mass spectrometry for fatty acid methyl ester (GC-MS-FAME) analysis (Agilent 7890A-5975C, Agilent Technologies, Santa Clara, CA, USA) according to the method used in our previous study [18]. In brief, fatty acid methyl ester (FAME) formation was initialized after heating and hydrolysis of samples with KOH, MeOH and H_2SO_4. The FAME fractions were collected after hexane extraction. FA composition of the FAME fraction was determined by capillary GC on a SP-2560 capillary column (Supelco, Bellefonte, PA, USA) operated using similar parameters as described previously [18]. FAs were identified by comparing their specific retention time and *m/z* model with a fatty acid methyl ester standard (Supelco 37 Component FAME mix, Sigma-Aldrich, Steinheim, Germany). After automated peak integration, the concentrations of FAs were calculated using calibration curves fitted by a linear regression model and finally expressed as mg/100 g [18].

2.8. Analysis of Non-Volatile Polar Metabolite Profiles in Sow Colostrum and Milk

Non-volatile polar metabolite composition in sow colostrum and milk samples were characterized using non-targeted proton nuclear magnetic resonance (^{1}H-NMR) metabolomics, according to the method used in our previous work [18]. In brief, the pH of samples was adjusted to 6.0. Lipid and large protein fractions were removed by dichloromethane extraction and ultra-centrifugation (74200× *g* for 60 min at 4 °C), respectively. The supernatant serum was then ultra-filtrated through a Pall Nanosep® centrifugal device with 3 kDa molecular weight cutoffs (Pall Life Sciences, Ann Arbor, MI, USA). Finally, the clear milk serum was mixed 1:1 (v/v) with a phosphate buffer pH 6.0 consisting of 1 mM 3-(Trimethylsilyl) propionic-2, 2, 3, 3-d4 acid sodium salt (TSP) (Merck, Darmstadt, Germany) as an internal standard. Samples were then subjected to a 500 MHz NOESY-GPPR-1D-^{1}H-NMR spectrometer (Bruker, Rheinstetten, Germany) operated with similar parameters as described in our previous study [18]. ^{1}H-NMR spectra were corrected, pre-treated, and segmented using a binning technique. Metabolite identification was performed by consulting the Chenomx NMR suite 8.2 library (Chenomx Inc., Edmonton, AB, Canada), Livestock Metabolome Database (www.lmdb.ca; accessed on 12 August 2023), and literature sources [16,18,25,26]. The sum of signal intensity corresponding to respective metabolites was expressed in arbitrary units. The ^{1}H-NMR signal intensities of respective compounds were expressed as $\log_{10}$ transformed [arbitrary unit] and introduced as variables in the statistical analysis [18].

2.9. Statistical Analysis

Descriptive statistics (i.e., mean, standard deviation and range) were performed using SAS version 9.4 (SAS Institute Inc., Cary, NC, USA). Sow performances were analyzed via multiple analyses of variance, using the general linear model procedure of SAS. Piglet performance (i.e., body weight at 0, 1, 3 and 21 days of age) were analyzed by using the general linear mixed model procedure of SAS. For all analyses, the statistical models included the fixed effect of the group (control and treatment group), parity classes (1, 2–4 and 5–7) and the interaction between group and parity classes. Sow identity was included in the models as a random effect. The effect of the probiotic feed additive on the incidence of piglets' diarrhea on each day was analyzed by the Wilcoxon Rank Sum test. A general linear model procedure was used to analyze the effects of the group (control and treatment group), parity classes (1, 2–4 and 5–7) and the interaction between group and parity classes on colostrum and milk composition on days 3 and 17 (on each day). Least-squares means were obtained from each class of the parity. The probability at $p < 0.05$ was regarded to be statistically significant.

GC-derived lipidomic and ^{1}H-NMR-derived metabolomic data were pretreated, normalized and subjected to multivariate statistical analysis in MetaboAnalyst 5.0 software (www.metaboanalyst.ca; accessed on 7 October 2023). Partial least-squares discriminant analysis (PLS-DA) was applied to visualize distinctive fatty acid and non-volatile polar metabolite patterns between control and treatment samples with important statistical parameters; i.e., % prediction accuracy, R^2 and Q^2 values, and variable importance in projection (VIP) scores, as described in our previous work [20].

3. Results

3.1. Sow Reproductive Performances

On average, the sow parity number was 2.6 ± 1.8. The number of total piglets born/litter and the number of live-born piglets/litter were 12.5 ± 2.9 and 10.8 ± 2.9 piglets/litter, respectively. The effects of dietary *C. butyricum* probiotic additive on sow reproductive performance are presented in Table 2. The results showed that there was no significant influence of *C. butyricum* probiotics on the reproductive performance of the sows in this study ($p > 0.05$).

Table 2. Effect of dietary *C. butyricum* probiotic additive on sow reproductive performances ($n = 64$ sows). Values are the least-squares means of samples ± SEM.

Parameters	Control	Treatment	*p* Value
Number of sows, sows	35	29	
Sow parity number	3.2 ± 0.1	3.1 ± 0.1	0.726
Number of totals born piglet/litter, piglets	12.3 ± 0.5	12.9 ± 0.6	0.501
Number of live born piglet/litter, piglets	10.8 ± 0.5	11.3 ± 0.6	0.488
Percentage of stillborn piglet/litter, %	7.1 ± 2.6	8.4 ± 2.9	0.724
Percentage of mummified fetus/litter, %	4.0 ± 1.8	2.8 ± 2.0	0.668
Number of weaned piglet/litter, piglets	7.1 ± 0.6	6.9 ± 0.6	0.757
Litter weight at weaning, kg	47.7 ± 3.8	46.5 ± 3.6	0.802
Backfat thickness at farrowing, mm	21.4 ± 1.0	20.9 ± 1.1	0.732
Backfat thickness at weaning, mm	18.6 ± 0.7	18.1 ± 0.8	0.612
Backfat loss during lactation (%)	9.6 ± 2.4	5.0 ± 2.8	0.215
Milk yield between Day 3 and 10 of lactation, kg	7.6 ± 0.4	7.5 ± 0.4	0.822
Milk yield between Day 10 and 17 of lactation, kg	7.8 ± 0.5	7.5 ± 0.4	0.651
Litter weight gain between Day 3 and 10 of lactation, kg	1.5 ± 0.2	1.1 ± 0.2	0.078
Litter weight gain between Day 10 and 17 of lactation, kg	0.8 ± 0.2	0.8 ± 0.2	0.817
Piglet preweaning mortality between Day 3–21, %	28.9 ± 4.7	27.6 ± 4.6	0.846

When the sow parity number was considered (Table 3), a significant difference in relative backfat loss was observed between primiparous sows in the control (11.6%) and treatment groups (−2.3%, $p = 0.019$). Sows in parity 2–4 that received the probiotic additive tended to deliver a higher litter weight at weaning ($p = 0.149$), with a lower relative backfat loss ($p = 0.167$) compared to those in the control group (Table 2). However, primiparous sows fed with the *C. butyricum* probiotics tended to yield a higher milk quantity (7.8 kg/day) compared to the control group (6.4 kg/day) ($p = 0.181$).

Table 3. Effect of dietary *C. butyricum* probiotic additive on sow reproductive performances by parity. Values are the least-squares means of samples.

Parameters	Parity					
	1		2–4		5–7	
	Control	Treatment	Control	Treatment	Control	Treatment
Number of total born piglet/litter, piglets	12.1	13.7	12.3	11.9	12.6	13.0
Number of live born piglet/litter, piglets	9.4	11.4	11.1	10.4	11.9	12.2
Percentage of stillborn piglet/litter, %	12.1	7.1	7.0	12.7	2.1	5.5
Percentage of mummified fetus/litter, %	7.0	7.9	1.7	0.6	3.2	0.1
Number of weaned piglet/litter, piglets	6.3	5.2	6.9	8.2	8.2	7.2
Litter weight at weaning, kg	41.7	32.9	45.3	55.3	56.2	51.2
Backfat loss during lactation (%)	11.6^{b}	-2.3^{a}	12.0	5.0	5.2	12.3
Milk yield between Day 3 and 10 of lactation, kg	6.4	7.8	8.1	7.6	8.2	7.0
Milk yield between Day 10 and 17 of lactation, kg	7.0	7.5	7.4	7.8	8.9	7.1
Litter weight gain between Day 3 and 10 of lactation, kg/day	1.1	0.6	1.7	1.6	1.8	1.0
Litter weight gain between Day 10 and 17 of lactation, kg/day	0.8	0.8	0.9	0.8	0.9	0.7
Piglet preweaning mortality between Day 3–21, %	30.6	36.7	28.1	20.1	28.1	26.0

^{a,b} Different superscript letters within the same parity class indicate a significant difference at $p < 0.05$.

3.2. Pre-Weaning Piglet Performances

The average piglet's birth weight and piglet's weight at 24 h after birth were 1.48 ± 0.41 and 1.55 ± 0.41 kg, respectively. No effect of *C. butyricum* probiotic additive on pre-weaning piglet performance was found in this study (Table 4). In sow parity 5–7, the piglet weight at day 21 of lactation was higher in the treatment than in the control group (4.55 vs. 3.29 kg, $p = 0.026$) (Table 5). The incidence of diarrhea in piglets throughout the entire lactation period is shown in Figure 1. The figure revealed that piglets belonging to sows fed with the *C. butyricum* probiotics had significantly lower diarrhea scores compared to those from mothers in the control group ($p < 0.05$).

Table 4. Pre-weaning piglet performances in control (n = 426) and treatment (n = 368) groups and at different parity classes of the sows. Values are the least-squares means of samples.

Parameters	Group				Parity				
	Control	Treatment	SEM	*p* Value	1	2–4	5–7	SEM	*p* Value
Piglet birth weight, kg	1.45	1.43	0.07	0.818	1.18 ^{b}	1.59 ^{a}	1.56 ^{a}	0.08	<0.001
Piglet weight at 24 h after birth, kg	1.56	1.50	0.06	0.506	1.30 ^{b}	1.68 ^{a}	1.61 ^{a}	0.08	<0.001
Piglet weight at Day 3, kg	1.79	1.80	0.06	0.914	1.64 ^{b}	1.92 ^{a}	1.82 ^{ab}	0.09	0.006
Piglet weight at Day 21, kg	3.87	4.32	0.19	0.085	3.84 ^{b}	4.54 ^{a}	3.92 ^{ab}	0.29	0.023

^{a,b} Different superscript letters within the same row indicate a significant difference at $p < 0.05$. SEM = standard error of mean.

Table 5. Pre-weaning piglet performances in control (n = 426) and treatment (n = 368) groups by different parity classes of the sows. Values are the least-squares means of samples.

Parameters	Parity 1			Parity 2–4			Parity 5–7		
	Control	Treatment	SEM	Control	Treatment	SEM	Control	Treatment	SEM
Piglet birth weight, kg	1.29	1.08	0.13	1.62	1.55	0.11	1.45	1.67	0.11
Piglet weight at 24 h after birth, kg	1.37	1.22	0.11	1.72	1.64	0.11	1.59	1.64	0.11
Piglet weight at Day 3, kg	1.70	1.57	0.10	1.90	1.94	0.08	1.76	1.88	0.14
Piglet weight at Day 21, kg	3.84	3.84	0.31	4.49	4.58	0.25	3.29 ^{b}	4.55 ^{a}	0.40

^{a,b} Different superscript letters within the same row and the same class of parity indicate a significant difference at $p < 0.05$. SEM = standard error of mean.

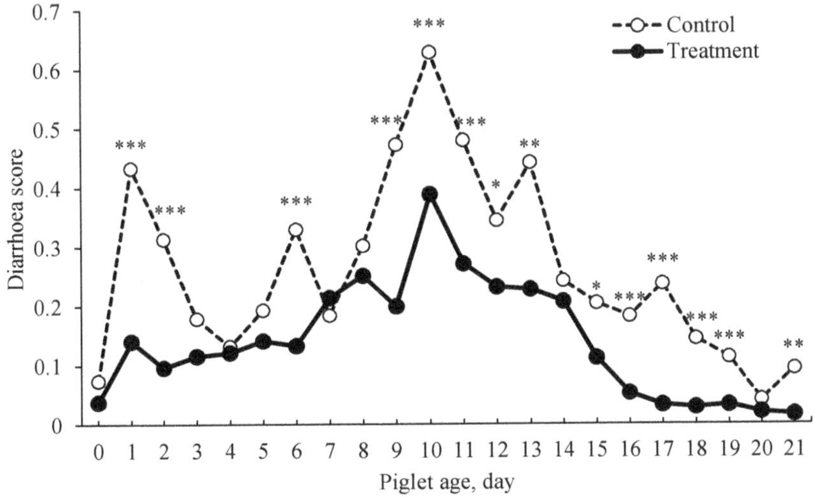

Figure 1. Incidence of diarrhea in piglets, from birth to 21 days in in the control and treatment groups. Level of significant difference at * $p < 0.05$, ** $p < 0.01$ and *** $p < 0.001$.

3.3. Major Chemical Composition of Sow Colostrum and Milk

There was no significant effect of *C. butyricum* probiotic additive on changes in fat, protein, lactose, dry matter and casein concentration of colostrum and milk samples obtained from sows in the treatment group of this study ($p > 0.05$) (Table 6). Similar results were found for antibody levels, IgG and IgA, in colostrum samples (Table 6).

Table 6. Effect of dietary *C. butyricum* probiotic additive on the variations in major chemical composition of sow colostrum (day 0), transient milk (day 3) and mature milk (day 17). Values are the least-squares means of samples.

Composition	Day 0				Day 3				Day 17			
	Con	Treat	SEM	*p* Value	Con	Treat	SEM	*p* Value	Con	Treat	SEM	*p* Value
Fat, g/100 g	5.99	5.53	0.49	0.492	11.13	9.97	0.64	0.222	8.13	7.62	0.26	0.159
Protein, g/100 g	15.80	17.10	0.12	0.118	5.73	5.90	0.31	0.722	5.20	4.96	0.16	0.273
Lactose, g/100 g	2.437	2.366	0.67	0.666	4.25	4.39	0.08	0.260	4.46	4.71	0.10	0.074
DM, g/100 g	24.98	25.66	0.51	0.506	22.80	21.86	0.75	0.398	19.56	19.13	0.32	0.335
Casein, g/100 g	12.58	13.70	0.10	0.103	4.169	4.26	0.16	0.695	4.04	4.07	0.08	0.813
IgG, mg/mL	42.02	40.80	0.76	0.763	NA	NA	NA	NA	NA	NA	NA	NA
IgA, mg/mL	10.23	10.71	0.78	0.779	NA	NA	NA	NA	NA	NA	NA	NA

Con = control group, Treat = treatment group, SEM = standard error of mean. NA = not applicable.

3.4. Changes in Fatty Acid Profiles of Sow Colostrum and Milk

Data from our previous work have demonstrated that the day after farrowing provided a significant impact on the biomolecular profile; i.e., fatty acids and non-volatile polar metabolites, of sow colostrum and milk [20]. Therefore, the influence of dietary *C. butyricum* probiotic additive on the FA profiles of milk samples were evaluated for colostrum (day 0), transient (day 3) and mature milk (day 17), independently. Three separate PLS-DAs were performed for the comparison of samples collected within the same day (Figure 2). Regarding colostrum, PLS-DA demonstrated a good distinction pattern between the control and *C. butyricum* treatment group, with a prediction accuracy of 69.71%, $R^2 = 0.623$ and $Q^2 = 0.511$ (Figure 2A). VIP scores with a value greater than 1.0 were used to screen out the discriminant FAs. Results indicated that variations in the concentration of capric (C10:0), lauric (C12:0), eicosatrienoic (C20:3n3), palmitoleic (C16:1), docosatetraenoic (C22:4), caprylic (C8:0), eicosadienoic (C20:2n6) and paullinic (C20:1n7) acid were responsible for the discrimination (Figure 2B). Continuing with transient milk: a good distinction pattern between the control and *C. butyricum* treatment group was also shown by PLS-DA with a prediction accuracy of 64.29%, $R^2 = 0.699$ and $Q^2 = 0.372$ (Figure 2C). VIP scores indicated that variations in the concentration of lauric (C12:0), palmitoleic (C16:1), myristoleic (C14:1), DPA (22:5n3), capric (C10:0), behenic (C22:0) and caprylic (C8:0) acid were responsible for the discrimination (Figure 2D). In matured milk, a distinct discriminative pattern between the control and *C. butyricum* treatment group was still observed by PLS-DA with a prediction accuracy of 57.14%, $R^2 = 0.652$ and $Q^2 = 0.449$ (Figure 2E). VIP scores indicated that variations in the concentration of linolenic (C18:3n3), DPA (22:5n3), docosahexaenoic (C22:6n3), paullinic (C20:1n7), myristoleic (C14:1) and arachidic (C20:0) acid were responsible for the discrimination (Figure 2F). The statistically significant levels of indicative FAs for discrimination between the control and *C. butyricum* treatment group in colostrum, transient and mature milk are demonstrated in Table 7. Based on chemometric results, a significant impact of *C. butyricum* probiotic feed additive on the variation in milk FA profiles was continually remarkable throughout the entire lactation period of the sows.

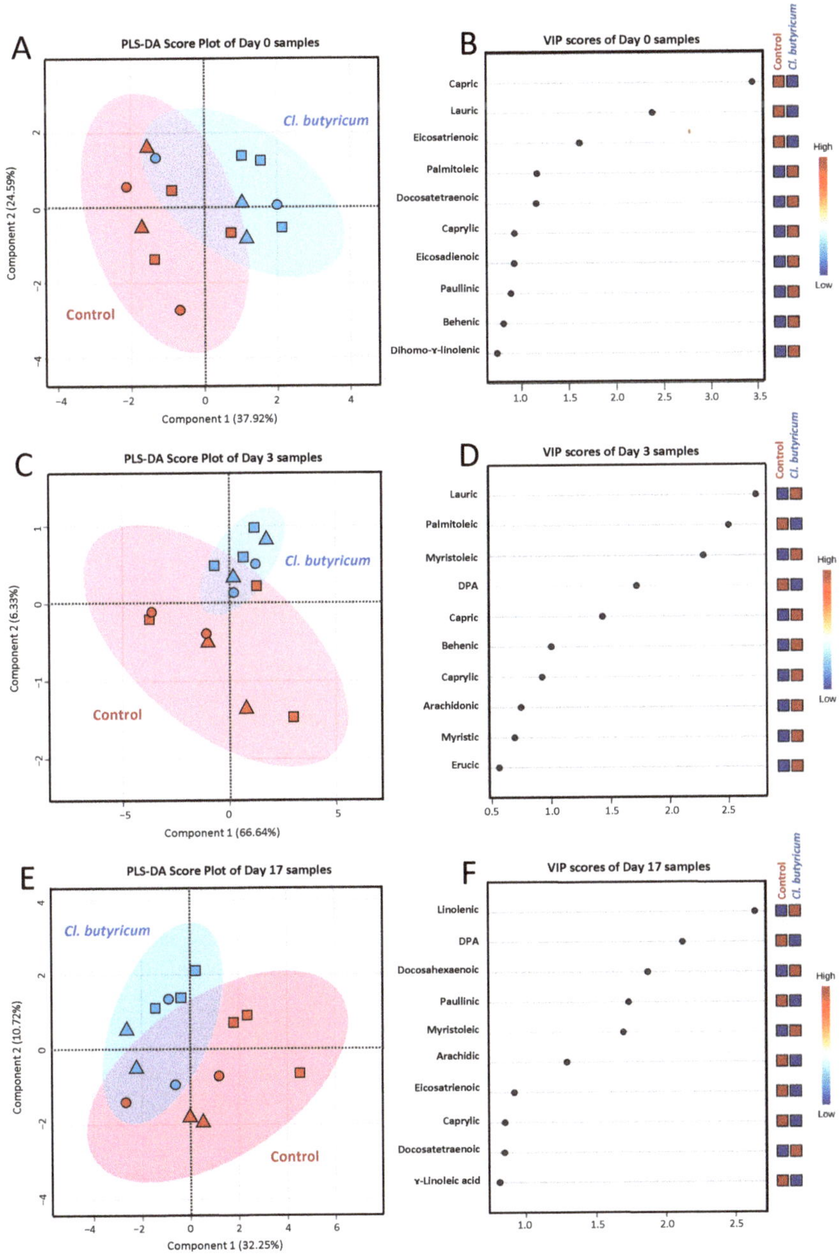

Figure 2. PLS-DA score plots for the comparison of fatty acid profiles of colostrum (day 0; panel (**A**)), transient milk (day 3; panel (**C**)) and mature milk (day 17; panel (**E**)) samples collected from sows in the control (red color) and *C. butyricum* treatment (blue color) groups. Samples from animals in parity 1 (■), parity 2–4 (▲) and parity 5–7 (•) are differently symbolized. VIP scores derived from the comparison among samples of the same day postpartum and indicative fatty acids accountable for the discrimination are visualized in panel (**B**), (**D**) and (**F**), respectively.

Table 7. Effect of dietary *C. butyricum* probiotic additive on the variations of indicative fatty acids in sow colostrum (day 0), transient milk (day 3) and mature milk (day 17). Values are the least-squares means of samples.

Fatty Acid	Day 0				Day 3				Day 17			
	Con	Treat	SEM	*p* Value	Con	Treat	SEM	*p* Value	Con	Treat	SEM	*p* Value
Caprylic acid	0.011 *	0.014	0.005	0.708	0.013	0.013	0.004	0.967	0.029	0.033	0.004	0.512
Capric acid	0.122	0.017	0.062	0.265	0.060	0.070	0.027	0.816	0.182	0.199	0.028	0.673
Lauric acid	1.521	1.018	0.321	0.300	1.642	1.770	0.467	0.851	3.067	3.185	0.504	0.874
Myristoleic acid	-	0.033	0.007	0.672	0.104	0.113	0.035	0.973	0.284	0.333	0.052	0.520
Palmitoleic acid	1.649	1.958	0.100	0.059	2.625 [a]	1.254 [b]	0.260	0.006	1.642	0.580	0.091	0.641
Cis-10-heptadecarnoic acid	0.114	0.132	0.009	0.156	0.177	0.192	0.018	0.570	0.156	0.187	0.014	0.148
Linoleic acid	26.42	24.46	0.863	0.147	19.07	19.88	0.574	0.345	19.410	18.652	0.869	0.555
Arachidic acid	0.167	0.165	0.018	0.928	0.196	0.182	0.010	0.346	0.190	0.166	0.011	0.164
Paullinic acid	0.300	0.366	0.030	0.121	0.627	0.548	0.065	0.414	0.655	0.408	0.109	0.147
Linolenic acid	1.804	1.735	0.160	0.769	1.377	1.386	0.097	0.944	1.300	1.475	0.147	0.426
Eicosadienoic acid	0.519	0.592	0.045	0.284	0.646	0.585	0.051	0.424	0.463	0.400	0.053	0.427
Eicosatrienoic acid	0.322	0.124	0.122	0.285	0.114	0.113	0.010	0.946	0.084	0.104	0.012	0.264
Docosatetraenoic acid	0.244	0.285	0.022	0.230	0.217	0.202	0.025	0.688	0.220	0.186	0.030	0.457
Docosapentaenoic acid	0.357	0.367	0.03	0.788	0.378	0.272	0.073	0.334	0.450	0.246	0.098	0.181

Con = control group, Treat = treatment group, SEM = standard error of mean. * Fatty acid contents are expressed as mg/100 g. [a,b] Different superscript letters within the same row indicate a significant difference at $p < 0.05$.

3.5. Changes in Non-Volatile Polar Metabolite Profiles of Sow Colostrum and Milk

Non-volatile polar metabolites including amino acids, carbohydrates, alcohols, organic acids and lipid derivatives in colostrum and milk samples were identified by a non-targeted ^{1}H-NMR analysis. As mentioned in FA profile analysis, three separate PLS-DAs were performed for comparison of non-volatile polar metabolite profiles of colostrum (day 0), transient (day 3) and mature milk (day 17) samples (Figure 3). Regarding colostrum, PLS-DA demonstrated a good distinction pattern between the control and *C. butyricum* treatment group, with a prediction accuracy of 78.57%, R^2 = 0.604 and Q^2 = 0.561 (Figure 3A). VIP scores indicated that variations in the concentration of ribose, lactose, carnitine, threonine, lactate, choline and o-phosphocholine were responsible for the discrimination (Figure 3B). In the case of transient milk, it was remarkable that the distinction between the two groups of samples disappeared during this transition period (Figure 3C). Interestingly, however, a clear distinction between the control and *C. butyricum* treatment group returned to be remarkable in mature milk again. This change in metabolite pattern was demonstrated by PLS-DA with a prediction accuracy of 71.42%, R^2 = 0.721 and Q^2 = 0.540 (Figure 3D). VIP scores indicated that variations in the concentration of uracil, UDP-galactose, UDP-glucose, UDP-*N*-Acetylglucosamine, lactate, *N*-Acetylglucosamine, *N*-acetylglutamate, threonine, acetate, UMP, adenine, hypoxantine, alanine, dimethylamine, glycero-3-P-choline, carnitine, O-acetylcholine and creatinine were responsible for the discrimination in mature milk from the control and *C. butyricum* treatment groups (Figure 3E). Variations in the concentrations of indicative non-volatile polar metabolites in sow colostrum, transient and mature milk—with their statistically significant levels—are demonstrated in Table 8. Based on the overall metabolite pattern recognition, a significant impact of dietary *C. butyricum* probiotic additive was only observed in colostrum and mature milk.

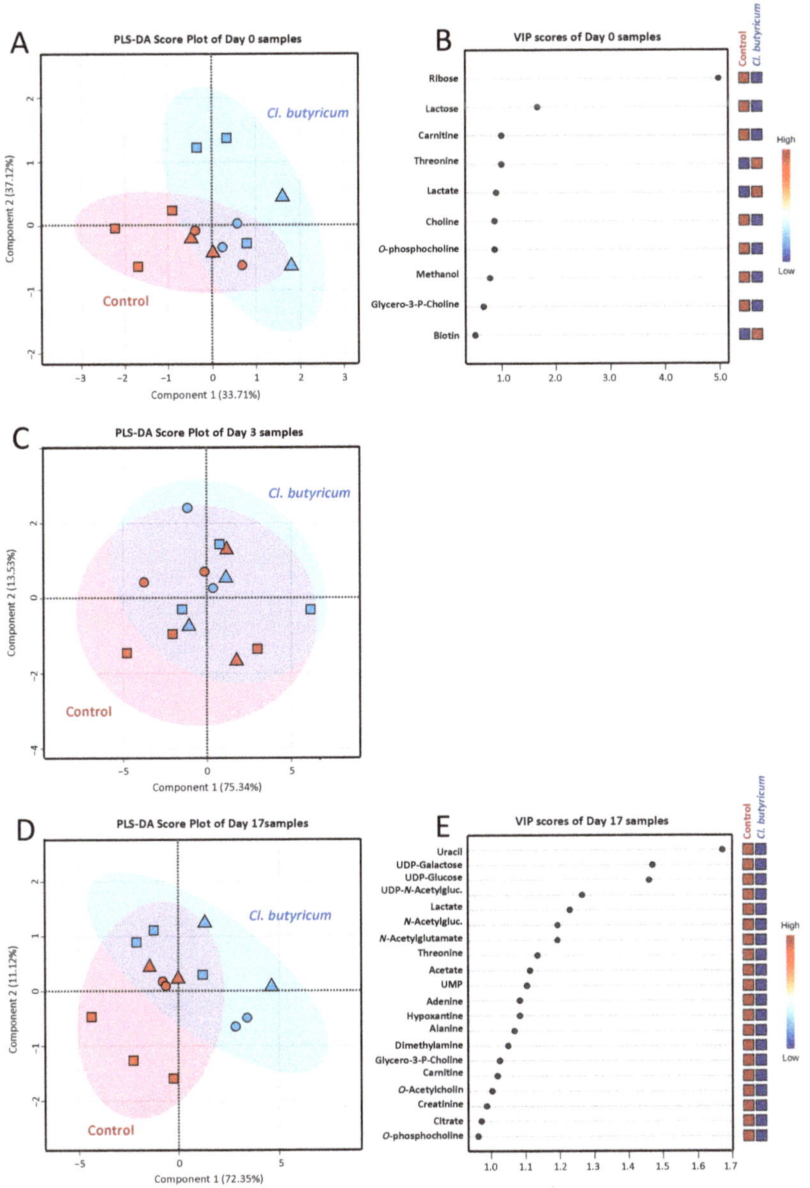

Figure 3. PLS-DA score plots for the comparison of non-volatile polar metabolite profiles of colostrum (day 0; panel (**A**)), transient milk (day 3; panel (**C**)) and mature milk (day 17; panel (**D**)) samples collected from sows in the control (red color) and *C. butyricum* treatment (blue color) group. Samples from animals in parity 1 (■), parity 2–4 (▲) and parity 5–7 (•) are differently symbolized. VIP scores derived from the comparison among samples of the same day postpartum and indicative metabolites accountable for the discrimination are visualized in panel (**B**) and (**E**), respectively.

Table 8. Effect of dietary *C. butyricum* probiotic additive on the variations of indicative non-volatile polar metabolites in sow colostrum (day 0), transient milk (day 3) and mature milk (day 17). Values are the least-squares means of samples.

Metabolite	Day 0				Day 3				Day 17			
	Con	Treat	SEM	*p* Value	Con	Treat	SEM	*p* Value	Con	Treat	SEM	*p* Value
Carnitine	9.260 *	9.175	0.038	0.154	8.527	8.523	0.062	0.967	8.016	7.862	0.047	0.052
Choline	9.301	9.225	0.037	0.178	8.571	8.562	0.065	0.925	8.054	7.908	0.046	0.055
Dimethylamine	8.611	8.624	0.033	0.774	7.809	7.773	0.068	0.713	7.336 [a]	7.175 [b]	0.048	0.045
Lactate	8.809	8.864	0.030	0.227	7.897	7.832	0.095	0.643	7.442	7.258	0.059	0.057
Lactose	9.837	9.714	0.043	0.081	10.169	10.209	0.052	0.601	10.602	10.899	0.040	0.567
N-Acetylglucosamine	9.418	9.425	0.033	0.896	8.626	8.575	0.077	0.650	8.122	7.945	0.055	0.053
N-Acetylglutamate	9.283	9.287	0.037	0.941	8.531	8.490	0.078	0.720	8.065 [a]	7.889 [b]	0.052	0.044
O-Acetylcholine	9.436	9.389	0.037	0.384	8.676	8.665	0.065	0.912	8.185 [a]	8.034 [b]	0.034	0.041
O-Phosphocholine	9.301	9.225	0.037	0.178	8.571	8.562	0.065	0.925	8.054	7.907	0.046	0.055
Ribose	6.339 [a]	5.982 [b]	0.089	0.022	6.536	6.397	0.237	0.688	7.448	7.410	0.056	0.644
sn-Glycero-3-phosphocoline	9.466	9.405	0.038	0.284	8.715	8.702	0.066	0.888	8.208 [a]	8.056 [b]	0.046	0.045
Threonine	8.617	8.680	0.029	0.169	7.687	7.624	0.094	0.654	7.237	7.066	0.060	0.078
UDP-Galactose	8.359	8.336	0.037	0.669	7.627	7.550	0.084	0.533	7.186 [a]	6.966 [b]	0.057	0.026
UDP-Glucose	8.454	8.435	0.037	0.725	7.693	7.630	0.083	0.607	7.265 [a]	7.047 [b]	0.055	0.023
UDP-*N*-Acetylglucosamine	9.059	9.063	0.034	0.937	8.359	8.322	0.073	0.731	7.915 [a]	7.723 [b]	0.050	0.026
Uracil	8.170	8.139	0.042	0.616	7.394	7.328	0.108	0.677	7.057 [a]	6.817 [b]	0.054	0.014

Con = control group, Treat = treatment group, SEM = standard error of mean. * Non-volatile polar metabolite contents are expressed as $\log_{10}$ transformed [arbitrary unit]. [a,b] Different superscript letters within the same row indicate a significant difference at $p < 0.05$.

4. Discussion

The *C. butyricum* has benefits for animal health; i.e., improving gut microbiota and intestinal barrier functions leading to increase the growth performances of both sows and piglets [4,5]. Relative backfat loss serves as an important parameter that indicates sow energy mobilization. It should be noted that backfat thickness is associated with a variety of reproductive parameters. For example, loss of backfat thickness could increase the number of stillborn piglets and decrease litter size [27]. Moreover, excessive backfat loss during lactation is associated with prolonged weaning-to-estrus intervals and reduced farrowing rates [28,29]. It is recognized that 15–20% of primiparous sows are often culled due to reproductive problems [30–32]. The present study found that the application of dietary *C. butyricum* probiotic additive during late gestation throughout the entire lactating period of the sows could reduce relative backfat loss, especially in primiparous sows. It has been documented that probiotics enhance intestinal barrier function and enzymatic production, which leads to improved nutrient digestion, health, and reproductive performances of the sows [4,33,34]. *C. butyricum* probiotics are noted to have a capacity to produce short-chain fatty acids (SCFAs) such as butyrate from carbohydrate metabolism, that could provide energy for epithelial cells and promote intestinal barrier function [35,36]. Furthermore, Niu et al. [37] demonstrated the association between a high abundance of *Clostridium* spp. in the gastrointestinal tract and higher backfat thickness. Therefore, feeding sows with *C. butyricum* probiotic additives may affect energy metabolization and digestion. Indeed, our result was consistent with the work of Konieczka et al. [38], who also found that feed formulation with probiotic *Bacillus subtilis* and *Bacillus amyloliquefaciens* could reduce backfat thickness loss in sows during lactation.

In the present study, the average body weight of all piglets at birth and during lactation was not affected by the *C. butyricum* probiotic additive to the sow's diet during late gestation to lactation. Nevertheless, the piglets of sows fed with *C. butyricum* incorporated in their diets had higher body weights on days 21 than those of the control group. *C. butyricum* probiotics have been found to enhance the growth performance of suckling piglets by improving milk quality and increasing the lactose and protein content in milk [39]. In contrast, in our study, the *C. butyricum* probiotic additive only tended to improve the lactose content in mature milk (day 17; $p = 0.074$), which is a source of energy for piglet metabolism.

Direct feeding of *C. butyricum* probiotics has been shown to promote the growth performance of piglets by improving enterocyte morphology, increasing villus height, improving the villus height–cell depth ratio and strength of the intestinal mucosa cell wall, enabling better nutrient absorption and a reduced diarrhea score [40,41]. Moreover, López et al. (2021) found that *C. butyricum* increased intestinal butyric acid levels, thereby improving the intestinal wellness and health status [42]. In agreement with the present study, enhanced growth coinciding with the reduction of diarrhea scores in pre-weaning piglets was also observed. However, it could be an indirect effect of lactating sows fed with probiotic *C. butyricum* incorporated diets. The results showed that piglets born from *C. butyricum*-fed sows had high body weights at day 21 when compared to the control group. Moreover, feeding *C. butyricum* probiotics to sows reduced the incidence of diarrhea in piglets throughout the lactation period. Pre-weaning piglet mortality is primarily caused by crushing, low viability and diarrhea [43,44]. Therefore, preventing diarrhea in suckling piglets is essential to minimize these problems. Probiotic additives have been shown to reduce piglet diarrhea both before weaning [45,46] and after weaning [41,42]. In general, the *C. butyricum*—a butyric acid-producing bacterium—lowers gut pH, which enhances antibacterial effects [47,48]. Cao et al. [39] indicated that *C. butyricum* probiotic additive promotes an intestinal microecology that supports beneficial bacteria subpopulations, such as *Bacteroidetes* and *Prevotellaceae* spp., while reducing harmful bacteria, including *Streptococcus*, *Escherichia*, and *Shigella*, in pre-weaning piglets. Tang et al. [5] further indicated that supplementing with this probiotic strain significantly reduces the colonization of harmful bacteria in the intestinal tract and enhances the expression of tight junctions (TJs) to improve intestinal barrier function. Furthermore, a reduction in serum lipopolysaccharides endotoxin concentrations, the major factors that induce inflammation and disrupt TJ protein, was found in sows fed with *C. butyricum* probiotics on day 21 of lactation [39]. This is in agreement with Kong et al. [49] who reported that consumption of *C. butyricum* probiotics benefited the gastrointestinal tract microbiome by increasing the beneficial bacteria and reducing harmful pathogens in the intestines of children [50]. Although *C. butyricum* probiotic additive in the present study was applied in the diets of sows, significant improvement in diarrhea incidence was remarkable in piglets. This might be due to the bacteria from the sows which colonized the mammary teats and were transferred to the piglets during suckling a few days after birth [34,41]. Therefore, the addition of *C. butyricum* probiotics additive to the diet of sows may increase the concentration of butyric acid in the gastrointestinal tract of the piglets. As a result, piglets from sows receiving probiotics have a lower incidence of diarrhea and gain a higher body weight. This improvement in digestive function and absorptive capacity of the intestine contributes to the overall health and growth of the piglets.

The significant effect of dietary *C. butyricum* probiotic additive was neither observed on gross chemical composition nor immunological quality of colostrum and milk of the sows in the probiotic-treated group in this study. This was in accordance with two studies which previously reported no significant influence of probiotic administration—i.e., *C. butyricum* [34] and *Bacillus licheniformis* and *Bacillus subtilis* [51]—on the contents of fat, protein, and lactose in sow colostrum. However, both studies found significant changes in fat, protein and lactose content in the mature milk of sows fed with probiotics. Indeed, many studies have reported significant impacts of probiotic administration—i.e., *Bacillus subtilis* and *Bacillus licheniformis* [8,13], *Saccharomyces boulardii* [7], *Saccharomyces cerevisiae* [14]—on the variations in fat, protein, lactose and dry matter content in both colostrum and mature milk of sows. Regarding immunological parameters, significant rises in IgG, IgA and IgM level in sow colostrum were found after probiotic administration in many studies [8,14,45]. In the present study, the effect of dietary *C. butyricum* probiotic additive on the colostral immunoglobulin contents could not be observed. It should be mentioned that inconsistent and very divergent results have been documented regarding the scientific benefits of probiotics when applied in the field. This can be related to the variations of probiotic strains, dosage and delivery methods; sow individuality and herd; animal housing and farming practices, as well as other environmental factors linked to specific swine production systems [4]. In addition to there being no significant effect on gross chemical and immune components, fatty acid and non-volatile polar metabolite profiling was performed to provide more insights on the impact of dietary *C. butyricum* probiotic additive on the metabolome of sow colostrum, transient and mature milk in this study.

It is well recognized that nutritional strategies during late gestation and the lactation period can induce changes in FA composition in sow colostrum and milk [52]. Alterations in milk FA profiles linked to probiotic administration have been reported in goats [53], ewes [54] and dairy cows [55]. However, information regarding the effect of probiotic administration on FA modification in sow milk is rather limited. Our results demonstrated a substantial impact of dietary *C. butyricum* probiotic additive on sow milk FA profiles. A continued distinction pattern of FA profiles between the control and *C. butyricum* treatment groups was observed in the colostrum, transient and mature milk of the sows. PLS-DA derived VIP scores suggested that variations in the concentration of: (i) saturated fatty acids (SFA) including caprylic (C8:0), capric (C10:0), lauric (C12:0), arachidic (C20:0) and behenic (C22:0) acid; (ii) monounsaturated fatty acids (MUFA) including myristoleic (C14:1), palmitoleic (C16:1) and paullinic (C20:1n7) acid; and (iii) polyunsaturated fatty acids (PUFA) including linolenic (C18:3n3), eicosadienoic (C20:2n6), eicosatrienoic (C20:3n3), docosatetraenoic (C22:4), DPA (22:5n3) and docosahexaenoic (C22:6n3) acid were accountable for the discrimination. Although chemometric analysis revealed a good distinction pattern in colostrum and milk FA profiles associated with probiotic *C. butyricum* consumption, the p values of individual FAs were not statistically significant ($p > 0.05$). The most prominent increase in palmitoleic acid (C16:1) was detected in colostrum ($p = 0.059$) and transient milk ($p = 0.006$) of the sows in the treatment group. Also, the concentrations of lauric (C12:0), myristoleic (C14:1) and linolenic (C18:3n3) acid tended to increase in the transient and mature milk of the sows in the *C. butyricum* treatment group. The higher levels in the milk FA could be attributed to improved nutrient digestion and absorption of the sows induced by the *C. butyricum* probiotics. Increasing trends of FAs in sow milk were also observed after probiotic yeast intake in the study of Domingos et al. [56]. It has been documented that certain medium- and long-chain FAs have promising antibacterial activities along with bioactivities to enhance epithelial barrier functions and gut health [57,58]. Therefore, a higher abundance of these FAs in milk might be linked to the reduction of diarrhea scores in pre-weaned piglets belonging to the sows in the probiotics treatment group of this study.

The application of non-targeted ^{1}H-NMR metabolomics is well acknowledged in lactation research. The advantage is to provide comprehensive characterization of overall metabolites present in milk and their modifications under different conditions [59]. Information regarding changes in milk metabolite profiles after probiotic administration has been reported in livestock such as dairy cows [60], donkeys [61] and sows [6–8,19,45,56]. Additionally, parity numbers and lactation stages were found to be the main factors influencing the composition of metabolomics in colostrum and milk in both monogastric and non-monogastric animals [18,62]. Our results demonstrated a substantial impact of dietary *C. butyricum* probiotic additive on the metabolite profiles of sow milk. A good distinction between the control and *C. butyricum* treatment groups was notably observed in colostrum and mature milk. PLS-DA-derived VIP scores suggested that variations in the concentration of certain carbohydrates, amino acids, amines, organic acids as well as their derivatives were accountable for the discrimination. It should be noted that the concentration of most indicative metabolites significantly decreased ($p < 0.05$)—i.e., ribose, dimethylamine, *N*-acetylglutamate, O-acetylcholine, sn-glycero-3-phosphocoline, UDP-galactose, UDP-glucose, UDP-*N*-acetylglucosamine and uracil—or tended to change ($p < 0.10$)—i.e., lactose, lactate, carnitine, choline, *N*-acetylglucosamine, *O*-phosphocholine and threonine—in the milk of sows in the probiotics treatment group. The influence of the dietary probiotic *Bacillus subtilis* and *Bacillus amyloliquefaciens* applications on sow milk metabolome has been reported by Saladrigas-García et al. [19]. In their study, variations in milk metabolites—especially lactose, UDP-*N*-acetylglucosamine, creatine phosphate, UDP-galactose and glycoprotein—were found to be associated with administration of the tested probiotic *Bacillus* strains. In this study, similar indicative metabolites—i.e., lactose, UDP-N-acetylglucosamine, and UDP-galactose—were observed in association with the use of the probiotic, *C. butyricum*. Another recent study focusing on the impact of multi-species probiotics (SLAB51)—consisting of *Streptococcus thermophilus, Bifidobacterium lactis, Lactobacillus brevis, Lactobacillus plantarum, Lactobacillus paracasei, Lactobacillus acidophilus,* and *Lactobacillus helveticus*—administration on donkey milk metabolome found significant changes in the concentration of 12 metabolites—i.e., lactose, *O*-phosphocholine, sn-glycero-3-phosphocholine, 4-pyridoxate, caprylate, isovalerate, butyrate, 2-oxoisocaproate, glucose, glucose-1-phosphate, glutamine, and 4-hydroxyphenilacetate—in donkey milk ($p < 0.05$) induced by administration of the probiotic mixture SLAB51 [60]. Moreover, this study found a decreasing trend in the concentration of lactate ($p = 0.16$) and threonine ($p = 0.14$) in donkey milk induced by SLAB51 probiotics. Alterations in lactose, lactate, threonine, *O*-phosphocholine and sn-Glycero-3-phosphocholine contents were also linked to the application of the *C. butyricum* probiotics in our work. Nevertheless, it should be mentioned that the beneficial effects of dietary probiotic administration on milk production and compositional changes reported in various livestock species are very case-specific and still inconclusive. Inconsistent findings and great variability in results could be due to probiotic-specific factors—e.g., probiotic strains and mixed formulation, dosage, mode and duration of administration to the subject animals—as well as host-specific factors; e.g., animal species and breed, health and physiological status, digestive system and gut health, diet composition and feeding regime [60]. Modifications in milk composition could be mediated by changes in digestive efficiency, nutrient absorption and metabolic response of the sows induced by probiotics and other beneficial microbes in their gut microbiota [19]. Therefore, further research is needed to better understand the mechanisms by which probiotic administration can impact milk composition, and to determine the optimal dosages and feeding durations for various probiotic strains. Moreover, the impact of probiotics on the fecal microbiota of sows and piglets is another point of interest that requires further investigation.

5. Conclusions

The present study demonstrated the impacts of *C. butyricum* probiotic feed additive on sow and piglet performances, together with the alterations in lipidomic and metabolomic profiles of sow milk. The results revealed that diets with *C. butyricum* probiotic additive provided no significant impact on the overall reproductive performance of sows. However, it was remarkable that *C. butyricum* probiotic additive resulted in a significantly lower backfat loss in primiparous sows and a significant increase in piglet weight at day 21 of lactation in parity 5–7 sows. In addition, the piglets from sows fed with probiotic *C. butyricum*-added diets experienced significantly lower diarrhea scores throughout the lactation period. In addition to animal performances, *C. butyricum* probiotic additive also induced notable changes in lipidomic and metabolomics profiles of sow colostrum and mature milk. Significant variations in the concentration of certain indicative fatty acids and metabolite compounds indicated a notable impact on the nutritional profile of sow milk. In conclusion, the use of *C. butyricum* probiotics in sows may improve sow body condition and reduce diarrhea incidence in piglets, with underlying changes in milk composition that warrant further investigation. These findings support the potential of *C. butyricum* probiotics as a beneficial feed additive in swine production.

Author Contributions: Conceptualization, J.R., K.K., W.V.d.V., S.S. and M.N.; methodology, J.R., S.P., T.F., K.K., W.V.d.V., S.S. and M.N.; investigation, J.R., J.S., K.H. and W.M.; resources, K.K. and W.V.d.V.; data curation, J.R., J.S., K.H., W.M., S.S. and M.N.; Formal analysis, S.P. and T.F., writing—original draft preparation, J.R. and S.P.; writing—review and editing, T.F., S.S., S.P. and M.N.; supervision, S.S. and M.N.; project administration, M.N., funding acquisition, M.N. All authors have read and agreed to the published version of the manuscript.

Funding: The PhD scholarship of J. Ruampatana is supported by the Second Century Fund (C2F), Chulalongkorn University and the 90th Anniversary of Chulalongkorn University Scholarship (Batch 55, 2024).

Institutional Review Board Statement: All animal related protocols in this study were approved by the Institutional Animal Care and Use Committee (IACUC) at the Faculty of Veterinary Science, Chulalongkorn University (Approval number 2031056) and followed the guidelines documented in "The ethical Principles and Guidelines for the Use of Animals for Scientific Purposes" edited by the National Research Council of Thailand.

Informed Consent Statement: Not applicable.

Data Availability Statement: Data produced in this study are available from the corresponding authors on reasonable request.

Acknowledgments: Technical contributions from undergraduate students in FTCU Dairy Research group were highly appreciated.

Conflicts of Interest: K. Kanjanavaikoon and W. Van der Veken are employed by Huvepharma Co., Ltd. The remaining authors declare that the research was conducted in the absence of any commercial or financial relationships that could be construed as a potential conflict of interest.

References

1. Jacobson, M. On the infectious causes of neonatal piglet diarrhoea—A review. *Vet. Sci.* **2022**, *9*, 422. [CrossRef] [PubMed]
2. Ruiz, V.L.A.; Bersano, J.G.; Carvalho, A.F.; Catroxo, M.H.B.; Chiebao, D.P.; Gregori, F.; Miyashiro, S.; Nassar, A.F.C.; Oliveira, T.M.F.S.; Ogata, R.A.; et al. Case–control study of pathogens involved in piglet diarrhea. *BMC Res. Notes* **2016**, *9*, 22. [CrossRef] [PubMed]
3. FAO/WHO. Guidelines for the Evaluation of Probiotics in Food: Report of a Joint FAO (Food and Agriculture Organization of the United Nations)/WHO (World Health Organization) Working Group on Drafting Guidelines for the Evaluation of Probiotics in Food. Available online: https://openknowledge.fao.org/server/api/core/bitstreams/382476b3-4d54-4175-803f-2f26f3526256/content (accessed on 31 May 2023).
4. Barba-Vidal, E.; Martín-Orúe, S.M.; Castillejos, L. Practical aspects of the use of probiotics in pig production: A review. *Livest. Sci.* **2019**, *223*, 84–96. [CrossRef]
5. Tang, X. Probiotic roles of Clostridium butyricum in piglets: Considering aspects of intestinal barrier function. *Animals* **2024**, *14*, 1069. [CrossRef] [PubMed]
6. Jarosz, Ł.; Ciszewski, A.; Marek, A.; Grądzki, Z.; Kaczmarek, B.; Rysiak, A. The effect of feed supplementation with EM Bokashi® multimicrobial probiotic preparation on selected parameters of sow colostrum and milk as indicators of the specific and nonspecific immune response. *Probiotics Antimicrob. Proteins* **2022**, *14*, 1029–1041. [CrossRef] [PubMed]
7. Rocha, V.P.; Araújo, L.R.S.; de Mendonça, I.B.; Martins, L.P.; de Alcântara Araújo, G.G.; Watanabe, P.H.; Andrade, T.S.; Evangelista, J.N.B. Effects of Saccharomyces cerevisiae var. boulardii CNCM I-1079 on performance, colostrum and milk composition, and litter performance of mixed-parity sows in a tropical humid climate. *Trop. Anim. Health Prod.* **2022**, *54*, 41. [CrossRef]
8. Gu, X.L.; Song, Z.H.; Li, H.; Wu, S.; Wu, S.S.; Ding, Y.N.; He, X.; Yin, Y.L.; Fan, Z.Y. Effects of dietary isomaltooligosaccharide and *Bacillus* spp. supplementation during perinatal period on lactational performance, blood metabolites, and milk composition of sows. *J. Sci. Food Agric.* **2019**, *99*, 5646–5653. [CrossRef] [PubMed]
9. Nuntapaitoon, M.; Muns, R.; Theil, P.K.; Tummaruk, P. Factors influencing colostrum consumption by piglets and their relationship with survival and growth in tropical climates. *Livest. Sci.* **2019**, *224*, 31–39. [CrossRef]
10. Wiecek, J.; Rekiel, A.; Bartosik, J.; Glogowski, R.; Kuczynska, B. Colostrum and milk quality of sows fed different diets during mid-pregnancy. *J. Anim. Feed Sci.* **2018**, *27*, 248–254. [CrossRef]
11. Hasan, S.; Orro, T.; Valros, A.; Junnikkala, S.; Peltoniemi, O.; Oliviero, C. Factors affecting sow colostrum yield and composition, and their impact on piglet growth and health. *Livest. Sci.* **2019**, *227*, 60–67. [CrossRef]
12. Holen, J.P.; Woodworth, J.C.; Tokach, M.D.; Goodband, R.D.; DeRouchey, J.M.; Gebhardt, J.T.; DeDecker, A.E.; Martinez, X. Evaluation of essential fatty acids in lactating sow diets on sow reproductive performance, colostrum and milk composition, and piglet survivability. *J. Anim. Sci.* **2022**, *100*, skac167. [CrossRef]
13. Han, L.; Azad, M.A.K.; Huang, P.; Wang, W.; Zhang, W.; Blachier, F.; Kong, X. Maternal supplementation with different probiotic mixture from late pregnancy to day 21 postpartum: Consequences for litter size, plasma and colostrum parameters, and fecal microbiota and metabolites in sows. *Front. Vet. Sci.* **2022**, *9*, 726276. [CrossRef]
14. Zhao, Y.; Wang, Q.; Zhou, P.; Li, Z.; Zhong, W.; Zhuo, Y.; Che, L.; Xu, S.; Fang, Z.; Jiang, X.; et al. Effects of yeast culture supplementation from late gestation to weaning on performance of lactating sows and growth of nursing piglets. *Animal* **2022**, *16*, 100526. [CrossRef]
15. Suh, J.H. Critical review: Metabolomics in dairy science—Evaluation of milk and milk product quality. *Food Res. Int.* **2022**, *154*, 110984. [CrossRef]
16. Picone, G.; Zappaterra, M.; Luise, D.; Trimigno, A.; Capozzi, F.; Motta, V.; Davoli, R.; Nanni Costa, L.; Bosi, P.; Trevisi, P. Metabolomics characterization of colostrum in three sow breeds and its influences on piglets' survival and litter growth rates. *J. Anim. Sci. Biotechnol.* **2018**, *9*, 23. [CrossRef]

17. Rocchetti, G.; O'Callaghan, T.F. Application of metabolomics to assess milk quality and traceability. *Curr. Opin. Food Sci.* **2021**, *40*, 168–178. [CrossRef]
18. Settachaimongkon, S.; Homyog, K.; Mekboonsonglarp, W.; Soonoue, P.; Lerdamnuaylarp, T.; Prayoonpeeraput, P.; Theil, P.K.; Nuntapaitoon, M. Dynamics of fatty acid and non-volatile polar metabolite profiles in colostrum and milk depending on the lactation stage and parity number of sows. *Sci. Rep.* **2023**, *13*, 1989. [CrossRef]
19. Saladrigas-García, M.; Solà-Oriol, D.; López-Verg, S.; D'Angelo, M.; Collado, M.C.; Nielsen, B.; Faldyna, M.; Prez, J.F.; Martín-Orúe, S.M. Potential effect of two *Bacillus* probiotic strains on performance and fecal microbiota of breeding sows and their piglets. *J. Anim. Sci.* **2022**, *100*, skac163. [CrossRef]
20. Eur-Lex, Commission Implementing Regulation (EU) No 2021/1411 of 27 August 2021 Concerning the Renewal of the Authorisation of Clostridium Butyricum FERM BP-2789 as a Feed Additive for Chickens Reared for Laying, Turkeys for Fattening, Turkeys Reared for Breeding, Minor Avian Species (Excluding Laying Birds), Weaned Piglets and Weaned Minor Porcine Species, Its Authorisation for Chickens for Fattening, Suckling Piglets and Suckling Minor Porcine Species, and Repealing Implementing Regulations (EU) No 373/2011, (EU) No 374/2013 and (EU) No 1108/2014 (Holder of Authorisation Miyarisan Pharmaceutical Co. Ltd. Represented by Huvepharma NV Belgium)Text with EEA Relevance 2021 (OJ L 104/11, 30.8.2021, p. 1–3). Available online: https://eur-lex.europa.eu/legal-content/EN/TXT/PDF/?uri=CELEX:32021R1411 (accessed on 10 June 2024).
21. Committee on Nutrient Requirements of Swine; National Research Council. *Nutrient Requirements of Swine: 11th Revised Edition*; National Academies Press: Washington, DC, USA, 2012.
22. Theil, P.K.; Flummer, C.; Hurley, W.L.; Kristensen, N.B.; Labouriau, R.L.; Sørensen, M.T. Mechanistic model to predict colostrum intake based on deuterium oxide dilution technique data and impact of gestation and prefarrowing diets on piglet intake and sow yield of colostrum1. *J. Anim. Sci.* **2014**, *92*, 5507–5519. [CrossRef]
23. Hansen, A.V.; Strathe, A.B.; Kebreab, E.; France, J.; Theil, P.K. Predicting milk yield and composition in lactating sows: A Bayesian approach. *J. Anim. Sci.* **2012**, *90*, 2285–2298. [CrossRef]
24. Nuntapaitoon, M.; Suwimonteerabutr, J.; Am-in, N.; Tienthai, P.; Chuesiri, P.; Kedkovid, R.; Tummaruk, P. Impact of parity and housing conditions on concentration of immunoglobulin G in sow colostrum. *Trop. Anim. Health Prod.* **2019**, *51*, 1239–1246. [CrossRef]
25. Luangwilai, M.; Duangmal, K.; Chantaprasarn, N.; Settachaimongkon, S. Comparative metabolite profiling of raw milk from subclinical and clinical mastitis cows using ^{1}H-NMR combined with chemometric analysis. *Int. J. Food Sci. Technol.* **2021**, *56*, 493–503. [CrossRef]
26. Luise, D.; Picone, G.; Balzani, A.; Capozzi, F.; Bertocchi, M.; Salvarani, C.; Bosi, P.; Edwards, S.; Trevisi, P. Investigation of the defatted colostrum1H-NMR metabolomics profile of gilts and multiparous sows and its relationship with litter performance. *Animals* **2020**, *10*, 154. [CrossRef]
27. Thongkhuy, S.; Chuaychu, S.B.; Burarnrak, P.; Ruangjoy, P.; Juthamanee, P.; Nuntapaitoon, M.; Tummaruk, P. Effect of backfat thickness during late gestation on farrowing duration, piglet birth weight, colostrum yield, milk yield and reproductive performance of sows. *Livest. Sci.* **2020**, *234*, 103983. [CrossRef]
28. De Rensis, F.; Gherpelli, M.; Superchi, P.; Kirkwood, R.N. Relationships between backfat depth and plasma leptin during lactation and sow reproductive performance after weaning. *Anim. Reprod. Sci.* **2005**, *90*, 95–100. [CrossRef]
29. Serenius, T.; Stalder, K.J.; Baas, T.J.; Mabry, J.W.; Goodwin, R.N.; Johnson, R.K.; Robison, O.W.; Tokach, M.; Miller, R.K. National Pork Producers Council Maternal Line National Genetic Evaluation Program: A comparison of sow longevity and trait associations with sow longevity. *J. Anim. Sci.* **2006**, *84*, 2590–2595. [CrossRef]
30. Tummaruk, P.; Kesdangsakonwut, S.; Kunavongkrit, A. Relationships among specific reasons for culling, reproductive data, and gross morphology of the genital tracts in gilts culled due to reproductive failure in Thailand. *Theriogenology* **2009**, *71*, 369–375. [CrossRef]
31. Heinonen, M.; Leppävuori, A.; Pyörälä, S. Evaluation of reproductive failure of female pigs based on slaughterhouse material and herd record survey. *Anim. Reprod. Sci.* **1998**, *52*, 235–244. [CrossRef]
32. López-Serrano, M.; Reinsch, N.; Looft, H.; Kalm, E. Genetic correlations of growth, backfat thickness and exterior with stayability in large white and landrace sows. *Livest. Prod. Sci.* **2000**, *64*, 121–131. [CrossRef]
33. van der Aar, P.J.; Molist, F.; van der Klis, J.D. The central role of intestinal health on the effect of feed additives on feed intake in swine and poultry. *Anim. Feed. Sci. Technol.* **2017**, *233*, 64–75. [CrossRef]
34. Cho, J.H.; Zhao, P.Y.; Kim, I.H. Probiotics as a dietary additive for pigs: A review. *J. Anim. Vet. Adv.* **2011**, *10*, 2127–2134. [CrossRef]
35. Jørgensen, H.; Larsen, T.; Zhao, X.Q.; Eggum, B.O. The energy value of short-chain fatty acids infused into the caecum of pigs. *Br. J. Nutr.* **1997**, *77*, 745–756. [CrossRef] [PubMed]
36. van Hoffen, E.; Ruiter, B.; Faber, J.; M'Rabet, L.; Knol, E.F.; Stahl, B.; Arslanoglu, S.; Moro, G.; Boehm, G.; Garssen, J. A specific mixture of short-chain galacto-oligosaccharides and long-chain fructo-oligosaccharides induces a beneficial immunoglobulin profile in infants at high risk for allergy. *Allergy* **2009**, *64*, 484–487. [CrossRef] [PubMed]
37. Niu, Q.; Li, P.; Hao, S.; Zhang, Y.; Kim, S.W.; Li, H.; Ma, X.; Gao, S.; He, L.; Wu, W.; et al. Dynamic Distribution of the Gut Microbiota and the Relationship with Apparent Crude Fiber Digestibility and Growth Stages in Pigs. *Sci. Rep.* **2015**, *5*, 9938. [CrossRef] [PubMed]

38. Konieczka, P.; Ferenc, K.; Jørgensen, J.N.; Hansen, L.H.B.; Zabielski, R.; Olszewski, J.; Gajewski, Z.; Mazur-Kuśnirek, M.; Szkopek, D.; Szyryńska, N.; et al. Feeding Bacillus-based probiotics to gestating and lactating sows is an efficient method for improving immunity, gut functional status and biofilm formation by probiotic bacteria in piglets at weaning. *Anim. Nutr.* **2023**, *13*, 361–372. [CrossRef] [PubMed]
39. Cao, M.; Li, Y.; Wu, Q.J.; Zhang, P.; Li, W.T.; Mao, Z.Y.; Wu, D.M.; Jiang, X.M.; Zhuo, Y.; Fang, Z.F.; et al. Effects of dietary *Clostridium butyricum* addition to sows in late gestation and lactation on reproductive performance and intestinal microbiota. *J. Anim. Sci.* **2019**, *97*, 3426–3439. [CrossRef] [PubMed]
40. Hamer, H.M.; Jonkers, D.; Venema, K.; Vanhoutvin, S.; Troost, F.J.; Brummer, R.J. Review article: The role of butyrate on colonic function. *Aliment. Pharmacol. Ther.* **2008**, *27*, 104–119. [CrossRef] [PubMed]
41. Chen, L.; Li, S.; Zheng, J.; Li, W.; Jiang, X.; Zhao, X.; Li, J.; Che, L.; Lin, Y.; Xu, S.; et al. Effects of dietary *Clostridium butyricum* supplementation on growth performance, intestinal development, and immune response of weaned piglets challenged with lipopolysaccharide. *J. Anim. Sci. Biotechnol.* **2018**, *9*, 62. [CrossRef]
42. López, M.; Madrid, J.; Hernández, F.; Ros, M.A.; Segura, J.C.; López, M.J.; Pallarés, F.J.; Sánchez, C.J.; Martínez-Miró, S. Effect of Feed Supplementation with Clostridium butyricum, Alone or in Combination with Carob Meal or Citrus Pulp, on Digestive and Metabolic Status of Piglets. *Animals* **2021**, *11*, 2924. [CrossRef]
43. Vaillancourt, J.P.; Stein, T.E.; Marsh, W.E.; Leman, A.D.; Dial, G.D. Validation of producer-recorded causes of preweaning mortality in swine. *Prev. Vet. Med.* **1990**, *10*, 119–130. [CrossRef]
44. Koketsu, Y.; Takenobu, S.; Nakamura, R. Preweaning mortality risks and recorded causes of death associated with production factors in swine breeding herds in Japan. *J. Vet. Med. Sci.* **2006**, *68*, 821–826. [CrossRef] [PubMed]
45. Laskowska, E.; Jarosz, Ł.; Grądzki, Z. Effect of multi-microbial probiotic formulation bokashi on pro- and anti-inflammatory cytokines profile in the serum, colostrum and milk of sows, and in a culture of polymorphonuclear cells isolated from colostrum. *Probiotics Antimicrob. Proteins* **2019**, *11*, 220–232. [CrossRef] [PubMed]
46. Gao, Q.; Qi, L.; Wu, T.; Wang, J. *Clostridium butyricum* activates TLR2-mediated MyD88-independent signaling pathway in HT-29 cells. *Mol. Cell. Biochem.* **2012**, *361*, 31–37. [CrossRef] [PubMed]
47. Lyberg, K.; Lundh, T.; Pedersen, C.; Lindberg, J.E. Influence of soaking, fermentation and phytase supplementation on nutrient digestibility in pigs offered a grower diet based on wheat and barley. *Anim. Sci.* **2006**, *82*, 853–858. [CrossRef]
48. Long, S.F.; Xu, Y.T.; Pan, L.; Wang, Q.Q.; Wang, C.L.; Wu, J.Y.; Wu, Y.Y.; Han, Y.M.; Yun, C.H.; Piao, X.S. Mixed organic acids as antibiotic substitutes improve performance, serum immunity, intestinal morphology and microbiota for weaned piglets. *Anim. Feed. Sci. Technol.* **2018**, *235*, 23–32. [CrossRef]
49. Kong, Q.; He, G.Q.; Jia, J.L.; Zhu, Q.L.; Ruan, H. Oral administration of *Clostridium butyricum* for modulating gastrointestinal microflora in mice. *Curr. Microbiol.* **2011**, *62*, 512–517. [CrossRef] [PubMed]
50. Seki, H.; Shiohara, M.; Matsumura, T.; Miyagawa, N.; Tanaka, M.; Komiyama, A.; Kurata, S. Prevention of antibiotic-associated diarrhea in children by *Clostridium butyricum* MIYAIRI. *Pediatr. Int.* **2003**, *45*, 86–90. [CrossRef] [PubMed]
51. Alexopoulos, C.; Georgoulakis, I.E.; Tzivara, A.; Kritas, S.K.; Siochu, A.; Kyriakis, S.C. Field evaluation of the efficacy of a probiotic containing *Bacillus licheniformis* and *Bacillus subtilis* spores, on the health status and performance of sows and their litters. *J. Anim. Physiol. Anim. Nutr.* **2004**, *88*, 381–392. [CrossRef] [PubMed]
52. Zhang, S.; Chen, F.; Zhang, Y.; Lv, Y.; Heng, J.; Min, T.; Li, L.; Guan, W. Recent progress of porcine milk components and mammary gland function. *J. Anim. Sci. Biotechnol.* **2018**, *9*, 77. [CrossRef]
53. Apás, A.L.; Arena, M.E.; Colombo, S.; González, S.N. Probiotic administration modifies the milk fatty acid profile, intestinal morphology, and intestinal fatty acid profile of goats. *J. Dairy Sci.* **2015**, *98*, 47–54. [CrossRef]
54. Payandeh, S.; Kafilzadeh, F.; Juárez, M.; De La Fuente, M.A.; Ghadimi, D.; Martínez Marín, A.L. Probiotic supplementation effects on milk fatty acid profile in ewes. *J. Dairy Res.* **2017**, *84*, 128–131. [CrossRef] [PubMed]
55. Izadi, B.; Mohebbi-Fani, M.; Hosseinzadeh, S.; Shekarforoush, S.S.; Nazifi, S.; Rasooli, A. Alteration of fatty acid profile of milk in Holstein cows fed Bacillus coagulans as probiotic: A field study. *Iran. J. Vet. Res.* **2021**, *22*, 100–106. [CrossRef] [PubMed]
56. Domingos, R.L.; Silva, B.A.N.; Bravo de Laguna, F.; Araujo, W.A.G.; Gonçalves, M.F.; Rebordões, F.I.G.; Evangelista, R.P.; de Alkmim, T.C.C.; Miranda, H.A.F.; Cardoso, H.M.C.; et al. *Saccharomyces cerevisiae* var. *Boulardii* CNCM I-1079 during late gestation and lactation improves voluntary feed intake, milk production and litter performance of mixed-parity sows in a tropical humid climate. *Anim. Feed. Sci. Technol.* **2021**, *272*, 114785. [CrossRef]
57. Lauridsen, C. Effects of dietary fatty acids on gut health and function of pigs pre- And post-weaning. *J. Anim. Sci.* **2020**, *98*, skaa086. [CrossRef]
58. Mitchell, M.K.; Ellermann, M. Long chain fatty acids and virulence repression in intestinal bacterial pathogens. *Front. Cell. Infect. Microbiol.* **2022**, *12*, 928503. [CrossRef]
59. Scano, P.; Cusano, E.; Caboni, P.; Consonni, R. NMR metabolite profiles of dairy: A review. *Int. Dairy J.* **2019**, *90*, 56–67. [CrossRef]
60. Nalla, K.; Manda, N.K.; Dhillon, H.S.; Kanade, S.R.; Rokana, N.; Hess, M.; Puniya, A.K. Impact of probiotics on dairy production efficiency. *Front. Microbiol.* **2022**, *13*, 805963. [CrossRef] [PubMed]

61. Laus, F.; Laghi, L.; Bazzano, M.; Cifone, M.G.; Cinque, B.; Yang, Y.; Marchegiani, A. Donkey colostrum and milk: How dietary probiotics can affect metabolomic profile, alkaline sphingomyelinase and alkaline phosphatase activity. *Metabolites* **2023**, *13*, 622. [CrossRef]
62. Vo, T.T.P.; Buranakar, C.; Chamsuwan, S.; Thammacharoen, S.; Ratchakom, P.; Nuntapaitoon, M.; Settachaimongkon, S. Changes in the metabolomic profiles of mammary secretion in relation to dam litter size and parity number in Black Bengal goats. *Vet. World* **2024**, *17*, 1469–1481. [CrossRef]

animals

MDPI

Article

The Effects of Microbial Additive Supplementation on Growth Performance, Blood Metabolites, Fecal Microflora, and Carcass Characteristics of Growing–Finishing Pigs

Hyuk-Jun Lee [1,†], Bu-Gil Choi [1,†], Young-Ho Joo [1], Chang-Hyun Baeg [1], Ji-Yoon Kim [1], Dong-Hyeon Kim [2], Seong-Shin Lee [3] and Sam-Churl Kim [1,*]

1 Division of Applied Life Science (BK21Four, Institute of Agriculture and Life Science), Gyeongsang National University, Jinju 52828, Republic of Korea; hyukjun0209@gmail.com (H.-J.L.); vet00161@gmail.com (B.-G.C.); wn5886@gmail.com (Y.-H.J.); back5254@naver.com (C.-H.B.); in42869@naver.com (J.-Y.K.)

2 Dairy Science Division, National Institute of Animal Science, Rural Development Administration, Cheonan 31000, Republic of Korea; kimdh3465@korea.kr

3 Animal Nutrition and Physiology Division, National Institute of Animal Science, Rural Development Administration, Wanju 55356, Republic of Korea; shin7398@korea.kr

* Correspondence: kimsc@gnu.ac.kr; Tel.: +82-55-772-1947; Fax: +82-55-772-1949

† These authors contributed equally to this work.

Simple Summary: Dietary probiotic strategies that benefit animal performance by producing antibacterial substances in the intestine, competing with harmful gut flora, and stimulating the immune system should be developed. Thus, this study examined optimum levels of growing–finishing pigs using a mixture of microbial additives producing antimicrobial substances and digestive enzymes to improve growth performance, blood metabolites, fecal microflora, and carcass characteristics. Three treatments were used: 0, 0.5, and 1.0% microbial additives in the basal diet, which led to a higher average daily gain and feed efficiency in growing–finishing pigs but not in the initial and final weights. Supplementation of pig diets with microbial additives has been demonstrated to be an effective strategy for improving the conent of immunoglobulin G (IgG) as a blood metabolite, increasing fecal lactic acid bacteria count, and reducing *Escherichia coli* (*E. coli*) count in pig manure. However, the use of microbial additives to improve carcass characteristics has been questioned due to their lack of influence on pigs. Consequently, 1.0% microbial additive could be optimal for growing–finishing pigs to improve growth performance, IgG content, and the fecal microflora environment.

Abstract: This study aimed to assess the effects of microbial additives that produce antimicrobial and digestive enzymes on the growth performance, blood metabolites, fecal microflora, and carcass characteristics of growing–finishing pigs. A total of 180 growing–finishing pigs (Landrace × Yorkshire × Duroc; mixed sex; 14 weeks of age; 58.0 ± 1.00 kg) were then assigned to one of three groups with three repetitions (20 pigs) per treatment for 60 days of adaptation and 7 days of collection. Dietary treatments included 0, 0.5, and 1.0% microbial additives in the basal diet. For growth performance, no significant differences in the initial and final weights were observed among the dietary microbial additive treatments, except for the average daily feed intake, average daily gain, and feed efficiency. In terms of blood metabolites and fecal microflora, immunoglobulin G (IgG), blood urea nitrogen, blood glucose, and fecal lactic acid bacteria count increased linearly, and fecal *E. coli* counts decreased linearly with increasing levels of microbial additives but not growth hormones and *Salmonella*. Carcass quality grade was improved by the microbial additive. In addition, carcass characteristics were not influenced by dietary microbial additives. In conclusion, dietary supplementation with 1.0% microbial additive improved average daily gain, feed efficiency, IgG content, and fecal microflora in growing–finishing pigs.

Keywords: blood metabolite; carcass characteristic; growth performance; fecal microflora; microbial additive; pig

1. Introduction

Over the past 50 years, antibiotic-based growth promoters have been used in farm animal production in several countries. Dietary supplementation of farm animals with antibiotics improves growth productivity, disease prevention, and farm income [1,2]. However, the excessive use of antibiotics has notably impacted livestock and public health owing to problems such as residual livestock products, antibiotic resistance when ingesting residual livestock products, and the multiplication of pathogenic harmful bacteria. Therefore, growth-promoting antibiotics have been prohibited in animal feed in Europe since 2006 and in South Korea since 2011 [3–5]. Consequently, as there is increasing awareness of the potential negative effects of animal diets, there has been increased interest in producing livestock without using antibiotic growth promoters [6]. Many livestock producers have suggested the use of various antibiotics to enhance animal performance, disease prevention, health, and meat quality. One of the most effective strategies that has been successfully used to control these problems is microbial additives. Microbial additives are preparations or products containing defined concentrations of live microorganisms that are sufficient to alter the intestinal microflora of the host and exert beneficial health effects [7]. Microbial additive supplementation has been suggested to improve growth performance [8–10], the immune system [11,12], and fecal microflora [13,14]. Multi-strain or multi-species microbial additives have been found to be more effective than mono-strain or single-species additives [15]. For example, considering non-antibiotic feed additives in pig diets, the additives that are available for improving growth performance or the gut and fecal environment through inclusion in diets and include the believed mechanisms for each additive are classified into six primary categories: (i) acidifier, (ii) mineral, (iii) prebiotics, (iv) probiotics (direct-fed microbials, DFM), (v) nucleotides, and (vi) plant extracts, as described by Liu et al. [6]. More recently, probiotics and prebiotics have been used successfully in pig diets for several years. Firstly, probiotics are commonly known as direct-fed microbials and are considered "live microorganisms that confer a health benefit on the host when administered in adequate amounts [16]". Prebiotics are non-digestible, fermented food substrates that stimulate growth, change the composition and activity of gut microorganisms, and improve host health [17]. These positive effects of probiotic and prebiotic supplementation may be worthwhile as feed additives for animals to be used in a way that has more benefits for animal health and performance, especially in growing to fattening phase situations and animals exposed to greater pathogenic loads [18]. Currently, the aim of the pig industry has focused on the accumulation of scientific evidence with respect to microbial additives, such as probiotics and prebiotics, and their effect on the growth, production, and health of pigs, as well as their effect on the immune system, digestive tract, and blood metabolites. In response to San Andres et al. [18], these changes were aimed at improving the ability of pigs to prevent pathogenic bacteria from colonizing the intestinal system, which can be accomplished via mechanisms that reduce the damaging effects of pathogens on the host. Microbial additives containing *Bacillus* spores improved the weight gain, feed conversion ratio, and carcass quality of the pigs [8]. In contrast, probiotics with the *Lactobacillus* strain can increase the gut immunoglobulin A (IgA) immune response and promote the gut immunological barrier [12], while *Saccharomyces* supplementation increased fecal lactic acid bacteria counts and pathogenic bacteria such as *Escherichia coli* (*E. coli*) decreased in number in pigs [13]. However, few attempts have been made to develop multi-microbe microbial additive products, and reports on the effects of multi-species microbial additives on growing–finishing pigs are limited.

Therefore, we hypothesized a positive influence of multi-microbe microbial additive products on the production and immune response of the blood metabolites, microflora, and meat quality of pigs. The objective of this study was to investigate the effect of microbial additives producing antimicrobial substances and digestive enzymes on the growth performance, blood metabolites, fecal microflora, and carcass characteristics of growing to finishing pigs.

2. Materials and Methods

The animal experiments were conducted at the Goseong Pig Farm (Gyeongnam, Republic of Korea) and were approved by the Animal Care and Use Committee of Gyeongsang National University, Jinju, Republic of Korea (GNU-200608-P0034).

2.1. Probiotics

The microbes included *Lactobacillus plantarum* SK3121 (>9.0 log10 colony-forming units (CFU)/g), *Bacillus subtilis* SK877 (>9.0 log10 CFU/g), *B. amyloliquefaciens* BBG-B5 (>9.0 log10 CFU/g), and *Saccharomyces cerevisiae* SK3587 (>9.0 log10 CFU/g), which were used as the seedstocks. *L. plantarum* SK3121 and *B. subtilis* SK877 were isolated based on their antimicrobial activity in Kimchi (Korean traditional fermented cabbage) and digestive enzyme activity in corn silage, respectively [4]. *B. amyloliquefaciens* BBG-B5 and *S. cerevisiae* SK3589 were isolated from pig feces based on their digestive enzyme activity, nutrients, and growth factors. The microbial additive used in this study, in which the seedstocks were applied into the grain mixtures at a 2% (as-fed basis) and ensiled at 30 °C for 7 days, was purchased from Big Biogen (Anseong, Republic of Korea). The counts of the microbial additive are presented in Table 1.

Table 1. Microbial counts of dietary additives used in this study (log10 CFU/g).

Item	Microbial Additive
Lactic acid bacteria	7.98
Bacillus subtilis	7.94
Yeast	8.09

2.2. Animal Management

A total of 180 growing–finishing pigs (Landrace × Yorkshire × Duroc; mixed sex; 14 weeks of age; 58.0 ± 1.00 kg) were randomly divided into three treatments with three repetitions (20 pigs) per treatment for 60 days of adaptation and 7 days of collection. The dietary treatments consisted of 0% (basal diet), 0.5% (basal diet + 0.5% microbial additive), and 1.0% (basal diet + 1.0% microbial additive). The basal diet was used throughout the experimental period (Table 2). The pens were fully slatted with concrete panels, and the light and temperature conditions were automatically controlled. Pigs were fed ad libitum using a one-hole feeder in each pen. The diet was delivered twice daily at 09:00 h and 17:00 h, and water was provided ad libitum per pen via nipples.

Table 2. Ingredients and chemical compositions of basal diets (DM basis).

Item	Basal Diet
Ingredient, %	
Corn	48.5
Soybean meal	31.9
Rice bran	5.00
Tallow	4.80
Lupine	3.20
Molasses	3.00
Calcium phosphate	1.60
Lysine	0.50
Methionine	0.50
Sodium chloride	0.30
Mineral premix [1]	0.60
Vitamin primix [2]	0.10
Total	100.0

Table 2. *Cont.*

Item	Basal Diet
Chemical compositions [3]	
ME, kcal/kg	3100
Dry matter, %	87.4
Crude protein, %	18.8
Ether extract, %	9.59
Crude ash, %	7.62

[1] One kilogram of the diet contained the following: Fe, 70 mg; Cu, 50 mg; Zn, 25 mg; Mn, 30 mg; I, 0.7 mg; Co, 0.5 mg; Se, 0.26 mg. [2] One kilogram of the diet contained the following: vitamin A, 16,000 IU; vitamin D_3, 3000 IU; vitamin E, 40 IU; vitamin B1, 2.5 mg; vitamin B2, 20 mg; vitamin B6, 4 mg; vitamin B12, 0.076 mg; vitamin K3, 2.5 mg; panthothenic acid, 40 mg; niacin, 75 mg; biotin 0.15 mg; folic acid, 0.65 mg; ethoxyquin, 12 mg. [3] Values represent the results of three samples, each assayed in triplicate.

2.3. *Analysis*

2.3.1. Diet Chemical Composition

The feed (1 kg) was dried at 65 °C for 48 h in a forced-air oven and ground using a cutting mill to pass through a 1 mm screen (Shinmyung Electric Co., Ltd., Gimpo, Republic of Korea). The metabolizable energy in the feed was calculated using the energy values of the ingredients obtained from the NRC [19]. The dry matter concentration was determined using a forced-air drying oven at 105 °C for 24 h. Crude protein and ether extracts were measured according to the Kjeldahl method (method number 984.13; AOAC, [20]) and the Soxhlet method (method number 920.39; AOAC, [20]), respectively. Crude ash was determined via incineration at 550 °C for 4 h in a muffle furnace.

2.3.2. Microbial Counts

The microbial additive sample (20 g) was placed in 180 mL of distilled water and processed in a blender for 30 s. The extract was filtered through two layers of cheesecloth and diluted (10^{-6} to 10^{-8}) to determine the microbial counts for lactic acid bacteria (LAB), bacilli, and yeast [21,22]. Microbial counts were measured via plate counting on Lactobacilli Man Rogosa Sharpe agar (MRS; Difco, Detroit, MI, USA) for LAB, Luria-Bertani agar (LB; Difco, Detroit, MI, USA) for *Bacillus* and potato dextrose agar (PDA; Difco, Detroit, MI, USA) for yeast. The MRS agar plates were maintained in a CO_2 incubator (Thermo Scientific, Waltham, MA, USA) at 30 °C for 48 h. The LB agar and PDA plates were incubated for 48 h at 30 °C under an aerobic incubator (Johnsam Corp., Boocheon, Republic of Korea) [23]. Visible colonies on the plates were counted and expressed as colony-forming units (log10 CFU/g of the sample).

2.3.3. Growth Performance

To analyze growth performance, each pig was weighed at the beginning (day 1) and end (day 60) of the experimental period to calculate the average daily gain (ADG). Feed intake was measured for each individual pen, and feed efficiency was determined by dividing the ADG by the average daily feed intake (ADFI) over 60 days (gain/intake). Additionally, ADFI was calculated by subtracting the feed remaining in the feeder from the feed offered.

2.3.4. Blood Metabolites

At 60 days, blood samples were collected from 10 mL vacuum tubes containing K_3EDTA (Becton Dickinson Vacutainer Systems, Franklin Lakes, NJ, USA) and then centrifuged at 3000× *g* for 15 min to separate the serum. Serum immunoglobulin G (IgG) and growth hormone levels were determined using commercial enzyme-linked immune sorbent assay (ELISA) kits. The plasma concentration of blood urea nitrogen (BUN) was determined using a UREA/BUN kit (Roche, Basel, Switzerland). An enzymatic kinetic assay was used to determine the plasma glucose concentration (GLU Kit; Roche, Mannheim, Germany).

2.3.5. Fecal Microflora

To measure LAB, *Salmonella enterica*, and *E. coli* loads, fecal samples (200 g) were collected monthly from each pen at five random locations and immediately analyzed. Each fecal sample (10 g) was weighed and placed in a stomacher bag containing 90 mL of sterile saline (0.9%) at a dilution of 1:10. Fecal samples were then plated on Difco MRS agar (Difco, Detroit, MI, USA), DifcoTM SS agar (Becton, Dickinson and Company, Sparks, MD, USA), and DifcoTM Violet Red Bile agar (Becton, Dickinson and Company, Sparks, MD, USA). The MRS agar plates were incubated in a CO_2 incubator (Thermo Scientific, Waltham, MA, USA) at 30 °C for 48 h, whereas the SS agar and Violet Red Bile agar plates were incubated for 48 h at 37 °C in an aerobic incubator (Johnsam Corp., Boocheon, Republic of Korea). Visible colonies from the plates were counted, and the number of CFU/g of fecal extract at weeks 0, 30, and 60 was calculated. Microbiological data were transformed to log10.

2.3.6. Carcass Characteristics

At the end of the feeding trial, all animals were moved to the Goryeong Nonghyup Meat Processing Facility, Goryeong, Republic of Korea, and slaughtered as approved by the Ministry of Agriculture, Food, and Rural Affairs after 24 h rest. Subsequently, all cold carcasses were chilled at 2 °C for 24 h, and then, carcass characteristics (carcass weight, back fat thickness, and carcass quality grade) were measured according to the guidelines of the Animal Products Grading Service, Republic of Korea [24].

2.4. Statistical Analysis

Data were analyzed using analysis of variance (ANOVA) in the generalized linear model (GLM) procedure of SAS (Statistical Analysis System, version 8.2, [25]), followed by Tukey's test to identify differences among the treatments. Significant effects were set at $p < 0.05$ and < 0.1 as tendencies. The IML procedure in SAS was used to generate linear and quadratic orthogonal polynomial coefficients for the unequally spaced data in the experiment. When a polynomial contrast (linear and quadratic effects) was significant, the effects of increasing the microbial additive supplementation levels were used.

3. Results

3.1. Growth Performance

During the 60-day experimental period, the ADFI decreased linearly ($p = 0.017$) in the microbial additive supplementation groups, which was lower than in the control group (Table 3). In addition, ADG and feed efficiency increased linearly ($p = 0.011$ and 0.015, respectively) with increasing levels of microbial additives ($p < 0.05$). No significant differences in the initial and final weights ($p > 0.05$) were observed among the treatments.

Table 3. Effects of microbial additive supplementation on the growth performance of growing–finishing pigs.

Item	Supplement, % [1]			SEM	*p*-Value	Contrast	
	0	0.5	1.0			Linear	Quadratic
Initial weight, kg	58.5	59.0	58.0	1.322	0.640	0.595	0.425
Final weight, kg	100.6	104.0	103.5	3.589	0.440	0.286	0.558
Average daily feed intake, kg/d	1.84 [a]	1.77 [b]	1.73 [b]	0.057	0.046	0.017	0.044
Average daily gain, kg/d	0.70 [b]	0.75 [a]	0.76 [a]	0.041	0.011	0.011	0.343
Feed efficiency (Gain:intake)	0.38 [b]	0.42 [a]	0.44 [a]	0.032	0.033	0.015	0.131

[1] Supplemented microbial additive at 0, 0.5, and 1.0% of basal diet. [a,b] Means in the same row with different superscripts differ significantly ($p < 0.05$).

3.2. Blood Metabolites

Regarding blood metabolites, the blood glucose concentration was the highest with 1.0% supplementation ($p = 0.046$, Table 4). In addition, IgG, BUN, and blood glucose levels increased linearly with increasing levels of microbial additives (p = 0.031, 0.049, and 0.003, respectively). No significant differences in the concentration of growth hormone were observed among the treatments (p = 0.212).

Table 4. Effects of microbial additive supplementation on the blood metabolites of growing–finishing pigs.

Item	Supplement, % [1]			SEM [2]	p-Value	Contrast	
	0	0.5	1.0			Linear	Quadratic
IgG, mg/mL	21.5	22.4	23.9	2.415	0.854	0.031	0.265
Growth hormone, ng/mL	0.22	0.23	0.24	0.097	0.212	0.858	0.626
Blood urea nitrogen, mg/dL	16.0	16.4	17.5	1.683	0.331	0.049	0.144
Blood glucose, mg/dL	63.1 ^{b}	63.2 ^{b}	67.9 ^{a}	2.356	0.046	0.003	0.101

[1] Supplemented microbial additive at 0, 0.5, and 1.0% of basal diet. [2] SEM, standard error of the mean. ^{a,b} Means in the same row with different superscripts differ significantly ($p < 0.05$).

3.3. Fecal Microflora

Fecal *Salmonella* was not detected in any of the treatments during the 60-day experimental period (Table 5). On days 30 and 60, fecal LAB counts increased linearly (p = 0.015 and 0.036, respectively) with increasing levels of microbial additives, whereas fecal *E. coli* counts decreased linearly (p = 0.048 and 0.039, respectively) with increasing levels of microbial additives.

Table 5. Effects of microbial additive supplementation on the fecal microflora of growing–finishing pigs.

Day	Microflora	Supplement, % [1]			SEM [2]	p-Values	Contrast	
		0	0.5	1.0			Linear	Quadratic
0 day	LAB [3]	6.21	6.28	6.38	0.303	0.807	0.071	0.584
	Salmonella	ND [4]	ND	ND	N/A [5]	N/A	N/A	N/A
	E. coli	4.18	4.08	3.97	0.589	0.494	0.068	0.777
30 day	LAB	6.68 ^{b}	7.03 ^{a}	7.15 ^{a}	0.124	0.026	0.015	0.584
	Salmonella	ND	ND	ND	N/A	N/A	N/A	N/A
	E. coli	4.13 ^{a}	3.90 ^{b}	3.74 ^{c}	0.089	0.044	0.048	0.777
60 day	LAB	6.72 ^{b}	6.88 ^{ab}	7.09 ^{a}	0.203	0.046	0.036	0.909
	Salmonella	ND	ND	ND	N/A	N/A	N/A	N/A
	E. coli	4.07 ^{a}	3.92 ^{b}	3.86 ^{b}	0.069	0.039	0.039	0.163

[1] Supplemented microbial additive at 0, 0.5, and 1.0% of basal diet. [2] SEM, standard error of the mean. [3] LAB, lactic acid bacteria. [4] ND, not detected. [5] N/A, not applicable. $^{a–c}$ Means in the same row with different superscripts differ significantly ($p < 0.05$).

3.4. Carcass Characteristics

Regarding carcass characteristics, we observed that the "1+" carcass quality grade was higher in the microbial additive supplementation groups (0.5% and 1%) than in the control group (Table 6). In addition, there were no significant differences in carcass weight or back-fat thickness among the treatments at 60 d (p = 0.637).

Table 6. Effects of microbial additive supplementation on the carcass characteristics of growing–finishing pigs.

Item	Supplement, % [1]			SEM [2]	*p*-Value	Contrast	
	0	0.5	1.0			Linear	Quadratic
Carcass weight, kg	78.2	78.3	79.7	4.801	0.756	0.160	0.474
Back-fat thickness, mm	19.5	19.1	20.5	3.478	0.637	0.082	0.157
Carcass quality grade, % (1+:1:2)	7:15:78	13:17:70	15:27:58	N/A [3]	N/A	N/A	N/A

[1] Supplemented microbial additive at 0, 0.5, and 1.0% of basal diet. [2] SEM, standard error of the mean. [3] N/A, not applicable.

4. Discussion

ADG, ADFI, and F:G ratio are vital parameters for assessing performance during the pig growth phase [26]. In this study, the use of microbial additives was shown to have significant effects on pig growth performance, suggesting the beneficial effects of currently used probiotic formulations. In other words, growing–finishing pigs supplemented with probiotics demonstrated greater body weight, ADG, and feed efficiency or lower ADFI than pigs in the control group who were not supplemented with probiotics. Chen et al. [11] reported increased ADG in growing pigs fed diets supplemented with 0.2% bacillus-based probiotics. Similarly, Jeon et al. [27] reported increased ADG and feed efficiency in growing pigs fed a probiotic-supplemented diet. According to several studies, complex probiotics positively affect the growth performance of growing–finishing pigs [8,10]. Our results are consistent with those of San Andres et al. [18] and Hong et al. [28], who reported that pigs fed multi-species microbial additive diets had significantly increased ADG, and during days 28 to 35 after weaning, the use of prebiotic mixtures improved the growth performance of nursery pigs. Giang et al. [14] reported that adding a mixture of probiotics (LAB complex, *Bacillus*, and *Saccharomyces*) increased ADG and improved feed efficiency compared with the control. Notably, the above-mentioned microbial complex also has probiotic potential in growing to finishing pigs. Thus, microbial additives improve daily gain and feed efficiency owing to the digestive enzymes and growth factors derived from probiotics. For example, the addition of direct-fed microbes, commonly known as probiotics, to swine feed can improve gut health by changing the microflora environment that suppresses pathogens. Additionally, it results in increased nutrient digestibility, improved health status, and the improved growth performance of pigs [14,29,30]. The beneficial effects of prebiotics in pigs have been linked to their increased fermentability. This occurs due to apoptosis in the small intestine, which leads to increased intestinal cell proliferation, subsequently improving digestive and absorptive capacities [31,32]. The growth of weaning pigs depends on the abundance of LAB and *Bifidobacteria* [33,34]. These bacteria and their fermentation products (short-chain fatty acids and polyamines) represent the energy supply for colonic epithelial cells, aiding absorption [33–35]. As mentioned above (probiotics), beneficial microbes (such as LAB) in prebiotics can produce bacteriocins, lactic acid, and other compounds that improve the intestinal environment and may inhibit the growth of certain pathogens [36]. According to Liu et al. [37], using 100 or 200 mg/kg of chito-oligosaccharide (derived from chitosan) in diets improved the growth performance and digestibility of dietary nutrients in weaning pigs. However, these positive effects of probiotics may be attributed to differences in the bacterial species used in the microbial additive preparations and pig genotypes [38].

Notably, the addition of the 1% microbial extract resulted in the highest IgG concentration. This plays a major role in antibody-mediated defense mechanisms [39,40] and suggests that IgG is more important for development than the other blood metabolites in this study. Probiotics control the production of lymphocyte cytokines and exert a major effect on the immune system [41]. Cho et al. [42] observed that microbial supplementation directly added to pig diets may also cause a decrease in immune stimulation by reducing pro-inflammatory cytokines in enterocytes. Therefore, an immune change can shift the energy utilized in excessive immune stimulation toward growth and improve feed efficiency. Furthermore, our results are well supported by those of a previous study in which growing-to-finishing pigs that received supplementation with *B. subtilis* had a positive impact on the evident increase in the effect of the probiotic on IgG [43]. However, there were no differences in serum IgA and immunoglobulin M levels between the groups. Similarly, Wang et al. [44] showed that a combination of *B. subtilis* and *Enterococcus faecium* in sow diets increased serum IgG levels. This implies that an increasing IgG concentration results in a better immune response and health in growing to finishing pigs. However, the growth hormones in this study did not produce the expected results because their content in all treatments was similar, suggesting no considerable effect on the growth hormones of pigs during the growing to finishing period. Growth hormone is an important factor that primarily regulates animal growth through related receptors and downstream pathways [45]. Significant correlations between growth levels and increased weight have been reported based on animal data [46]. In this study, an increase in BUN and blood glucose values with microbial addition compared to the controls was not observed in growing pigs. Higher BUN levels represent lower nitrogen absorption efficiency, indicating an increase in lean body mass [47]. BUN levels generally decrease when the protein mass and absorption are reduced [47]. Duan et al. [48] reported that the grower phase, the control group (0%), had significantly lower BUN values than the 0.1% and 0.3% *Lactobacillus lactis* groups, whereas no difference was observed in the BUN values among the three groups for the finisher phase. For example, in the digestive tract, probiotics increase ammonia fixation and alleviate decreases in amino acid availability, which can be reduced by increasing the concentration of blood urea [49]. One observation with supplementary microbial additives at the 1% level was an increase in blood glucose concentrations. Thus, higher blood glucose levels might be explained by the activity of digestive enzymes from the microbes used or a response to increased energy absorption in the intestine [50,51]. Balasubramanian et al. [1] and Devi and Kim [52] found that microbial additive supplementation (0.1 or 0.2 g/kg multi-species probiotic, 0.2% medium-chain fatty acids, and 0.1% probiotic) had a significant effect on pig blood glucose concentrations. In contrast, Chen et al. [9] reported that feeding pigs with microbial additives (0.1 and 0.2% complex probiotics) did not affect their blood traits. However, this was not the case in the present study. At present, the mechanisms underlying these blood parameters remain unclear. In addition, blood glucose and BUN levels were within the reference ranges [53]. In terms of the effect of microbial additives on gastrointestinal health, enhancement in the ability of growing–finishing pigs to digest and ferment nutrients may correspond to an increase in the growth performance associated with immune system stimulation, including a decrease in pathogenic bacteria [44].

In this study, we determined the effects of microbial supplementation on the fecal microflora of pigs (Table 5). The increased fecal LAB or reduced fecal *E. coli* after microbial additive supplementation compared to the control is in line with the findings of Balasubramanian et al. [13], who suggested that a microbial additive containing 0.01 and 0.02% *Bacillus* spp. in basal diets affected fecal LAB counts and inhibited fecal *E. coli* counts. This may be partly explained by the presence of LAB, which are excellent antibacterial agents that suppress the growth of pathogenic microorganisms. Similar findings were reported by Lu et al. [54], who noted that supplementing the diet with a probiotic complex altered the bacterial community in the feces of weaned piglets. A study on the inclusion of multi-probiotics was reported by Giang et al. [14] in that the results of increased fecal LAB count and decreased fecal *E. coli* count in growing pigs owed to the inhibition of pathogenic microbial growth and activity by the probiotic characteristics. In addition, it has been reported that probiotics with *Bacillus* strains can not only change intestinal bacteria through colonization but can also produce specific bacteriocins by inhibiting the widest range of pathogenic bacteria [55]. In general, *Lactobacillus* spp. in probiotics can induce beneficial enzyme activities, such as sucrase, lactase, and tripeptidase in the pig small intestine and thereby promote the growth of "good" bacteria through their functions that help the absorption of nutrients and keep the balance of the intestinal or fecal microbiota [56]. Based on this information, this could be a probable reason to support our results on fecal microbes. Surprisingly, no fecal *Salmonella* in pig manure was detected in any of the treatments despite the antibacterial activity linked to the pig gut.

Furthermore, microbial supplementation resulted in no significant differences in carcass characteristics, indicating that no noticeable changes in carcass characteristics were observed during the 60-day experimental period. Exceptionally, pigs supplemented with microbial additives tended to have slight increases in carcass weight and back-fat thickness at the 1% level compared to the other groups. Junka et al. [57] and Ganeshkumar et al. [58] observed a significantly increased carcass weight in pigs that received probiotic supplementation. Chu et al. [59] reported that the carcass weight decreased in pigs fed diets supplemented with microbial additives. Because of this back-fat thickness, our observations were not in accordance with those of previous studies. Grela et al. [60] found an effect of prebiotics on back-fat thickness, which was lower in pigs fed dried Jerusalem artichokes. Other results reported by Chang et al. [61] stated that probiotic treatment groups had no significant effect on backfat thickness in pigs. Consequently, the outcomes of these studies may have been attributed to the different concentrations of microbial additives used or various important factors, such as the composition and form of the feed, interactions with probiotics, or probiotic strains [62]. Among the carcass characteristics, our data showed a higher "1+" carcass quality grade by increasing the microbial additive amount. These findings are consistent with those of a previous study, which reported that supplementing growing–finishing pig diets with *Bacillus* spp. probiotics increased meat carcass quality grade [13]. Min et al. [63] observed no beneficial effects on carcass quality grade in growing–finishing pigs fed a dietary mixture of proteases and probiotics. The discrepancies between the results of our study and those of previous studies may be due to differences in microbial abilities. However, further studies are required to evaluate the exact mechanisms of microbial action on carcass grade.

5. Conclusions

In conclusion, this study provides an extensive investigation of the growth performance, blood metabolites, fecal microflora, and carcass characteristics of growing to finishing pigs fed diets supplemented with microbial additives. The results show that dietary supplementation with 1.0% microbial additive effectively improved the growth performance (ADG and feed efficiency) and IgG content of the growing to finishing pigs. In addition, the 1.0% dietary microbial additives boosted the fecal microflora environment by increasing fecal LAB levels and decreasing fecal *E. coli* counts. In particular, among the carcass characteristics, these results gained a higher "1+" carcass quality grade by increasing the microbial additive, which may be due to differences in the ability of the microbials used. This study contributes to our knowledge of sustainable manure management techniques by offering valuable insights into the optimization of microbial additive levels.

Author Contributions: Conceptualization, H.-J.L., B.-G.C. and S.-C.K.; methodology, H.-J.L., D.-H.K. and S.-C.K.; software, B.-G.C., Y.-H.J. and S.-S.L.; validation, Y.-H.J., D.-H.K., S.-S.L. and S.-C.K.; formal analysis, H.-J.L., B.-G.C., C.-H.B. and J.-Y.K.; investigation, H.-J.L., B.-G.C., C.-H.B. and J.-Y.K.; resources, H.-J.L. and B.-G.C.; data curation, H.-J.L. and B.-G.C.; writing—original draft preparation, H.-J.L. and B.-G.C.; writing—review and editing, D.-H.K. and S.-C.K.; visualization, H.-J.L. and B.-G.C.; supervision, D.-H.K. and S.-C.K.; project administration, S.-C.K.; funding acquisition, H.-J.L., B.-G.C. and S.-C.K. All authors have read and agreed to the published version of the manuscript.

Funding: This research was supported by the IPET (Korea Institute of Planning and Evaluation for Technology in Food, Agriculture, Forestry and Fisheries; Project No. 315017-05-2-SB030) and the Ministry of Agriculture, Food and Rural Affairs, Korea.

Institutional Review Board Statement: The animal experiments were conducted at the Goseong Pig Farm (Gyeongnam, South Korea) and were approved by the Animal Care and Use Committee of Gyeongsang National University, Jinju, South Korea (GNU-200608-P0034).

Informed Consent Statement: Not applicable.

Data Availability Statement: The data presented in this study are available from the corresponding author upon request.

Conflicts of Interest: The authors declare no conflicts of interest.

References

1. Balasubramanian, B.; Lee, S.I.; Kim, I.H. Inclusion of dietary multi-species probiotic on growth performance, nutrient digestibility, meat quality traits, faecal microbiota and diarrhoea score in growing-finishing pigs. *Ital. J. Anim. Sci.* **2018**, *17*, 100–106. [CrossRef]
2. Kong, X.F.; Yin, F.G.; He, Q.H.; Liu, H.J.; Li, T.J.; Huang, R.L.; Fan, M.Z.; Liu, Y.L.; Hou, Y.Q.; Li, P.; et al. *Acanthopanax senticosus* extract as a dietary additive enhances the apparent ileal digestibility of amino acids in weaned piglets. *Livest. Sci.* **2009**, *123*, 261–267. [CrossRef]
3. Kim, Y.J.; Chung, T.H.; Choi, I.H. Influence of supplemental *Schisandra chinensis* powder on growth performance, serum cholesterol, and meat quality of broilers. *Acta. Agric. Scand. Sect. A—Anim. Sci.* **2013**, *63*, 175–182. [CrossRef]
4. Lee, H.J.; Cho, S.L.; Kim, D.H.; Amanullah, S.M.; Joo, Y.H.; Yang, H.S.; Choi, L.H.; Moon, S.S.; Kim, S.C. Effects of *Chrysanthemum boreale* probiotic on growth performance and meat quality of Hanwoo Steer. *J. Agric. Life Sci.* **2015**, *48*, 85–94. [CrossRef]
5. Mathur, S.; Singh, R. Antibiotic resistance in food lactic acid bacteria—A review. *Int. J. Food Microbiol.* **2005**, *105*, 281–295. [CrossRef]
6. Liu, Y.; Espinosa, C.D.; Abelilla, J.J.; Casas, G.A.; Vanessa Lagos, L.; Lee, S.A.; Kwon, W.B.; Mathai, J.K.; Navarro, D.M.D.L.; Jaworski, N.W.; et al. Non-antibiotic feed additives in diets for pigs: A review. *Anim. Nutr.* **2018**, *4*, 113–125. [CrossRef]
7. Schrezenmeir, J.; de Vrese, M. Probiotics, probiotics and synbiotics approaching a definition. *Am. J. Clin. Nutr.* **2001**, *73*, 361s–364s. [CrossRef]
8. Alexopoulos, C.; Georgoulakis, I.E.; Tzivara, A.; Kyriakis, C.S.; Govaris, A.; Kyriakis, S.C. Field evaluation of the effect of a probiotic-containing *Bacillus licheniformis* and *Bacillus subtilis* spores on the health status, performance, and carcass quality of grower and finisher pigs. *J. Vet. Med. Ser. A* **2004**, *51*, 306–312. [CrossRef]
9. Chen, Y.J.; Son, K.S.; Min, B.J.; Cho, J.H.; Kwon, O.S.; Kim, I.H. Effects of dietary probiotic on growth performance, nutrients digestibility, blood characteristics and fecal noxious gas content in growing pigs. *Asian-Australas. J. Anim. Sci.* **2005**, *18*, 1464–1468. [CrossRef]
10. Shon, K.S.; Hong, J.W.; Kwon, O.S.; Min, B.J.; Lee, W.B.; Kim, I.H.; Park, Y.H.; Lee, I.S. Effects of *Lactobacillus reuteri* based direct-fed microbial supplementation for growing-finishing pigs. *Asian-Australas. J. Anim. Sci.* **2005**, *18*, 370–374. [CrossRef]
11. Chen, Y.J.; Min, B.J.; Cho, J.H.; Kwon, O.S.; Son, K.S.; Kim, H.J.; Kim, I.H. Effects of dietary *Bacillus*-based probiotic on growth performance, nutrients digestibility, blood characteristics and fecal noxious gas content in finishing pigs. *Asian-Australas. J. Anim. Sci.* **2006**, *19*, 587–592. [CrossRef]
12. Malin, M.; Suomalainen, H.; Saxelin, M.; Isolauri, E. Promotion of IgA immune response in patients with Crohn's disease by oral bacteriotherapy with *Lactobacillus GG. Ann. Nutr. Metab.* **1996**, *40*, 137–145. [CrossRef]
13. Balasubramanian, B.; Li, T.; Kim, I.H. Effects of supplementing growing-finishing pig diets with *Bacillus* spp. probiotic on growth performance and meat-carcass grade quality traits. *Rev. Bras. Zootec.* **2016**, *45*, 93–100. [CrossRef]
14. Giang, H.H.; Viet, T.Q.; Ogle, B.; Lindberg, J.E. Effects of supplementation of probiotics on the performance, nutrient digestibility and faecal microflora in growing-finishing pigs. *Asian-Australas. J. Anim. Sci.* **2011**, *24*, 655–661. [CrossRef]
15. Timmerman, H.M.; Koning, C.J.M.; Mulder, L.; Rombouts, F.M.; Beynen, A.C. Monostrain, multistrain and multispecies probiotics. A comparison of functionality and efficacy. *Int. J. Food Microbiol.* **2004**, *96*, 219–233. [CrossRef]
16. FAO/WHO (Food and Agriculture Organization/World Health Organization). Health and nutritional properties of probiotics in food including powder milk with live lactic acid bacteria. In *Report of a Joint FAO/WHO Expert Consultation on Evaluation of Health and Nutritional Properties of Probiotics in Food Including Powder Milk with Live Lactic Acid Bacteria*; FAO: Cordoba, Argentina, 2001; pp. 1–33.
17. Slavin, J. Fiber and prebiotics: Mechanisms and health benefits. *Nutrients* **2013**, *5*, 1417–1435. [CrossRef]
18. San Andres, J.V.; Mastromano, G.A.; Li, Y.; Tran, H.; Bundy, J.W.; Miller, P.S.; Burkey, T.E. The effects of prebiotics on growth performance and in vitro immune biomarkers in weaned pigs. *Transl. Anim. Sci.* **2019**, *3*, 1315–1325. [CrossRef]
19. NRC. *Nutrient Requirements of Swine*, 11th ed.; National Academy Press: Washington, DC, USA, 2012.
20. Association of Official Analytical Chemists. *Official Methods of Analysis*, 18th ed.; AOAC International: Washington, DC, USA, 2005.
21. Paradhipta, D.H.V.; Lee, S.S.; Kang, B.; Joo, Y.H.; Lee, H.J.; Lee, Y.; Kim, J.; Kim, S.C. Dual-purpose inoculants and their effects on corn silage. *Microorganisms* **2020**, *8*, 765. [CrossRef]
22. Paradhipta, D.H.V.; Joo, Y.H.; Lee, H.J.; Lee, S.S.; Kwak, Y.G.; Han, O.K.; Kim, D.H.; Kim, S.C. Effects of wild or mutated inoculants on rye silage and its rumen fermentation indices. *Asian-Australas. J. Anim. Sci.* **2020**, *33*, 949–956. [CrossRef]
23. Paradhipta, D.H.V.; Joo, Y.H.; Lee, H.J.; Lee, S.S.; Kim, D.H.; Kim, J.D.; Kim, S.C. Effects of inoculant application on fermentation quality and rumen digestibility of high moisture sorghum-sudangrass silage. *J. Appl. Anim. Res.* **2019**, *47*, 486–491. [CrossRef]
24. APGS. Animal Products Grading Service. Report of business for animal products grading. In *Animal Products Grading System*; National Livestock Co-Operative Federation: Seoul, Republic of Korea, 2007.
25. SAS Institute Inc. *SAS/STAT User's Guide, Version 9*; SAS Institute Inc.: Cary, NC, USA, 2002.

26. Sarkar, G.; Mondal, S.; Bhattacharya, D.; Ponraj, P.; Sawhney, S.; Bala, P.; Chakraborty, D.; Sunder, J.; De, A.K. Effect of a Multi-Strain Probiotic on Growth Performance, Lipid Panel, Antioxidant Profile, and Immune Response in Andaman Local Piglets at Weaning. *Fermentation* **2023**, *9*, 970. [CrossRef]
27. Jeon, B.S.; Kwag, J.H.; Yoo, Y.H.; Cha, J.O.; Park, H.S. Effects of feeding enzyme, probiotics or yucca power on pig growth and odor-generating substances in feces. *Korean J. Anim. Sci.* **1996**, *38*, 52–58.
28. Hong, J.W.; Kim, I.H.; Kwon, O.S.; Kim, J.H.; Min, B.J.; Lee, W.B. Effects of dietary probiotics supplementation on growth performance and fecal gas emission in nursing and finishing pigs. *Korean J. Anim. Sci. Technol.* **2002**, *44*, 305–314. [CrossRef]
29. Cromwell, G.L. Feed additives in swine diets. In *Sustainable Swine Nutrition*, 1st ed.; Chiba, L.I., Ed.; John Wiley Sons Inc.: Ames, IA, USA, 2013; pp. 302–304.
30. Kenny, M.; Smidt, H.; Mengheri, E.; Miller, B. Probiotics—Do They Have A Role in the Pig Industry? *Animal* **2011**, *5*, 462–470. [CrossRef]
31. Lindberg, J.E. Fiber effects in nutrition and gut health in pigs. *J. Anim. Sci. Biotechnol.* **2014**, *5*, 15. [CrossRef]
32. van der Aar, P.J.; Molist, F.; van der Klis, J.D. The central role of intestinal health on the effect of feed additives on feed intake in swine and poultry. *Anim. Feed Sci. Technol.* **2017**, *233*, 64–75. [CrossRef]
33. Abe, F.; Ishibashi, N.; Shimamura, S. Effect of administration of bifidobacteria and lactic acid bacteria to newborn calves and piglets. *J. Dairy Sci.* **1995**, *78*, 2838–2846. [CrossRef]
34. Modesto, M.; D'Aimmo, M.R.; Stefanini, I.; Trevisi, P.; De Filippi, S.; Casini, L.; Mazzoni, M.; Bosi, P.; Biavati, B. A novel strategy to select bifidobacterium strains and prebiotics as natural growth promoters in newly weaned pigs. *Livest. Sci.* **2009**, *122*, 248–258. [CrossRef]
35. Donohoe, D.R.; Garge, N.; Zhang, X.; Sun, W.; O'Connell, T.M.; Bunger, M.K.; Bultman, S.J. The microbiome and butyrate regulate energy metabolism and autophagy in the mammalian colon. *Cell Metab.* **2011**, *13*, 517–526. [CrossRef]
36. Ringø, E.; Schillinger, U.; Holzapfel, W. Antimicrobial activity of lactic acid bacteria isolated from aquatic animals and the use of lactic acid bacteria in aquaculture. In *Microbial Ecology of Growing Animals*; Saunders Ltd.: St. Louis, MO, USA, 2015; Volume 2, pp. 418–453.
37. Liu, P.; Piao, X.S.; Kim, S.W.; Wang, L.; Shen, Y.B.; Lee, H.S.; Li, S.Y. Effects of chito-oligosaccharide supplementation on the growth performance, nutrient digestibility, intestinal morphology, and fecal shedding of *Escherichia coli* and *Lactobacillus* in weaning pigs. *J. Anim. Sci.* **2008**, *86*, 2609–2618. [CrossRef]
38. Rekiel, A.; Wiecek, J.; Dziuba, M. Effect of feed additives on the results of fattening and selected slaughter and quality traits of pork meat of pigs with different genotypes. *Czech J. Anim. Sci.* **2005**, *50*, 561–567. [CrossRef]
39. Snoeck, V.; Peters, I.R.; Cox, E. The IgA system: A comparison of structure and function in different species. *Vet. Res.* **2006**, *37*, 455–467. [CrossRef]
40. Samolińska, W.; Kowalczuk-Vasilev, E.; Grela, E.R. Comparative effect of different dietary inulin sources and probiotics on growth performance and blood characteristics in growing-finishing pigs. *Arch. Anim. Nutr.* **2018**, *72*, 379–395. [CrossRef]
41. Mun, D.Y.; Kyoung, H.J.; Kong, M.H.; Ryu, S.D.; Jang, K.B.; Baek, J.R.; Park, K.I.; Song, M.H.; Kim, Y.H. Effects of *Bacillus*-based probiotics on growth performance, nutrient digestibility, and intestinal health of weaned pigs. *J. Anim. Sci. Technol.* **2021**, *63*, 1314–1327. [CrossRef]
42. Cho, J.H.; Zhao, P.Y.; Kim, I.H. Probiotics as a dietary additive for pigs: A review. *J. Anim. Vet. Adv.* **2011**, *10*, 2127–2134. [CrossRef]
43. Li, R.; Liu, J.; Liu, Y.; Cao, L.; Qiu, W.; Qin, M. Probiotic effects of *Bacillus subtilis* on growth performance and intestinal microecological balance of growing-to-finishing pigs. *J. Food Biochem.* **2023**, *2023*, 7150917. [CrossRef]
44. Wang, C.; Wei, S.; Xu, B.; Hao, L.; Su, W.; Jin, M.; Wang, Y. *Bacillus subtilis* and *Enterococcus faecium* co-fermented feed regulates lactating sow's performance, immune status and gut microbiota. *Microb. Biotechnol.* **2021**, *14*, 614–627. [CrossRef]
45. Renaville, R.; Hammadi, M.; Portetelle, D. Role of the somatotropic axis in the mammalian metabolism. *Domest. Anim. Endocrinol.* **2002**, *23*, 351–360. [CrossRef]
46. Owens, P.C.; Gatford, K.L.; Walton, P.E.; Morley, W.; Campbell, R.G. The relationship between endogenous insulin-like growth factors and growth in pigs. *J. Dairy Sci.* **1999**, *77*, 2098–2103. [CrossRef]
47. Lan, R.X.; Lee, S.I.; Kim, I.H. Effects of multistrain probiotics on growth performance, nutrient digestibility, blood profiles, faecal microbial shedding, faecal score and noxious gas emission in weaning pigs. *J. Anim. Physiol. Anim. Nutr.* **2016**, *100*, 1130–1138. [CrossRef]
48. Duan, H.; Lu, L.; Zhang, L.; Li, J.; Gu, X.; Li, J. Effects of *Lactobacillus Lactis* supplementation on growth performance, hematological parameters, meat quality and intestinal flora in growing-finishing pigs. *Animals* **2023**, *13*, 1247. [CrossRef]
49. Scheuemann, S.E. Effect of the probiotic Paciflor® (CIP 5832) on energy and protein metabolism in growing pigs. *Anim. Feed Sci. Technol.* **1993**, *41*, 181–189. [CrossRef]
50. Knecht, D.; Cholewińska, P.; Jankowska-Makosa, A.; Czyż, K. Development of swine's digestive tract microbiota and its relation to production indices—A review. *Animals* **2020**, *10*, 527. [CrossRef]
51. Lee, H.J.; Noh, H.T.; Paradhipta, D.H.V.; Joo, Y.H.; Lee, S.S.; Choi, J.S.; Kim, D.H.; Kim, S.K.; Kim, S.C. Impact of supplementary microbial additives producing antimicrobial substances and digestive enzymes on growth performance, blood metabolites, and fecal microflora of weaning pigs. *Animals* **2021**, *11*, 1217. [CrossRef]

52. Devi, S.M.; Kim, I.H. Effect of medium chain fatty acids (MCFA) and probiotic (*Enterococcus faecium*) supplementation on the growth performance, digestibility, and blood profiles in weaning pigs. *Vet. Med.* **2014**, *59*, 257–535. [CrossRef]
53. Mitruka, B.M.; Rawnsley, H.M. *Clinical Biochemistry and Hematological Reference Values in Normal Experimental Animals and Humans*, 2nd ed.; Masson: New York, NY, USA, 1977; pp. 215–237.
54. Lu, X.; Zhang, M.; Zhao, L.; Ge, K.; Wang, Z.; Jun, L.; Ren, F. Growth performance and post-weaning diarrhea in piglets fed a diet supplemented with probiotic complexes. *J. Microbiol. Biotechnol.* **2018**, *28*, 1791–1799. [CrossRef]
55. Dobson, A.; Cotter, P.D.; Ross, R.P.; Hill, C. Bacteriocin production: A probiotic trait? *Appl. Environ. Microbiol.* **2012**, *78*, 1–6. [CrossRef]
56. Collington, G.K.; Parker, D.S.; Armstrong, D.G. The influence of inclusion of either an antibiotic or a probiotic in the diet on the development of digestive enzyme activity in the pig. *Br. J. Nutr.* **1990**, *64*, 59–70. [CrossRef]
57. Jukna, Č.; Junka, V.; Šimkus, A. The effect of probiotics and phytobiotics on meat properties and quality in pigs. *Vet. Zootec.* **2005**, *29*, 80–84.
58. Ganeshkumar, S.G.; Gnanaraj, A.P.T.; Sivakumar, T.; Karthickeyan, S.M.K.; Murugan, M. Effect of probiotic supplementation on the carcass traits and sensory qualities of swill fed pork. *Tamilnadu J. Vet. Anim. Sci.* **2009**, *5*, 157–160.
59. Chu, G.M.; Yang, B.S.; Kim, H.Y.; Kim, J.H.; Ha, J.H.; Kim, C.H.; Lee, S.D.; Song, Y.M. Effects of supplemental fermented agro by-products diet on the growth performances, blood characteristics and carcass traits in fattening pigs. *Asian-Australas. J. Anim. Sci.* **2011**, *24*, 1464–1472. [CrossRef]
60. Grela, E.R.; Swiatkiewicz, M.; Florek, M.; Bakowski, M.; Skiba, G. Effect of inulin source and a probiotic supplement in pig diets on carcass traits, meat quality and fatty acid composition in finishing pigs. *Animals* **2021**, *11*, 2438. [CrossRef]
61. Chang, S.Y.; Belal, S.A.; Kang, D.R.; Choi, Y.I.; Kim, Y.H.; Choe, H.S.; Heo, J.Y.; Shim, K.S. Influence of probiotics-friendly pig production on meat quality and physicochemical characteristics. *Korean J. Food Sci. Anim. Resour.* **2018**, *38*, 403–416. [CrossRef]
62. Jørgensen, J.N.; Laguna, J.S.; Millán, C.; Casabuena, O.; Gracia, M.I. Effects of a Bacillus-based probiotic and dietary energy content on the performance and nutrient digestibility of wean to finish pigs. *Anim. Feed Sci. Technol.* **2016**, *221*, 54–61. [CrossRef]
63. Min, Y.J.; Choi, Y.H.; Choe, J.H.; Kim, Y.H.; Jeong, Y.D.; Kim, D.W.; Kim, J.E.; Jung, H.J.; Song, M.H. Effects of dietary mixture of protease and probiotics on growth performance, blood constituents, and carcass characteristics of growing-finishing pigs. *J. Anim. Sci. Technol.* **2019**, *61*, 272–277. [CrossRef] [PubMed]

animals

Article

Dietary 25 Hydroxyvitamin D3 Improved Serum Concentration Level and Alkaline Phosphatase Activity during Lactation but Had Meager Impact on Post-Farrowing Reproductive Performance in Sows

Prester C. John Okafor [1] and Nitipong Homwong [1,2,*]

1 Laboratory of Swine Science, Department of Animal Science, Faculty of Agriculture at Kamphaeng Saen, Kasetsart University, Kamphaeng Saen Campus, Nakhon Pathom 73140, Thailand; presterchukajohn.o@ku.th
2 National Swine Research and Training Center, Faculty of Agriculture at Kamphaeng Saen, Kasetsart University, Kamphaeng Saen Campus, Nakhon Pathom 73140, Thailand
* Correspondence: nitipong.h@ku.ac.th; Tel.: +66-942-914-063

Simple Summary: Vitamin D3 regulates many biological functions in mammals. One of its classical roles is to maintain calcium homeostasis and improve bone metabolism. The dietary requirement of this vitamin in sows often exceed that required for finishing pigs due to their role in mitigating osteomalacia, also known as soft bone. Commercial variants of this vitamin also exist as 25 hydroxyvitamin D3; however, their function in the reproductive performance of sows is not clearly understood. In this study, two commercial products were compared with regular Vitamin D3 in feeds administered to sows from gestation to lactation. Post-farrowing reproductive performance, serum alkaline phosphatase activity, and 25 hydroxyvitamin D3 concentration were compared. Feed intake, pre-weaning mortality, and the number of weaned piglets differed during lactation. Alkaline phosphatase activity and 25 hydroxyvitamin D3 concentration increased during lactation. This could be caused by an increase in metabolic demand for phosphorus and calcium during lactation. The current finding shows that the use of 25 hydroxyvitamin D3 in sow diets may improve other functions such as bone strength, calcium, and phosphorus homeostasis without necessarily affecting sow reproductive performance.

Abstract: Dietary 25 hydroxyvitamin D3 (25(OH)D3) promotes serum 25(OH)D3 concentration and alkaline phosphatase activity (ALP); however, post-farrowing reproductive performance of lactating sows fed with 14-epimer of 25(OH)D3 is uncertain. This study investigated post-farrowing reproductive performance, serum ALP activity, and serum 25(OH)D3 concentration in sows fed VD3, 25(OH)D3, or 14-epi 25(OH)D3. Weaned sows (n = 203) in parities 2 and 3 were blocked weekly and treated with 2000 IU/kg VD3 (T1), 25 µg/kg 25(OH)D3:14-epi 25(OH)D3 (T2), or 50 µg/kg 25(OH)D3 (T3) diets, all equilibrated to 2000 IU/kg as fed. Sow performance, treatment, and sampling period effects were analyzed. Environmental conditions were analyzed as covariates. The number of piglets weaned ($p = 0.029$), pre-weaning mortality ($p = 0.029$), sampling period ($p < 0.001$), and treatment and period interaction ($p = 0.028$) differed significantly. There was an increase in 25(OH)D3 during lactation due to physiological demands for milk calcium and milk production. Supplementing twice the concentration of 25(OH)D3 compared to its epimer, 25(OH)D3:14-epi 25(OH)D3, had no significant effect on the post-farrowing reproductive performance of lactating sows. The effect of 25(OH)D3 on post-farrowing reproductive performance and ALP activity in sows was influenced by metabolic demand for calcium due to physiological changes during lactation as well as epimer conformation.

Keywords: vitamin D3; 25-hydroxyvitamin D3; alkaline phosphatase; epimers; post-farrowing; reproduction; performance; sow; lactation

1. Introduction

The need for improved reproductive performance has been a front-line objective in swine production [1,2]. Some crucial indices for estimating post-farrowing reproductive performance include litter size, live birth, the milking ability of sows, weaned piglets, pre-weaning mortality, and stillborn. These factors are also used in benchmarking [1,2]. The need for dietary support using nutrient supplements such as vitamin D3 (VD3) has also been extensively studied across various stages of sow reproductive life [3,4]. Metabolites of VD3 such as 25-hydroxyvitamin D3, 25(OH)D3, and 1,25-dihydroxyvitamin D3, 1,25(OH)2D3, are known to play a critical role during breeding, implantation, placentation, gestation, parturition, and lactation. For instance, the metabolism of available Ca produced during lactation largely depends on the regulatory functions of these metabolites [5]. They have also been reported to enhance milking ability, reduce stillbirth, and improve litter size [6–8]. A report by Weber and colleagues showed that the birth and weaning weights of piglets were improved [3]. The distinction in functions of VD3 and its metabolites in reproducing sows is still not clearly understood. Despite the above-mentioned effects of 25(OH)D3 on sow performance, some scholars reported no difference in the reproductive performance of sows fed either of these forms [9]. In addition to reproductive performance, metabolites of VD3 are well known regulators of alkaline phosphatase (ALP) activity.

The ALP enzymes are glycoproteins expressed during bone calcification, collagen formation, growth, and differentiation of tissues, as well as in bone remodeling [10]. This enzyme is highly expressed in growing animals due to an increase in the activity of osteoblasts. However, hepatic cells also play a role in augmenting the enzyme levels in mature animals [11]. The activity of ALP is required to metabolize inorganic phosphates, reduce pyrophosphates (which are known inhibitors of mineralization), and increase phosphate localization in osteons [12]. Reports have also shown that the activity of ALP is also promoted in a VD3 and 25(OH)D3 replete state [13,14]. Due to the crucial role of ALP in bone formation and mineralization, the present study aimed to understand the effects of the epimeric form of 25(OH)D3 on the activity of ALP in reproducing sows.

Epimers of 25(OH)D3 are analogues with similar structures and molecular weights as parent compounds, but they differ in the stereochemical structure of their respective side chains. The orientation of the compound often affects its biological function [15]. Some biologically active analogues include 3-epi 25(OH)D3 and 14-epi 25(OH)D3 produced by sigma-tropic hydrogen shifts at the third and fourteenth carbons of 25(OH)D3. This function is catalyzed by epimerases [16,17]. The method of synthesizing and harvesting these products might contribute to product differentiation [18]. Though chemical synthesis has been a common practice, fermentation technology has also been explored for commercial 25(OH)D3 production [19]. The 14-epi-analogs have been reported to promote the activation of the vitamin D receptor, a transcription factor that regulates genetic mechanisms involved in mineralization and bone fortification [20]. However, no studies have yet compared the 14-epimer of 25(OH)D3 in diets of gestating or lactating sows at peak parity. There is also a paucity of information on its role in ALP activity at this stage. The present study postulates that half a dose of 14-epimer of dietary 25(OH)D3 has a similar effect to the regular commercial 25(OH)D3 variant on serum concentration, ALP activity, and the reproductive performance of sows. Therefore, the objective of this study was to investigate post-farrowing reproductive performance, serum ALP activity, and 25(OH)D3 concentration in peak-parity sows fed dietary VD3 or either of two epimeric conformations of 25(OH)D3.

2. Materials and Methods

2.1. Animals and Dietary Treatments

A total of 203 weaned crossbreed sows (50% Landrace × 50% Yorkshire) in parity 2 and 3 (peak parity), which averaged 81 and 102 weeks of age, respectively, were included in the study. Sows were blocked by weaning week, and, on average, the study sample increased by approximately 11 sows for 18 consecutive weeks. Sows were randomly assigned to three treatments of approximately 67 sows each. Three dietary treatments contained either 2000 IU/kg of regular VD3 (T1), 25 μg/kg of 25(OH)D3:14-epi 25(OH)D3 (T2), or 50 μg/kg of 25(OH)D3 (T3). All treatment concentrations were equilibrated to 2000 IU/kg of VD3 in a base diet and supplied as mash feed for 20 weeks from gestation through lactation until weaning (Table 1). The nutrient compositions of gestation and lactation diets were formulated using FeedLIVE® Version 1.61 (Live Informatics Co., Ltd., Nonthaburi, Thailand) and are shown in Table 2. Average feed intake was limited to 2.5 kg daily from day 1 to 84 (84 days) and was increased to 3.3 kg from days 85 to 109 (26 days) during gestation. During lactation, sows were fed ad libitum until weaned. Lactation length (LL, days) was managed by the producer, and hence was considered a covariate in this study. All suckling piglets were offered mash creep feed (VD3 composition, equivalent to 4000 IU/kg) from 10 days of age until weaning at approximately 25 days of age.

Table 1. Forms and concentrations of vitamin D3 used as a dietary supplement.

Treatments	Active Substance/Product	Mass (g)/Metric Ton of Feed	Dose per kg of Diet	Product Conc. IU/Metric Ton
T1 [a/]	500,000	4	2000 IU	2,000,000
T2 [b/]	69.7 mg	360	25 μg/kg	2,000,000
T3 [b/]	12.5 g	4	50 μg/kg	2,000,000

T1 = VD3 (1 mg of VD3 = 500 IU); T2 = 25(OH)D3:14-epi 25(OH)D3 (1 mg = 80,000 IU); T3 = 25(OH)D3 (1 mg = 40,000 IU); [a/] Regular form of cholecalciferol, VD3; [b/] Metabolite form of cholecalciferol: 25-hydroxycholecalciferol, 25(OH)D3.

Table 2. Feed ingredients and calculated nutrient composition of basal diets for gestating and lactating sows.

Ingredient, %	Gestating Sow	Lactating Sow
Broken rice	10.00	35.00
Tapioca meal (70%)	30.00	5.00
Rice barn	23.29	15.00
Wheat barn	15.00	12.00
Soybean oil	1.99	5.70
Soybean meal (45.5%)	15.78	22.44
L-lysine	0.15	0.65
DL-methionine	0.08	0.13
L-threonine	0.07	0.09
Monodicalcium phosphate	1.08	1.41
Calcium carbonate	1.79	1.80
Salt	0.25	0.27
Premix [1]	0.50	0.50
¶ Optiphose®	0.01	0.01
Calculated nutrient composition, %		
Metabolizable energy, MJ/kg	12.34	13.81
Crude protein	14.00	17.50
Crude fat	6.22	8.79
Crude fiber	5.97	4.71
Calcium	1.15	1.15
Total phosphorus	0.86	0.85
Available phosphorus	0.41	0.45
Sodium	0.32	0.32
Lysine	0.80	1.36
Methionine	0.27	0.38
Methionine + Cystine	0.48	0.63

Table 2. *Cont.*

Ingredient, %	Gestating Sow	Lactating Sow
Threonine	0.53	0.68
Tryptophan	0.17	0.21

[1] Dietary premix per kilograms of feed contented as following; without vitamin D_3, vitamin A 3000 IU, vitamin E 11 IU, vitamin K_3 1.00 mg, vitamin B_1 0.40 mg, vitamin B_2 1.20 mg, vitamin B_6 1.50 mg, vitamin B_{12} 0.01 mg, pantothenic acid 4.00 mg, niacin 4.00 mg, folic acid 0.30 mg, biotin 0.40 mg, choline chloride 60.00 mg, ferrous 40.00 mg, copper chelate 36 mg, manganese chelate 10.80 mg, zinc chelate 36.00 mg, cobalt 0.40 mg, iodine 0.40 mg, and selenium 0.06 mg. ¶ phytase enzyme used in diets.

2.2. Feed Quality Control Analysis

As a quality control measure, five-hundred grams of feed from each batch supplied to the sows was sampled for proximate analysis. Dry matter (method 930.15), crude protein (method 2001.11), ether extract (method 2003.05), crude ash (method 942.05), crude fiber (method 978.10), calcium (method 927.02), and phosphorus (method 965.17) were analyzed according to the AOAC protocol [21]. The gross energy was determined using a bombs calorimeter (method ISO 9831) [22]. The outcome of the analysis is presented in Table 3. To quantify feed 25(OH)D3 concentration in treatment diets, feed samples were sent to BIOVET® laboratory (Peshtera, Bulgaria). The result is presented in Table 4.

Table 3. Proximate analysis showing the respective treatment composition of various nutrients in diets.

Diet	Gestating Sow			Lactating Sow		
Treatment	T1 ‡	T2 ‡	T3 ‡	T1 ‡	T2 ‡	T3 ‡
Gross energy [1], MJ/kg	18.77	19.31	19.40	18.02	17.67	17.61
Crude protein, %	15.30	15.37	15.09	19.49	18.79	18.93
Ether extract, %	7.78	8.26	7.85	12.54	11.78	12.23
Crude fiber, %	4.32	4.67	4.31	3.91	3.46	3.98
Crude ash, %	7.46	7.07	7.12	7.50	7.19	7.11
Calcium [2], %	1.29	1.22	1.18	1.39	1.35	1.27
Phosphorus, %	0.81	0.81	0.80	0.96	0.89	0.87

[1] Gross energy was analyzed using a bombs calorimeter (method ISO 9831). [2] Calcium was analyzed by using Atomic Absorption Spectrophotometry (Shimadzu, AA-7000). ‡ T1: Basal diet with Vitamin D_3 2000 IU plus Optiphos® 0.1 g per kg diet. T2: Basal diet with 25 μg 14-epi 25(OH)D_3 plus Optiphos® 0.1 g per kg diet. T3: Basal diet with 50 μg 25(OH)D_3 per plus Optiphos® 0.1 g per kg diet.

Table 4. Outcome of quality control analysis for 25(OH)D3 level in gestating and lactating diets.

Diet	Level (ng/g)		
	T1	T2	T3
Gestation	<2.00	20.20	57.30
Lactation	<2.00	22.70	50.00

Dietary 25(OH)D3 assay was analyzed using HPLC by BIOVET® laboratory (Peshtera, Bulgaria); the limits of detection and limits of quantitation were 2 and 5 ng/g at 10% coefficient of variation. T1 (VD3), T2 (25(OH)D3:14-epi 25(OH)D3), and T3 (25(OH)D3) represent the experimental groups and their designated dietary VD3 or 25(OH)D3 compositions.

2.3. Environment, Housing, and Management

During gestation, sows were housed individually in stalls. From day 110, sows were moved to a farrowing barn. Individual farrowing stalls were equipped with heated crates and creep area. Sows and piglets had unlimited access to drinking water in their respective stalls by nipple. Additionally, water was supplied regularly in bowls for easy access to younger piglets. Where necessary, sows were assisted by injecting oxytocin during farrowing. Piglets were weighed within 24 h of farrowing. Cross-fostering to equalize litter size was carried out within 48 h after farrowing; however, this was restricted to sows farrowed within treatment groups. The gestation and lactation barns were equipped with evaporative cooling system.

2.4. Records and Measures

All performance records and inventory in the production facility were digitized in PigLIVE® Version 4.0 (Live Informatics Co., Ltd., Nonthaburi, Thailand). A live record of the insemination date was documented; however, sows were confirmed in-pig 30 days after by a standing reflex during boar exposure. Gestation and lactation length (days) were days from insemination to farrowing and farrow to wean, respectively. Records collected at farrow were farrowing time (hours), which included the time from farrow of first piglet to completion, oxytocin use (ml), and percent oxytocin use. Post-farrowing reproductive performance indices recorded onsite were total born (TB), born alive (BA), still born, mummified, piglets and litter body weight (BW, kg), pre-weaning mortality, weaned piglets, and lactation feed intake (LFI, kg/day). Sow body condition, mainly BW and backfat (BF, mm) before farrowing, and at weaning, as well as percent BW and BF loss during lactation were also measured. Fecal score was recorded daily by observing piglets' fecal droplets in each pen. Stools were assigned 0, 1, or 2 indicating lumpy (no diarrhea), pasty, or liquid (diarrhea) stool, respectively.

2.5. Blood Sample Collection and Biochemical Analysis

Blood samples were collected by periods on day 5 post-farrowing (AF5), 25 post-farrowing (AF25), and day 6 post-weaning (AW6). One piglet from each sow was sampled for blood collection at weaning. Approximately 3 mL of blood was obtained from each sow through the jugular vein, aptly transferred into coagulant blood collection tubes, and transported in an ice box to the Laboratory of Swine Science, Faculty of Agriculture at Kamphaeng Saen for processing. Centrifugation was carried out at 2500× *g* for 15 min at 4 °C and the resulting serum was collected into 1.5 mL microtubes and stored at −20 °C for further analyses of 25(OH)D3 concentration (ng/mL) and ALP activity (U/L).

2.6. Alkaline Phosphatase Assay

The in vitro test for the quantification of serum ALP activity was carried out using an ALP kit (Mindray®, Shenzhen, China). The reagents consist of the following: R1, magnesium acetate-zinc sulfate in a solution of 2-amino-2-methyl-1-propanol buffer (2.5, 1.2 and 435 mmol/L respectively); and R2, p-nitrophenyl phosphate (60 mmol/L). A multi sera calibrator and control were used for quality control assessment (reference range: 80.6–98.6 U/L). Distilled water was used as a blank. The reaction volume for samples and reagents (sample: R2:R1) was set to 1:12.5:50. The absorbance reading was obtained using a chemistry analyzer (Mindray BS-120®, Shenzhen, China) at a reaction temperature of 37 °C and wavelengths between 405 and 546 nm. The detection range for linearity was between 5 and 800 U/L. The equation for the reaction can be stated as follows:

$$4-\text{Nitrophenylphosphate}+\text{H}_2\text{O}+\text{ALP}+\text{Mg}^+ \rightarrow 4-\text{Nitrophenol}+\text{P}_\text{i}$$

ALP catalyzes the hydrolysis of 4-nitrophenyl phosphate, producing 4-nitrophenol and inorganic phosphate. The alkaline buffer also acts as a phosphate-group acceptor. The activity of ALP is directly proportional to the rate of formation of 4-nitrophenol in the sample.

2.7. Analysis of Serum 25(OH)D3 Concentration

Sample preparation: Serum samples and acetonitrile were added dropwise into a microtube at a ratio of 1:2 (*v*/*v* serum/acetonitrile) and mixed thoroughly via vortex. The preparation was centrifuged thrice at 5000× *g* for 10 min at 25 °C, and the supernatant carefully collected in clean microtubes, filtered through a nylon membrane (0.22 µm, Whatman®, Kent, UK) into an amber glass vial, and transferred to an autosampler for chromatographic analysis.

Chromatographic analysis: Reverse-phase symmetry C18 column, (5 µM 4.6 × 250 mm, Waters, San Ramon, CA, USA) was used in this study. The mobile phase contained 100%

acetonitrile, delivered at a flow rate of 1.2 mL/min. Sample injection volumes were set at 50 μL. Injector and column temperatures were set at 25 and 40 °C, respectively. Chromatographic separation occurred at a detection wavelength of 264 nm using a photodiode array (PDA) detector (Waters, San Ramon, CA, USA).

Calibration curve: Standards of VD3 and 25(OH)D3 (Ehrenstorfer® GmbH, Augsburg, Germany) were used to generate calibration curves. Using the mobile phase as a diluent, 128, 64, 32, 16, 8, and 4 ng/mL of VD3 and 25(OH)D3 were prepared and used as calibrators. The chromatogram is presented in Figure 1.

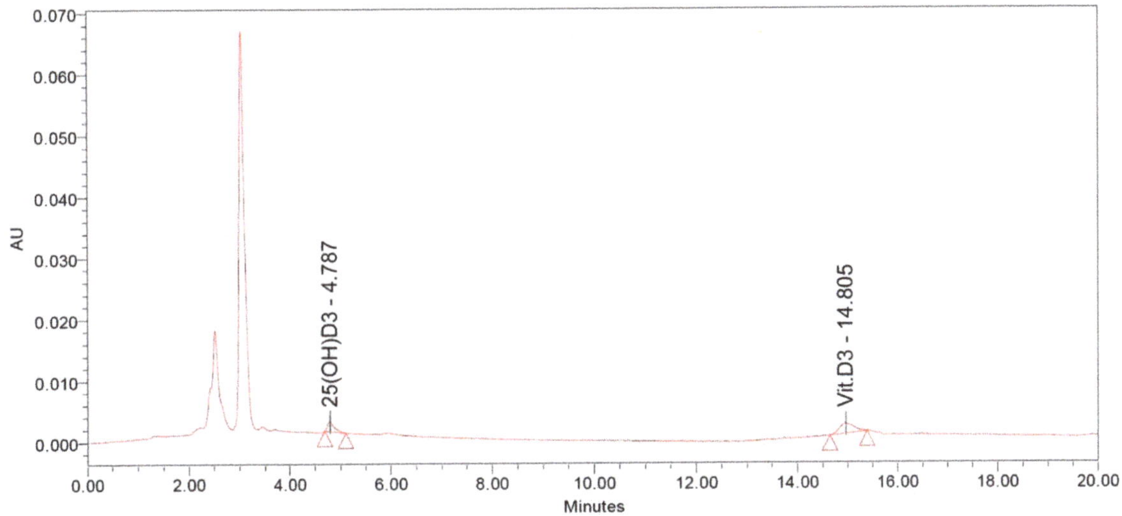

Figure 1. Chromatogram of vitamin D3 and 25-hydroxyvitamin D3 extraction using 100 percent acetonitrile.

2.8. Statistical Analysis

Gestation or lactation in days were analyzed using general linear models according to the statistical model: $Y_{ijk} = \mu + W_j + T_k + e_{ijk}$. The term, "μ" is the coefficient for grand mean; Y_{ijk} is the gestation or lactation length of ith sow in k^{th} treatment inseminated in a specific week j^{th}. Residual error in the model was denoted as e_{ijk}.

Post-farrowing reproductive performance was analyzed using general linear models. A statistical model was considered as follows: $Y_{ijkl} = \mu + W_j + LL_i + T_k + e_{ijkl}$, where Y_{ijkl} is the observed reproductive performance variables of sow_i with lactation length (LL_i) in the treatment group (T_k), from a lactation sow in weaned week j^{th} (W_j) with the experimental error, e_{ijkl}. Lactation length (LL_j) was considered as the model covariate when needed.

Piglet fecal scores from sows in treatments were analyzed using a non-parametric Kruskal–Wallis test.

The general linear mixed-effects model for an analysis of 25(OH)D3 concentration and ALP activity was as follows: $Y_{ijk} = \mu + T_j + P_k + T_j{:}P_k + SOW_i + e_{ijk}$, where Y_{ijk} = vector of response for i^{th} sow in j^{th} treatment at period k^{th} of blood collection; μ = grand mean; SOW_i was the random effect; T_j was treatment effects; P_k was period effects; $T_j{:}P_k$ was treatment–period interaction effects; and the term e is the coefficient for the residual error of all model terms.

Residual distributions from general linear models and general linear mixed-effects models were studied. Assumptions for normality, linearity, and heteroskedasticity were tested. In cases of non-normality, Box–Cox transformation was used to estimate transformation parameters [23]. Arcsine square root transformation was also applied in modes when deemed appropriate. Model selection was performed using Akaike Information Criteria

(AIC) for either nested or un-nested models. The lower the AIC, the better the model fit. Graphs were plotted using "ggplot 2" package [24], and treatment means were compared using the "emmeans" package [25]. All statistical analyses were carried out using R version 4.2 [26]. Statistical significance was set at $p < 0.05$.

3. Results

3.1. Post-Farrowing Reproductive Performance

The effect of dietary supplementation of 25(OH)D3 on post-farrowing reproductive performance is presented in Table 5. At a 95 percent confidence level, no significant difference was observed in gestation length; however, lactation length differed significantly between T1 and T2 (p = 0.039). Lactation length was designated as a covariate for all subsequent statistical models. Lactation feed intake also differed significantly between sows T2 and T3 (p = 0.023). Figure 2 shows the average daily feed intake during lactation. Neither of the three treatments showed a significant difference in body weight nor backfat within this period of lactation. Pre-weaning mortality as well as the number of piglets weaned were significantly different between treatments. The average pre-weaning mortality of piglets varied significantly between T1 and T2 (p = 0.029). Weaned piglets were also statistically significant between treatments (p = 0.029). The average fecal score for all treatment groups during lactation was less than two, indicating no severe sign of diarrhea.

Table 5. Post-farrowing reproductive performance of sows and piglets.

Parameter	T1	T2	T3	p-Value
	Mean ± SEM			
Gestation length, days	117.93 ± 0.20	118.04 ± 0.20	118.41 ± 0.20	1/ 0.168
Lactation length, days	23.74 ± 0.21 ^{b}	24.49 ± 0.22 ^{a}	24.30 ± 0.22 ^{ab}	1/ 0.039 *
LFI, kg/day	6.37 ± 0.15 ^{ab}	6.11 ± 0.14 ^{b}	6.58 ± 0.14 ^{a}	1/ 0.023 *
BW before farrowing, kg	297.60 ± 2.79	298.25 ± 2.77	299.46 ± 2.85	1/ 0.895
BW at weaning, kg	257.23 ± 3.15	252.89 ± 3.26	260.59 ± 3.27	1/ 0.250
BF before farrowing, kg	18.35 ± 0.25	18.80 ± 0.25	18.65 ± 0.25	1/ 0.414
BF at weaning, mm	17.78 ± 0.24	17.33 ± 0.25	17.65 ± 0.25	1/ 0.421
BW loss, % ¶	14.40 ± 0.57	14.84 ± 0.60	13.50 ± 0.60	1/ 0.104
BF loss, % ¶	10.32 ± 1.28	11.99 ± 1.33	8.37 ± 1.34	1/ 0.156
Total born, head	15.73 ± 0.38	16.45 ± 0.37	15.72± 0.38	1/ 0.278
Piglets BW, kg	1.47 ± 0.03	1.40 ± 0.03	1.41 ± 0.03	1/ 0.140
Piglets born alive, head	14.15 ± 0.36	14.94 ± 0.35	14.67 ± 0.36	1/ 0.286
Stillborn piglets, head	1.61 ± 0.15	1.39 ± 0.15	1.13 ± 0.16	1/ 0.100
Mummified piglets, head	0.48 ± 0.09	0.66 ± 0.09	0.69 ± 0.09	1/ 0.228
Pre-wean mortality, head	0.81 ± 0.13 ^{a}	1.26 ± 0.13 ^{b}	0.84 ± 0.13 ^{ab}	1/ 0.029 *
Weaned piglets, head	11.42 ± 0.13 ^{a}	10.97 ± 0.13 ^{b}	11.39 ± 0.13 ^{a}	1/ 0.029 *
Time of farrowing, hour	4.39 ± 0.28	4.15 ± 0.25	4.16 ± 0.31	1/ 0.086
Oxytocin dose, ml	0.86 ± 0.13	1.05 ± 0.13	1.13 ± 0.14	1/ 0.294
Litter fecal score, score	1.44 ± 0.13	1.70± 0.13	1.41 ± 0.13	2/ 0.399

1/ p-value obtained from a general linear model. 2/ p-value obtained from non-parametric test using Kruskal–Wallis test. ¶ Arcsine square root transformation applied to response variable. ab Different superscripts within the same row indicate statistical significance ($p < 0.05$). * $p < 0.05$. LFI, lactation feed intake; SEM, standard error of mean; BW, body weight; BF, back fat. Litter fecal score was as follows: 0 (no diarrhea/solid); 1 (pasty but not liquid); 2 (diarrhea/liquid). T1 (VD3), T2 (25(OH)D3:14-epi 25(OH)D3), and T3 (25(OH)D3) fed sows.

3.2. ALP

The activity of ALP was estimated on three levels, treatment, period, and interaction (treatment × period) (Table 6). The activity of ALP did not differ significantly at the treatment level. There was a significant difference in ALP activity at AF5, AF25, and AW6 ($p < 0.001$). The interaction between treatment and period was also statistically different (p = 0.028). At AF25, ALP activity was at its peak with estimated values of 100.80 ± 10.52, 77.56 ± 10.52, and 96.70 ± 10.87 U/L in T1, T2, and T3, respectively. The lowest estimated activity was recorded at AW6 (Figure 3A).

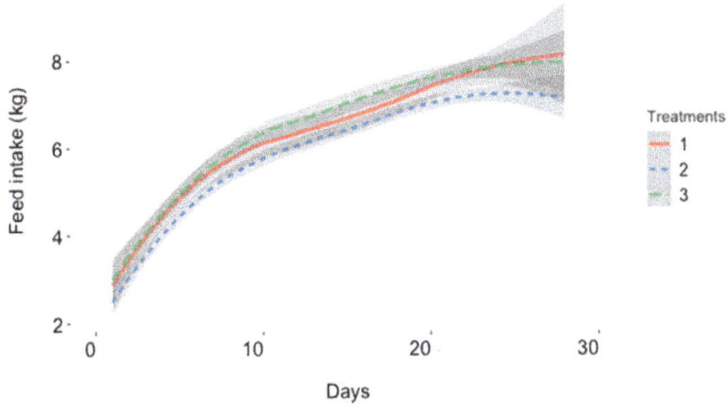

Figure 2. Estimate of average daily feed intake of multiparous sows during lactation.

Table 6. Effects of treatment and periods on ALP activity and 25(OH)D3 concentration in lactating sows.

Treatment Effects				
Parameters	**T1**	**T2**	**T3**	***p*-Value**
		Mean ± SEM		
ALP, U/L ‡	73.35 ± 10.02	61.49 ± 10.02	69.02 ± 10.36	1/ 0.820
ALP AF5, U/L	67.62 ± 6.81	55.78 ± 6.81	58.17 ± 7.04	2/ 0.527
ALP AF25, U/L	100.82 ± 10.86	77.58 ± 10.86	96.66 ± 11.24	2/ 0.865
ALP AW6, U/L	51.65 ± 4.70	51.15 ± 4.70	52.16 ± 4.86	2/ 0.805
25(OH)D3, ng/mL ‡	54.67 ± 9.37	61.96 ± 9.01	58.20 ± 8.98	1/ 0.880
25(OH)D3, AF5, ng/mL	54.90 ± 7.84	49.69 ± 7.09	40.23 ± 7.52	2/ 0.724
25(OH)D3, AF25, ng/mL	55.78 ± 11.83	80.86 ± 10.59	80.02 ± 10.53	2/ 0.219
25(OH)D3, AW6, ng/mL	60.62 ± 9.14	48.55 ± 8.86	52.09 ± 7.97	2/ 0.637
Period effects				
	AF5	AF25	AW6	
ALP, U/L	60.58 ± 6.14 [a]	91.58 ± 6.14 [b]	51.64 ± 6.14 [c]	1/ <0.001 ***
25(OH)D3, ng/mL	42.59 ± 5.76 [a]	72.33 ± 5.43 [b]	47.75 ± 5.73 [a]	1/ <0.001 ***
Treatment × Period				
ALP, U/L †	—	—	—	1/ 0.028 *
25(OH)D3, ng/mL †	—	—	—	1/ 0.146

1/ *p*-value obtained from general linear mixed-effect model. 2/ *p*-value obtained from general linear model. ‡ Total treatment effect adjusted for individual sow difference and sample collection periods. † Interaction plots shown in Figure 3. [abc] Different superscript within the same row indicate statistical significance ($p < 0.05$). * $p < 0.05$, *** $p < 0.001$. LFI, lactation feed intake; SEM, standard error of mean; BW, body weight; BF, back fat. T1 (VD3), T2 (25(OH)D3:14-epi 25(OH)D3), and T3 (25(OH)D3) fed sows.

3.3. 25(OH)D3

The concentration of 25(OH)D3 was also established at the treatment, period, and interaction levels. As shown in Table 6, there was no significant difference in treatment means; however, the period effect was shown to be statistically significant ($p < 0.001$). The interaction model plotted in Figure 3B showed that 25(OH)D3 concentration was at its peak at AF25 with estimated values of 55.78 ± 11.83, 80.86 ± 10.59, and 80.02 ± 10.53 ng/mL for T1, T2, and T3, respectively.

The cocktail used for standard preparation contained 50 ng/mL dry mass of VD3 and 25(OH)D3. Separation was carried out in a reverse phase symmetry C18 column. 25(OH)D3 had a shorter retention time compared to VD3.

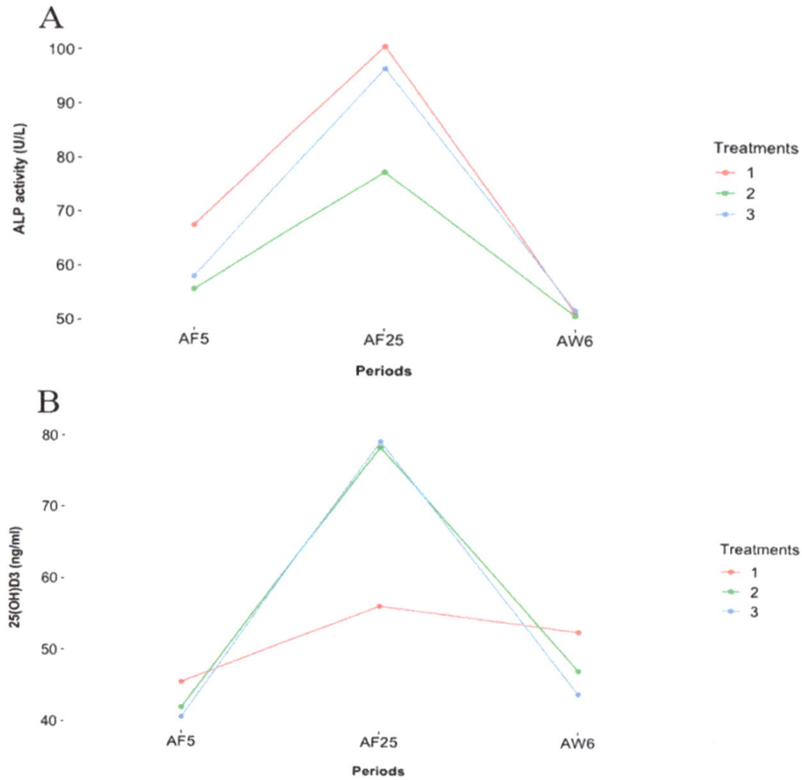

Figure 3. Interaction plots (treatment × sampled periods) depicting (**A**), alkaline phosphatase activity and (**B**), 25(OH)D3 concentration in lactating sows.

Feed intake increased non-linearly during lactation, with an average of 6.37, 6.11, and 6.58 in T1, T2, and T3, respectively. The average lactation length was 25 days. Treatments 1 (VD3), 2 (25(OH)D3:14-epi 25(OH)D3), and 3 (25(OH)D3) are diets fed to the experimental groups and their respective VD3 or 25(OH)D3 compositions.

Blood samples were collected at three periods: AF5—5 days post-farrowing; AF25—25 days post farrowing; and AW6—6 days post-weaning. There was an increase in alkaline phosphatase activity at AF25 in all treatments. Similarly, 25(OH)D3 concentration was also higher at AF25, however, with higher peaks in treatments 2 and 3 compared to the control. Treatments 1 (VD3), 2 (25(OH)D3:14-epi 25(OH)D3), and 3 (25(OH)D3) are diets fed to the experimental groups and their respective VD3 or 25(OH)D3 compositions.

4. Discussion

4.1. Post-Farrowing Reproductive Performance

To the best knowledge of the authors, this is the first study evaluating the effects of supplementing dietary 25(OH)D3 and its 14-epimer on reproductive performance, ALP activity, and serum 25(OH)D3 concentration in multiparous sows. Since the 14-epi-analogue of 25(OH)D3 purportedly has a higher potency than the regular form, a half-dose was used to provide the right empirical stance to substantiate or rescind the claim. The chromatographic analysis of feed samples showed a two-fold higher dietary concentration of 25(OH)D3 compared to 14-epi-25(OH)D3; however, this did not result in a two-fold increase in post-farrowing reproductive performance, indicating that the relationship between dietary concentration and reproductive performance may not be linear. Lactation feed intake, pre-weaning mortality, and the number of piglets weaned differed significantly between treatments. Because lactation length was strictly managed by the farm, it was considered as a covariate for subsequent analyses. The lengthy lactation culminated in a linear increase in feed intake. It was also shown that higher feed intake during lactation was associated with positive body reserve and improved estrus cycling [27]. This has a positive impact since lengthy lactation often resulted in improved milk production in sows, and, in turn, improved piglet vitality [28]. In other studies, sows with longer lactation length had a low wean-to-estrus interval in days [29,30]. Therefore, the practice in itself is intended to improve overall sow and piglet performance. The mortality rate reported herein was lower compared to that published previously [31]. This could be attributed to differences in farm management practices. According to Muns et al. (2016), pre-weaning mortality can be caused by the piglet, sow, or environment [32]. All housing facilities used for gestation and lactation sows in the present study were equipped with an evaporative cooling system, annulling the possibility of heat-induced mortality. A plausible explanation for the observed mortality rate in sows fed 14-epi 25(OH)D3 could be low piglet vitality and low feed intake, which are often associated with milking ability. The scope of this study did not incorporate the study of milking ability in sows; hence, empirical evidence was provided in this regard. Other studies have, however, demonstrated that piglet mortality and consequent low weaning proportion in sows fed 25(OH)D3 diets were linked to reduced milking [3,33]. With regard to feed intake, it is a known fact that sows consume more feed to replenish body reserve, which tends to derogate during lactation; therefore, adequate milking is often associated with reduced body reserve (often shown by reduction in backfat). In the current study, 14-epi 25(OH)D3-fed sows not only had the lowest feed intake during lactation, but comparatively had the lowest reduction in back fat. This could possibly result in low milking, low piglet vitality, and mortality. The current discussion so far has shown a notable distinction in lactation feed intake between the experimental groups. This finding agrees with a previous report stating that dietary 25(OH)D3 did not improve sow reproductive performance more than regular VD3 [9]. Both feed intake and the average number of weaned piglets differed significantly between experimental groups. These factors were mostly influenced by farm management practice and might not lead to any significant production loss or pejorative effect.

4.2. ALP

After an adjustment for treatment effects, the activity of ALP differed significantly with respect to the sampled periods. Higher activity was observed at AF25 compared to AF5 and AW6. The current observation further supports the need to consider the physiological stage of sows in studies involving ALP. Similar findings have also been reported in other studies [34,35]. The need for phosphorus (P) is the most probable driving factor for increased ALP activity during lactation. At the jejunum of the small intestine, dietary P is absorbed into the bloodstream and transported to organs where it mediates various metabolic functions, including those of fetal and mammary gland development during gestation and lactation, thereby contributing significantly to the total P level in the circulation [11]. In a P replete state, unbound P can be readily absorbed into the circulation; this process drives the activity of ALP [36]. Calcitriol, a hormonal form of VD3, is also one of the essential regulators of P homeostasis. A possible mechanism was explained in Kozai et al. where it was shown that 1-alpha hydroxylase is often expressed abundantly in a state of phosphate depletion [37]. Consequently, the depletion of phosphate will lead to an increased metabolic demand for P and, in turn, an increase in ALP activity.

4.3. 25(OH)D3

There was a significant difference in mean 25(OH)D3 concentrations at sampled periods, the highest being at AF25. The linear increase in dietary feed intake during lactation suggests that the observed changes in 25(OH)D3 concentration might partly be influenced by dietary levels. The current observation consents the finding that post-farrowing concentration increased linearly with dietary feed levels [38,39]. However, no statistical difference was observed at the treatment level. This shows that higher dietary concentration alone might not necessarily culminate in increased bioavailability. Hence, several biological factors could jointly affect serum concentration. As seen in previous reports, evidence has also shown that blood concentration is not only influenced by dietary levels but metabolic Ca demand as well [40,41]. Studies have shown that during lactation, available Ca and P are often elevated. The observed increase in the concentration of 25(OH)D3 at AF25 indicates an upsurge in the activity of 1-alpha hydroxylase, possibly in response to metabolic Ca and P demand [42]. Consistent with the current findings, the activities of 1-alpha hydroxylase and ALP are co-regulated in response to dietary Ca and P levels, which are the hormonal triggers for the assembly of their respective transcription machinery [35,42]. Though epimerization causes variation in the biological activity of 25(OH)D3, serum level is to a greater extent influenced by physiological changes as well as sows' nutrient demand during lactation.

5. Conclusions

The current study lucidly demonstrated the distinction in function of two 25(OH)D3 products differing in their respective epimeric conformations. The comparatively low lactation feed intake in 14-epi 25(OH)D3-fed sows led to a reduction in the average number of weaned piglets and slightly higher mortality. These results are a culmination of factors that are closely linked and can be improved by good farm management practices. Due to their subjective nature, their influence on the outcome of reproductive performance in this study can be considered meager. The current finding therefore supports the study hypothesis that half doses of 25(OH)D3:14-epi 25(OH)D3 and 25(OH)D3 have a similar effect on sows' reproductive performance. The finding that neither the activity of ALP nor 25(OH)D3 concentration were influenced by dietary treatments also assents the current null hypothesis. However, the major distinction observed at AF25 implies that bioavailable concentrations of 25(OH)D3 were also influenced by physiological changes as well as the metabolic demand for Ca and P, as seen during lactation. The 25(OH)D3:14-epi 25(OH)D3 is therefore a highly potent substitute for the regular 25(OH)D3 and VD3 variants in reproducing sows.

Author Contributions: Conceptualization, N.H.; methodology, N.H.; validation, P.C.J.O. and N.H.; formal analysis, P.C.J.O. and N.H.; investigation, N.H.; resources, N.H.; data curation, N.H.; writing—original draft preparation, P.C.J.O.; writing—review and editing, P.C.J.O. and N.H.; supervision, N.H.; project administration, N.H.; funding acquisition, N.H. All authors have read and agreed to the published version of the manuscript.

Funding: This research was made possible with financial support from the following establishments: Huvepharma Co-Project, Kasetsart University Graduate Scholarship Program, National Swine Research and Training Center, Kasetsart University and Thai Orchids Laboratories.

Institutional Review Board Statement: This study was conducted with the approval of the Institutional Animal Care and Use Committee of Kasetsart University, Thailand (ACKU62-AGK-006).

Data Availability Statement: The data are available on request from the corresponding author.

Acknowledgments: The technical assistance by Somnuk Promdang, Wannee Chewpreecha, Wunwinee Ahiwichai, Anchalee Khongpradit and Nattanit Jimongkolkul are gratefully acknowledged. The facility and animals for the research was provided by Papon Farms. The authors highly appreciate these contributors for their overwhelming support and kind gestures.

Conflicts of Interest: The authors declare no conflicts of interest, or significant financial support that could have influenced the outcome of this publication.

References

1. Foxcroft, G.; Patterson, J. Optimizing breeding management in a competitive world: Gilt and sow aspects. In Proceedings of the AASV 41st Annual Meeting, Omaha, NE, USA, 6–9 March 2010; pp. 6–9.
2. Koketsu, Y.; Tani, S.; Iida, R. Factors for improving reproductive performance of sows and herd productivity in commercial breeding herds. *Porc. Health Manag.* **2017**, *3*, 1. [CrossRef] [PubMed]
3. Weber, G.; Witschi, A.-K.; Wenk, C.; Martens, H. Triennial Growth Symposium—Effects of dietary 25-hydroxycholecalciferol and cholecalciferol on blood vitamin D and mineral status, bone turnover, milk composition, and reproductive performance of sows. *J. Anim. Sci.* **2014**, *92*, 899–909. [CrossRef] [PubMed]
4. Kogawa, M.; Anderson, P.H.; Findlay, D.M.; Morris, H.A.; Atkins, G.J. The metabolism of 25-(OH) vitamin D3 by osteoclasts and their precursors regulates the differentiation of osteoclasts. *J. Steroid Biochem. Mol. Biol.* **2010**, *121*, 277–280. [CrossRef] [PubMed]
5. Wuryastuti, H.; Stowe, H.D.; Miller, E.R. The influence of gestational dietary calcium on serum 1, 25-dihydroxycholecalciferol in sows and their pigs. *J. Anim. Sci.* **1991**, *69*, 734–739. [CrossRef] [PubMed]
6. Lauridsen, C. Trennial Growth Symposium—Establishment of the 2012 vitamin D requirements in swine with focus on dietary forms and levels of vitamin D. *J. Anim. Sci.* **2014**, *92*, 910–916. [CrossRef] [PubMed]
7. Lauridsen, C.; Halekoh, U.; Larsen, T.; Jensen, S.K. Reproductive performance and bone status markers of gilts and lactating sows supplemented with two different forms of vitamin D. *J. Anim. Sci.* **2010**, *88*, 202–213. [CrossRef]
8. Coffey, J.; Hines, E.; Starkey, J.; Starkey, C.; Chung, T. Feeding 25-hydroxycholecalciferol improves gilt reproductive performance and fetal vitamin D status. *J. Anim. Sci.* **2012**, *90*, 3783–3788. [CrossRef] [PubMed]
9. Flohr, J.R.; Tokach, M.D.; Dritz, S.S.; Derouchey, J.M.; Goodband, R.D.; Nelssen, J.L.; Henry, S.C.; Tokach, L.M.; Potter, M.L.; Goff, J.P.; et al. Effects of supplemental vitamin D3 on serum 25-hydroxycholecalciferol and growth of preweaning and nursery pigs. *J. Anim. Sci.* **2014**, *92*, 152–163. [CrossRef]
10. Shaheen, S.; Noor, S.S.; Barakzai, Q. Serum alkaline phosphatase screening for vitamin D deficiency states. *J. Coll. Physicians Surg. Pak.* **2012**, *22*, 424–427.
11. Khailova, L.; Robison, J.; Jaggers, J.; Ing, R.; Lawson, S.; Treece, A.; Soranno, D.; Osorio Lujan, S.; Davidson, J.A. Tissue alkaline phosphatase activity and expression in an experimental infant swine model of cardiopulmonary bypass with deep hypothermic circulatory arrest. *J. Inflamm.* **2020**, *17*, 27. [CrossRef]
12. Krawitt, E.L.; Korson, R. Effect of vitamin D on brush border alkaline phosphatase in the rat small intestine. *Enzyme* **1972**, *13*, 278–285. [CrossRef]
13. Kiran, B.; Prema, A.; Thilagavathi, R.; Rani, R.J. Serum 25-Hydroxy vitamin D, calcium, phosphorus and alkaline phosphatase levels in healthy adults above the age of 20 living in Potheri Village of Kancheepuram District, Tamilnadu. *J. Appl. Pharm. Sci.* **2014**, *4*, 030–034. [CrossRef]
14. McCarthy, R.A.; McCallion, N.; Molloy, E.J. Relationship Between Vitamin D and Alkaline Phosphatase Levels in Very Low Birth Weight (VLBW) Infants. *Arch. Dis. Child.* **2008**, *93*, ps315.
15. Kamao, M.; Tatematsu, S.; Hatakeyama, S.; Sakaki, T.; Sawada, N.; Inouye, K.; Ozono, K.; Kubodera, N.; Reddy, G.S.; Okano, T. C-3 epimerization of vitamin D3 metabolites and further metabolism of C-3 epimers: 25-hydroxyvitamin D3 is metabolized to 3-epi-25-hydroxyvitamin D3 and subsequently metabolized through C-1alpha or C-24 hydroxylation. *J. Biol. Chem.* **2004**, *279*, 15897–15907. [CrossRef] [PubMed]
16. Molnar, F.; Sigueiro, R.; Sato, Y.; Araujo, C.; Schuster, I.; Antony, P.; Peluso, J.; Muller, C.; Mourino, A.; Moras, D.; et al. 1alpha,25(OH)2-3-epi-vitamin D3, a natural physiological metabolite of vitamin D3: Its synthesis, biological activity and crystal structure with its receptor. *PLoS ONE* **2011**, *6*, e18124. [CrossRef] [PubMed]
17. Lensmeyer, G.; Poquette, M.; Wiebe, D.; Binkley, N. The C-3 Epimer of 25-Hydroxyvitamin D3 Is Present in Adult Serum. *J. Clin. Endocrinol. Metab.* **2012**, *97*, 163–168. [CrossRef] [PubMed]
18. Maynard, D.F.; Trankle, W.G.; Norman, A.W.; Okamura, W.H. 14-Epi Stereoisomers of 25-Hydroxy-and 1. alpha., 25-Dihydroxyvitamin D3: Synthesis, Isomerization to Previtamins, and Biological Studies. *J. Med. Chem.* **1994**, *37*, 2387–2393. [CrossRef] [PubMed]
19. Tang, D.; Liu, W.; Huang, L.; Cheng, L.; Xu, Z. Efficient biotransformation of vitamin D 3 to 25-hydroxyvitamin D 3 by a newly isolated Bacillus cereus strain. *Appl. Microbiol. Biotechnol.* **2020**, *104*, 765–774. [CrossRef] [PubMed]
20. Guy, E.; Lieve, V.; Natacha, R.; Frank, C.; Pierre De, C.; Maurits, V.; Giuseppe, T.-V.; Dino, M.; Roger, B.; Annemieke, V. Superagonistic Action of 14-epi-Analogs of 1,25-Dihydroxyvitamin D Explained by Vitamin D Receptor-Coactivator Interaction. *Mol. Pharmacol.* **2005**, *67*, 1566. [CrossRef]
21. Latimer, G.W.; International, A.O.A.C. *Official Methods of Analysis of AOAC International*, 21st ed.; AOAC International: Gaithersburg, MD, USA, 2019.
22. *ISO:9831*; Animal Feeding Stuffs, Animal Products, and Faeces or Urine—Determination of Gross Calorific Value—Bomb Calorimeter Method. International Organization for Standardization: Geneve, Switzerland, 1998. Available online: https://www.iso.org/obp/ui/#iso:std:iso:9831:ed-1:v1:en (accessed on 22 January 2024).
23. Cook, R.D.; Weisberg, S. *Applied Regression including Computing and Graphics*; John Wiley & Sons: New York, NY, USA, 2009.
24. Hadley, W. *ggplot2: Elegant Graphics for Data Analysis*; Springer: New York, NY, USA, 2016; p. 260. [CrossRef]
25. Russell, L. Emmeans: Estimated Marginal Means, Aka Least-Squares Means, R package version 1.4.2; An alternative to least squares means, Am Stat. 2019. Available online: https://github.com/rvlenth/emmeans (accessed on 24 October 2019).

26. R Core Team. *R: A Language and Environment for Statistical Computing;* R Core Team: Vienna, Austria, 2022.
27. Thekkoot, D.M.; Kemp, R.A.; Rothschild, M.F.; Plastow, G.S.; Dekkers, J.C. Estimation of genetic parameters for traits associated with reproduction, lactation, and efficiency in sows. *J. Anim. Sci.* **2016**, *94*, 4516–4529. [CrossRef]
28. Piles, M.; MartÌ, J.; Reixach, J.; S·nchez, J.P. Genetic parameters of sow feed efficiency during lactation and its underlying traits in a Duroc population. *Animal* **2020**, *14*, 889–898. [CrossRef]
29. Cabrera, R.A.; Boyd, R.D.; Jungst, S.B.; Wilson, E.R.; Johnston, M.E.; Vignes, J.L.; Odle, J. Impact of lactation length and piglet weaning weight on long-term growth and viability of progeny1,2. *J. Anim. Sci.* **2010**, *88*, 2265–2276. [CrossRef] [PubMed]
30. Hawe, S.J.; Scollan, N.; Gordon, A.; Magowan, E. Impact of sow lactation feed intake on the growth and suckling behavior of low and average birthweight pigs to 10 weeks of age. *Transl. Anim. Sci.* **2020**, *4*, 655–665. [CrossRef]
31. Kilbride, A.L.; Mendl, M.; Statham, P.; Held, S.; Harris, M.; Marchant-Forde, J.N.; Booth, H.; Green, L.E. Risks associated with preweaning mortality in 855 litters on 39 commercial outdoor pig farms in England. *Prev. Vet. Med.* **2014**, *117*, 189–199. [CrossRef]
32. Muns, R.; Nuntapaitoon, M.; Tummaruk, P. Non-infectious causes of pre-weaning mortality in piglets. *Livest. Sci.* **2016**, *184*, 46–57. [CrossRef]
33. Uddin, M.K.; Hasan, S.; Peltoniemi, O.; Oliviero, C. The effect of piglet vitality, birth order, and blood lactate on the piglet growth performances and preweaning survival. *Porc. Health Manag.* **2022**, *8*, 52. [CrossRef]
34. Di Martino, G.; Stefani, A.L.; Lippi, G.; Gagliazzo, L.; McCormick, W.; Gabai, G.; Bonfanti, L. The degree of acceptability of swine blood values at increasing levels of hemolysis evaluated through visual inspection versus automated quantification. *J. Vet. Diagn.* **2015**, *27*, 306–312. [CrossRef] [PubMed]
35. Wubuli, A.; Gerlinger, C.; Reyer, H.; Oster, M.; Muráni, E.; Trakooljul, N.; Ponsuksili, S.; Wolf, P.; Wimmers, K. Reduced phosphorus intake throughout gestation and lactation of sows is mitigated by transcriptional adaptations in kidney and intestine. *BMC Genom.* **2020**, *21*, 626. [CrossRef] [PubMed]
36. Golder, H.M.; McGrath, J.; Lean, I.J. Effect of 25-hydroxyvitamin D3 during prepartum transition and lactation on production, reproduction, and health of lactating dairy cows. *J. Dairy Sci.* **2021**, *104*, 5345–5374. [CrossRef]
37. Kozai, M.; Yamamoto, H.; Ishiguro, M.; Harada, N.; Masuda, M.; Kagawa, T.; Takei, Y.; Otani, A.; Nakahashi, O.; Ikeda, S.; et al. Thyroid Hormones Decrease Plasma 1α,25-Dihydroxyvitamin D Levels Through Transcriptional Repression of the Renal 25-Hydroxyvitamin D3 1α-Hydroxylase Gene (CYP27B1). *Endocrinology* **2013**, *154*, 609–622. [CrossRef]
38. Thayer, M.T.; Nelssen, J.L.; Langemeier, A.J.; Morton, J.M.; Gonzalez, J.M.; Kruger, S.R.; Ou, Z.; Makowski, A.J.; Bergstrom, J.R. The effects of maternal dietary supplementation of cholecalciferol (vitamin D3) and 25(OH)D3 on sow and progeny performance1. *Transl. Anim. Sci.* **2019**, *3*, 692–708. [CrossRef] [PubMed]
39. Zhou, H.; Chen, Y.; Lv, G.; Zhuo, Y.; Lin, Y.; Feng, B.; Fang, Z.; Che, L.; Li, J.; Xu, S.; et al. Improving maternal vitamin D status promotes prenatal and postnatal skeletal muscle development of pig offspring. *Nutrition* **2016**, *32*, 1144–1152. [CrossRef] [PubMed]
40. Jeong, J.H.; Hong, J.S.; Han, T.H.; Fang, L.H.; Chung, W.L.; Kim, Y.Y. Effects of dietary vitamin levels on physiological responses, blood profiles, and reproductive performance in gestating sows. *J. Anim. Sci. Technol.* **2019**, *61*, 294–303. [CrossRef] [PubMed]
41. Rodney, R.M.; Celi, P.; McGrath, J.J.; Golder, H.M.; Anderson, S.T.; McNeill, D.M.; Fraser, D.R.; Lean, I.J. Metabolic and production responses to calcidiol treatment in mid-lactation dairy cows. *Anim. Prod. Sci.* **2019**, *59*, 449–460. [CrossRef]
42. Grez-Capdeville, M.; Crenshaw, T.D. Estimation of phosphorus requirements of sows based on 24-h urinary phosphorus excretion during gestation and lactation. *Br. J. Nutr.* **2022**, *128*, 377–388. [CrossRef]

animals

MDPI

Article

Effects of Alternative Cassava and Taro Feed on the Carcass and Meat Quality of Fattening Pigs Reared under Ecuadorian Backyard Systems

Alfredo Valverde Lucio [1], Ana Gonzalez-Martínez [2], Julio Gabriel Ortega [1] and Evangelina Rodero Serrano [2,*]

1 Faculty of Natural Sciences and Agriculture, University of the South of Manabí (UNESUM), Jipijapa 130303, Manabí, Ecuador; yhonny.valverde@unesum.edu.ec (A.V.L.); julio.gabriel@unesum.edu.ec (J.G.O.)
2 Department of Animal Production, Faculty of Veterinary Sciences, University of Cordoba, AGR-134, ceiA3., 14071 Córdoba, Spain; agmartinez@uco.es
* Correspondence: pa1rosee@uco.es

Simple Summary: Pork is currently the cheapest protein source in the world. In the traditional rearing of backyard pigs in regions of Ecuador, cassava and taro crops are frequently used as replacement alternatives to corn in pig feed formulations. In this study, the quality and characteristics of the carcass and the behavior of the gastrointestinal tract (GIT) of 30 fattening pigs reared under the backyard production system were analyzed. The animals were fed with a conventional or alternative diet based on the addition of cassava and taro in doses of 32% and 42%. The results showed a higher effect of the geographical location than the feed administered to the animals. The morphological traits were those with lower changes between groups than the gastrointestinal tract measurements. The proportions of alternatives used in the formulations must be optimized, since this directly increases the amount of protein in the meat and the weight of the GIT, decreasing the degree of fattening of the carcass. In the production of backyard pigs in Ecuador based on the use of by-products and agricultural waste, it is necessary to promote the standardization of the type of pig that is raised, taking into account geographical location and promoting the use of local genetic resources.

Abstract: Ecuadorian small producers use crossbred animals with a low level of genetic improvement, which are fed with alternative feeds to decrease production costs. The objective of this study was to evaluate the effects of geographical location and three diets according to the amount of cassava and taro incorporated into the feed (T1 conventional feed; T2 and T3 with 32% and 42% of cassava and taro, respectively) in pigs reared under the backyard system. The results did not show many differences between the treatments for morphological traits; however, between geographical locations, significant differences were evidenced. The fat content from the first rib was higher in the T1 group. The intramuscular fat percentage was higher in the T1 group, contrary to the protein levels, which were higher in the T3 group in Esmeraldas and the T2 group in Ro Chico. In the gastrointestinal tract (GIT) and its attached organs, differences were found in the empty stomach weight, full and empty small intestine weight, liver weight, and total GIT weight, with the T2 and T3 groups having the largest and heaviest. Cassava and taro did not affect the morphometric behavior and quality of the carcass but increased the amount of protein in the meat and the weight of the GIT. Geographical location was also observed to have a significant effect.

Keywords: alternative diets; meat quality; morphometric traits; gastrointestinal tract

1. Introduction

Pork is currently the cheapest protein source in the world [1]. Its production reached 122 million tons in 2021, positioning it as the second-highest and the highest production and consumption worldwide, respectively [2]. In Ecuador, pork production in 2021 was

220,000 t, which was supported by backyard producers [3]. This backyard family production system traditionally uses agricultural feed alternatives generated from cultivation on their own farms, as well as cooking by-products to reduce production costs [4,5].

Small producers in Ecuador lack technical infrastructure as well as health plans. They use crossbred animals with a low level of genetic improvement, resulting from unplanned crossings between pure and crossbred animals, or between improved mixed breeds [6,7], according to the geographical location [5,8]; this directly affects the productivity, as it is a key factor for weight gain [9] and the quality of the carcass [10].

The quality of meat is defined by its palatability and consumer acceptance [11,12], and at the organoleptic level, it is measured by its color, smell, texture [13], and fat content, in addition to other technical aspects such as its pH, water retention capacity, and fatty acid and cholesterol profile [14]. It can also be defined by its health benefits (e.g., amounts of omega 3, vitamins, and amino acids) [15].

Production systems can interfere with obtaining a quality carcass [16,17]. In this sense, pigs raised in outdoor production systems, in which they consume pastures and complementary diets, grow healthier, and have better productivity [18], offer higher carcass yields and more tender meat, as well as a greater amount of intramuscular fat, unsaturated fatty acids, vitamin E, and antioxidants [19,20].

Pig production systems involve production costs in which feed accounts for at least 70% [21], with feed alternatives representing a sustainable and economical way of feeding [22], which, when dosed correctly, do not affect the quality of the pork [23].

The feed alternatives must be formulated considering the nutritional requirements of the animals to guarantee their productive performance [24], without disregarding the age of the animals, as well as the intestinal needs over a period of time between three and four weeks to adapt to the new diet [25].

The production of backyard pigs is represented by the social stratum [26] and constitutes an important source of income for the family economy, both as an accessible source of protein and as a tradable good in the market [27]. Currently, pig production and its derivatives are an important source of employment [28], which contributes to social development by guaranteeing food security [29] and supplying the needs of the population with quality meat [30].

The feed alternatives used in the breeding and fattening of backyard pigs in Ecuador include a diversity of feeds, among which cassava, taro, tagua, bananas, and squash stand out [5]. The use of cassava (*Manihot esculenta*) and taro (*Colocacia esculenta*) as corn substitutes lowers the production costs of backyard pigs [22]. In traditional diets, maize represents between 50% and 70% of the diet content, which considerably increases the costs of production [31]. The use of cassava and taro as a food alternative due to their high digestibility [32–34] provides acceptable results for production [12,22,35,36]. However, due to the content of antinutritional factors in both feed alternatives, it is necessary to subject them to prior cooking to reduce their negative effects [37–40].

There is no scientific evidence regarding the simultaneous use of cassava and taro in the quality of the carcass of backyard pigs. Therefore, the objective of this research was to evaluate the characteristics of the carcass, the quality of the meat, and the behavior of the gastrointestinal tract of fattening pigs fed with cassava and taro and raised under backyard production systems in Ecuador. Secondly, the effect of the amount of alternative feed in the diet was evaluated, as well as the geographical location of the animals, which is directly related to their genetic origin.

2. Materials and Methods

2.1. Selection and Preparation of Animals

A total of 42 castrated crossbred pigs (20 males and 22 females) were used, which were purchased at 60 days of age from producers in the study area. The animals used were Creole pigs mixed with the Pietrain breed coming from few litters to have greater homogeneity (two per geographical location). The experiments were carried out in two geographical locations (Figure 1), with the purpose of carrying out repetitions of the study. Fifteen pigs were raised in Quinindé, Esmeraldas province, and the remaining fifteen in Río Chico, Manabí province. Both geographical locations have a tropical climate in terms of their average annual temperatures, rainfall, and altitudes [22]. The experiments were conducted during July to November 2021.

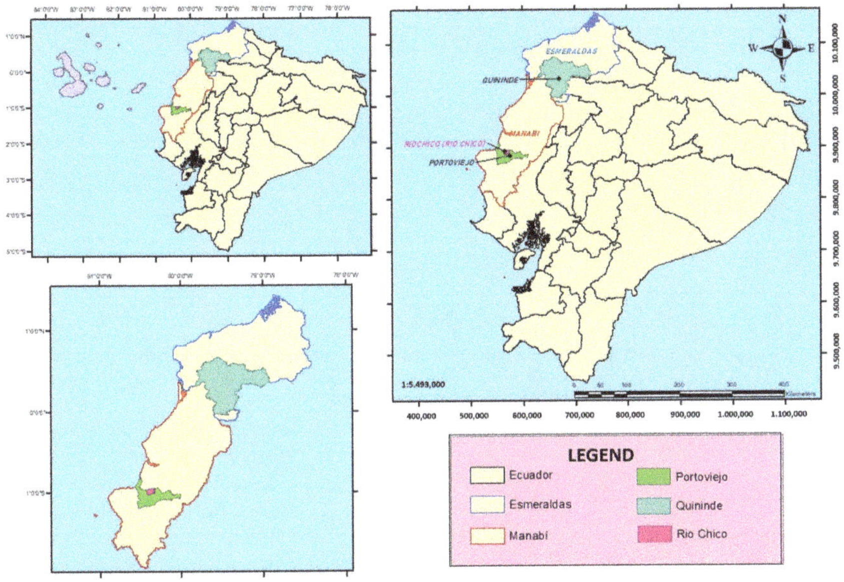

Figure 1. Geographic location of the sampling zones [22].

The animals from each geographical location were randomly distributed into three groups (four males and three females per group), and they were housed in traditional pig pens [22] with an area of 1.25 m^2 per animal. Before the experiment began, the pigs were given a ten-day period to adapt to the location, and a progressive change in their feed was carried out. It is worth noting that the volume of experimental feed was gradually increased every 5 days. The pigs were provided with water ad libitum through feeding bottles. Each group was given a different diet formulation (T1, T2, and T3, according to the following section). The feed was supplied twice a day at fixed times, at 8:00 a.m. and 3:00 p.m. Prior to the start of the experiment, the pigs had a period of ten days to adapt to the new feed, with the amount increasing progressively.

2.2. Preparation and Formulation of Diets

The feed alternative based on by-products of the cassava and taro processing industry were used. Both feed alternatives were administered to the animals after cooking to eliminate anti-nutritional components. Prior to cooking, the cassava and taro were weighed, washed, and chopped with the peel included; in addition, salt was added to increase the palatability of both feeds for the animals. After cooking, the feed was allowed to cool before mixing it with the rest of the components of the diet administered to the animals.

Three diets were formulated: one without the addition of cassava and taro (T1), and the remaining two with the addition of 32% (T2) and 42% (T3) of feed alternative in equal parts (Table 1). The protein and energy content of the diet was standardized according to the productive phase the pigs were in, which was growth or fattening. Thus, during the growth stage, the amount of protein was 18%, and in the fattening stage, it was 15%. The bromatological analysis of the feed alternatives used, as well as the formulas used for the experiment, were described in a previous study [22]. The animals received this diet during the 90 days that the trial lasted. All of the animals remained healthy throughout the experiment.

Table 1. Nutritional value and composition of the used diets [22].

Ingredients [1]	Phase					
	Growth			Fattening		
	T1 (Control)	T2 (32%)	T3 (42%)	T1 (Control)	T2 (32%)	T3 (42%)
Corn (kg)	23.64	9.55	4.55	23.64	9.55	6.36
Protein concentrate (kg) [2]	13.18	16.36	17.27	11.36	15.00	16.36
Rice powder (kg)	8.18	4.55	4.09	10.00	5.91	3.18
Cooked cassava (kg)		7.27	9.55		7.27	9.55
Cooked taro (kg)		7.27	9.55		7.27	9.55
Red palm oil (kg)	1	1	1	1	1	1
Salt (g)	0.25	0.25	0.25	0.25	0.25	0.25
Crude protein (%)	17.35	17.35	17.35	15.30	15.29	15.29
Gross energy (kcal/kg)	3098	3079	3077	3084	3079	3094

[1] Quantity per kilo of feed. [2] Soybean meal, rice by-products, banana meal, free fatty acids, molasses, calcium carbonate, mycotoxin binders, vitamin supplements (A, D3, E, and K3), riboflavin, niacin, thiamine mononitrate, cyanocobalamin, pyridoxine hydrochloride, biotin, trace mineral supplements, manganese sulfate, zinc sulfate, copper sulfate, ferrous sulfate, sodium, calcium iodide, methionine, lysine (such as hydrochloride and sulfate), threonine, choline chloride, antifungals, enzymes, antibiotics, and antioxidants.

2.3. Procedure for Obtaining the Data

The pigs were utilized in the development of two studies; one was recently published [22]. For the second piece of research, only 30 pigs were utilized, first choosing all the males in each lot and then females until there were 5 animals per group.

The animals, at 160 days and 77.84 ± 1.71 kg, were slaughtered after a ten-hour fasting period, in accordance with Ecuadorian regulations [41]. Each carcass was weighed and measured while hot, and morphometric measurements of the foreleg, leg, ham, and shank were conducted. The rest of the measurements were obtained 24 h after slaughter [42]. The quartering of the carcass was carried out according to the indications of Nieto et al. [43]. The head was removed by cutting at the occipito-atlas joint, and the feet by cutting at the carpus-metacarpal and tarsus-metatarsal joints. The carcass was split longitudinally and, finally, to prevent dehydration, kept at −20 °C in plastic bags. After 24 h since the slaughter had passed, the loin was separated by a cut that began just ventral to the ventral side of the scapula at the cranial end and followed the natural curvature of the vertebral column to the ventral edge of the psoas major at the caudal end of the loin. The ham was removed with a straight cut between the second and third sacral vertebrae, and then the foreleg was separated from the trunk. After the rib was separated from the vertebrae, measurements of the fat were taken at the first and last rib levels as well as haunch point. Once all of the parts (head, loin, ham, foreleg, ribs, and legs) had been separated, they were weighed and measured.

The following measurements were collected: (i) the weights of the head, loin, ribs, ham, foreleg, and legs; (ii) the length of the carcass, bone, and muscle of the foreleg and ham; (iii) the perimeters of the front shank and ham; (iv) the thickness of backfat (DBT) at the first and last rib levels and gluteus point; and (v) the loin and haunch fat. The measurement instruments used were a high-precision digital scale from Montero (Seattle, WA, USA), model TCS300JC61Z©, with a range of 300 kg to 2000 g (d = 100 g); a RexBeti Stainless Hardened © digital vernier caliper (measuring range: 5906 in. Precision: 0.1 inch) (Seattle, WA, USA); and a Jontex © brand digital scale (Seattle, WA, USA) with a maximum capacity of 40 kg and a minimum of 200 g (e = d = 5 g).

The digestive viscera of the gastrointestinal tract (GIT), stomach, liver, pancreas, small intestine, colon, cecum, and rectum were separated from the carcass to be individually measured and weighed, first full and then empty. The total weight of the viscera was calculated using the sum of the individual weights of each one of the parts, obtaining a weight for the total of the full GIT and another for the empty GIT [44].

The collection of weights and measures was carried out by the same technician for the two locations in order to reduce potential errors in obtaining the data [9].

For the meat quality analysis, a sample of 200 g of the longissimus lumbar muscle was taken at the level of the last rib 45 min after slaughter and was frozen at a temperature between −18 and −20 °C [42]. The bromatological analyses to determine the content of protein, fat, dry matter, moisture, ash, and pH were carried out in the Multianalityca S.A. laboratory (Quito—Ecuador) (certified SAE LEN 09-008). The reference methods of analysis were the following: moisture, Association of Official Agricultural Chemists, AOAC 925.10; crude protein, AOAC 2001.11; fat, AOAC 2003.06; ash, AOAC 923.03; and pH. Finally, the dry matter was estimated through the following calculation based on methods established by Maclean et al. [45]: dry matter = (initial weight − dry weight)/initial weight.

2.4. Statistical Analysis

IBM SPSS Statistics (version 26) software was used to perform the statistical analyses. All of the records were considered to be quantitative variables. After checking the normality and homogeneity of the variables, a mixed ANOVA with repeated measurements analysis was conducted. The statistical model included the fixed effects of treatment (T) and location (L) and their interaction (T × L). The repeated effect was location, and the subject of the repeated measurements was the animal nested within a group. When the fixed effects were significant, differences between the least squares means were assessed by a paired *t*-test at 5%. Moreover, Pearson correlations between carcass measurements were investigated in order to assess the relationship between the morphometric and compositional variables of the carcass and the GIT.

3. Results

3.1. Pig Carcass Morphology

Both of the effects considered (geographical location and feeding treatment) showed different results in terms of the morphological carcass characteristics (Table 2). The geographical location significantly affected ($p < 0.05$) most of the carcass characteristics of the backyard pigs in Ecuador, with the exceptions of the carcass yield, hot carcass weight, ham weight, rib weight, and leg weight. Meanwhile, the treatment only led to significant differences ($p < 0.05$) in ham perimeter, which showed the highest values in those pigs fed with conventional feed. However, the animals from Quinindé fed with 42% cassava and taro (T3_42%), as well as the animals from Río Chico that did not receive a feed alternative (T1_control), showed higher values for most of the parameters considered. The ham weight was significantly higher in pigs fattened in Quinindé with 42% of feed alternative. In general, the coefficients of variation were the lowest in Río Chico, and they showed different values between treatments.

Table 2. Morphological characteristics of carcasses (mean ± standard error (coefficient of variation)) of backyard pigs fed with different formulations of nutritional alternatives with cassava and taro from two locations in Ecuador (Quinindé and Río Chico).

Traits [1]	Quinindé			Río Chico			Location (L)	Treatment (T)	L × T
	Treatments [2]								
	T1 (Control)	T2 (32%)	T3 (42%)	T1 (Control)	T2 (32%)	T3 (42%)			
LV (kg)	66.98 ± 5.36 (17.87) [c]	73.37 ± 4.32 (13.16) [bc]	77.83 ± 3.29 (9.45) [abc]	87.91 ± 1.12 (2.85) [a]	79.58 ± 1.62 (4.57) [abc]	81.35 ± 1.27 (3.49) [ab]	$p < 0.01$	0.200	0.051
HCW (kg)	47.50 ± 4.75 (21.44)	51.55 ± 2.95 (12.78)	55.87 ± 3.73 (14.93)	60.21 ± 0.61 (2.27)	55.12 ± 2.44 (9.91)	53.66 ± 1.75 (7.28)	0.087	0.902	0.086
CY (%)	0.71 ± 0.2 (6.38)	0.70 ± 0.01 (2.58)	0.72 ± 0.03 (8.36)	0.69 ± 0.01 (1.96)	0.69 ± 0.03 (9.55)	0.66 ± 0.02 (6.06)	0.083	0.867	0.506
CL (cm)	62.00 ± 0.77 (2.79) [b]	63.00 ± 1.14 (4.05) [b]	63.00 ± 1.14 (4.05) [b]	75.72 ± 2.40 (7.15) [a]	74.72 ± 1.31 (3.89) [a]	73.72 ± 2.10 (6.37) [a]	$p < 0.01$	0.943	0.670

Table 2. *Cont.*

Traits [1]	Quinindé T1 (Control)	Quinindé T2 (32%)	Quinindé T3 (42%)	Río Chico T1 (Control)	Río Chico T2 (32%)	Río Chico T3 (42%)	Location (L)	Treatment (T)	L × T
	Treatments [2]								
FLL (cm)	29.00 ± 1.32 (9.90) ^{a}	29.80 ± 0.20 (4.05) ^{a}	31.20 ± 0.58 (4.18) ^{a}	28.97 ± 0.47 (3.66) ^{abc}	27.07 ± 0.37 (3.15) ^{bc}	27.72 ± 0.44 (3.45) ^{c}	$p < 0.01$	0.276	$p < 0.01$
LL (cm)	61.20 ± 1.71 (6.26) ^{ab}	63.80 ± 1.32 (4.62) ^{a}	63.00 ± 1.22 (4.35) ^{ab}	57.6 ± 0.42 (1.64) ^{bc}	54.96 ± 1.66 (6.76) ^{c}	54.04 ± 0.99 (4.11) ^{c}	$p < 0.01$	0.744	0.082
HL (cm)	43.60 ± 1.69 (8.67) ^{a}	45.40 ± 0.81 (4.00) ^{a}	43.40 ± 1.72 (8.86) ^{a}	35.11 ± 0.54 (3.45) ^{b}	33.58 ± 1.36 (9.05) ^{b}	33.64 ± 0.99 (6.61) ^{b}	$p < 0.01$	0.712	0.427
HP (cm)	75.40 ± 1.91 (5.67) ^{a}	69.60 ± 1.81 (5.80) ^{ab}	74.20 ± 1.69 (5.08) ^{ab}	74.13 ± 1.26 (3.79) ^{ab}	68.18 ± 2.42 (7.93) ^{ab}	67.08 ± 1.67 (5.57) ^{b}	$p < 0.05$	$p < 0.05$	0.209
FSP (cm)	14.60 ± 0.40 (6.13) ^{b}	14.40 ± 0.51 (9.92) ^{b}	14.80 ± 0.20 (3.02) ^{b}	15.82 ± 0.27 (3.78) ^{a}	15.28 ± 0.58 (8.48) ^{a}	14.84 ± 0.30 (4.54) ^{b}	$p < 0.05$	0.554	0.332
LW (kg)	5.46 ± 0.30 (12.48) ^{bc}	5.26 ± 0.44 (18.86) ^{c}	5.96 ± 0.40 (14.98) ^{abc}	6.74 ± 0.06 (2.00) ^{a}	6.65 ± 0.17 (5.61) ^{ab}	6.68 ± 0.12 (4.11) ^{ab}	$p < 0.01$	0.467	0.478
HW (kg)	13.89 ± 0.38 (6.15)	15.96 ± 1.5 (21.00)	17.34 ± 1.13 (14.59)	17.30 ± 0.31 (4.02)	15.54 ± 1.15 (16.53)	15.22 ± 0.70 (4.11)	0.712	0.760	$p < 0.05$
FW (kg)	10.66 ± 0.46 (9.64) ^{ab}	8.80 ± 0.68 (17.38) ^{b}	10.72 ± 0.61 (12.66) ^{ab}	12.09 ± 0.28 (5.22) ^{a}	11.25 ± 0.49 (9.80) ^{a}	10.98 ± 0.57 (11.56) ^{ab}	$p < 0.01$	0.054	0.141
RW (kg)	5.65 ± 0.35 (14.04)	4.86 ± 0.38 (17.46)	5.7 ± 0.6 (23.38)	5.66 ± 0.05 (1.91)	5.3 ± 0.22 (9.44)	5.18 ± 0.20 (8.82)	0.933	0.265	0.387
HDW (kg)	4.63 ± 0.10 (5.84) ^{b}	4.64 ± 0.12 (5.87) ^{b}	4.68 ± 0.29 (13.80) ^{b}	6.18 ± 0.28 (10.08) ^{a}	5.26 ± 0.50 (21.08) ^{ab}	5.02 ± 0.43 (19.24) ^{ab}	$p < 0.01$	0.201	0.162
FTW (kg)	1.11 ± 0.03 (5.84)	1.09 ± 0.06 (11.87)	1.11 ± 0.04 (8.06)	1.26 ± 0.07 (13.31)	1.16 ± 0.11 (21.49)	1.09 ± 0.06 (11.33)	0.262	0.425	0.472

[1] LV: live weight; HCW: hot carcass weight; CY: carcass yield; CL: carcass length; FLL: hand length; LL: leg length; HL: ham length; HP: ham perimeter; FSP: front shank perimeter; LW: loin weight; HW: ham weight; FW: foreleg weight; RW: rib weight; HDW: head weight; FTW: feet weight. [2] T1: conventional feed with no corn replacement; T2: corn replacement with 32% of cassava + taro; and T3: corn replacement with 42% of cassava + taro. In addition, $^{a, b, c}$ are for each control; least square means without a common superscript differ significantly ($p < 0.05$) between groups.

3.2. Fat Thickness and Content of Pig Carcass

The backfat thickness at the first rib level was significantly ($p < 0.05$) higher in the animals that did not receive a feed alternative (Río Chico = 2.30 cm; Quinindé = 2.07 cm) (Table 3). Meanwhile, the backfat thickness at the last rib level and haunch fat were significantly ($p < 0.05$) higher in the animals from Quinindé (T1 = 2.34 cm and 1.51 cm; T2 = 1.70 cm and 1.45 cm; T3 = 1.82 cm and 1.41 cm, respectively).

Table 3. Fat contents of the carcass (mean ± standard error (coefficient of variation)) of backyard pigs fed with different formulations of nutritional alternatives with cassava and taro from two locations in Ecuador (Quinindé and Río Chico).

Traits [1]	Quinindé T1 (Control)	Quinindé T2 (32%)	Quinindé T3 (42%)	Río Chico T1 (Control)	Río Chico T2 (32%)	Río Chico T3 (42%)	Location (L)	Treatment (T)	L × T
	Treatments [2]								
DBT1 (cm)	2.07 ± 0.24 (26.02)	1.23 ± 0.08 (15.21)	1.86 ± 0.23 (27.99)	2.30 ± 0.36 (34.71)	1.71 ± 0.23 (30.39)	1.81 ± 0.24 (29.89)	0.276	$p < 0.05$	0.569
DBT2 (cm)	2.34 ± 0.25 (23.46) ^{a}	1.70 ± 0.81 (40.57) ^{ab}	1.82 ± 0.23 (27.89) ^{a}	0.91 ± 0.03 (6.32) ^{bc}	0.88 ± 0.10 (24.66) ^{bc}	0.86 ± 0.08 (20.66) ^{c}	$p < 0.01$	0.199	0.267
DBT3 (cm)	1.52 ± 0.21 (31.13)	1.35 ± 0.29 (48.08)	1.54 ± 0.07 (9.66)	1.28 ± 0.14 (23.55)	1.16 ± 0.14 (27.47)	1.37 ± 0.13 (20.46)	0.176	0.494	0.980
LF (cm)	1.56 ± 0.26 (37.01)	1.26 ± 0.13 (22.51)	1.32 ± 0.06 (9.68)	0.94 ± 0.23 (55)	1.11 ± 0.3 (60.1)	1.09 ± 0.14 (29.42)	0.057	0.953	0.486
HF (cm)	1.51 ± 0.26 (38.45) ^{a}	1.45 ± 0.26 (40.03) ^{a}	1.41 ± 0.07 (10.43) ^{ab}	1.30 ± 0.25 (25.49) ^{ab}	0.73 ± 0.09 (25.9) ^{b}	0.92 ± 0.22 (53.87) ^{b}	$p < 0.01$	0.245	0.421

[1] DBT1: backfat thickness at first rib level; DBT2: backfat thickness at last rib level; DBT3: buttock fat; LF: loin fat; HF: haunch fat. [2] T1: conventional feed with no corn replacement; T2: corn replacement with 32% of cassava + taro; and T3: corn replacement with 42% of cassava + taro. In addition, $^{a, b, c}$ are for each control; least square means without a common superscript differ significantly ($p < 0.05$) between groups.

3.3. Pork Quality Analysis

The bromatological characters showed significant differences ($p < 0.05$) for the geographical location effect, except for the percentage of intramuscular fat (Table 4). The moisture content, protein, and ash were higher in animals raised in Río Chico, while the pH and percentage of dry matter were higher in pigs raised in Quinindé. On the contrary, the diet that the animals received only significantly affected ($p < 0.05$) the pH of the meat, this being higher in pigs fed with 42% of alternative feeds (Quinindé = 5.80; Río Chico = 5.57).

Table 4. Bromatological analysis of meat (mean ± standard error (coefficient of variation)) from backyard pigs fed with different formulations of nutritional alternatives with cassava and taro from two locations in Ecuador (Quinindé and Río Chico).

Traits [1]	Quinindé			Río Chico			Location (L)	Treatments (T)	L × T
	Treatments [2]								
	T1 (Control)	T2 (32%)	T3 (42%)	T1 (Control)	T2 (32%)	T3 (42%)			
H %	64.46 ± 2.74 (9.51) ^{bc}	62.56 ± 2.94 (10.51) ^{c}	69.34 ± 2.35 (7.57) ^{abc}	73.8 ± 0.45 (1.36) ^{a}	72.96 ± 0.38 (1.18) ^{ab}	76.41 ± 1.86 (5.44) ^{a}	$p < 0.01$	0.054	0.714
CP %	17.37 ± 0.81 (10.40) ^{c}	18.11 ± 0.66 (8.11) ^{c}	18.93 ± 0.22 (2.54) ^{bc}	21.51 ± 0.63 (6.54) ^{ab}	23.33 ± 0.77 (7.43) ^{a}	19.23 ± 1.16 (13.45) ^{bc}	$p < 0.01$	0.097	$p < 0.01$
IMF %	2.98 ± 0.60 (45.26)	2.71 ± 0.21 (17.7)	1.75 ± 0.11 (14.69)	3.13 ± 0.50 (35.75)	2.21 ± 0.66 (66.83)	2.88 ± 1.26 (97.73)	0.631	0.513	0.484
Ash %	0.86 ± 0.05 (12.12) ^{b}	0.82 ± 0.04 (10.77) ^{b}	0.88 ± 0.04 (10.88) ^{b}	1.42 ± 0.07 (10.48) ^{a}	1.50 ± 0.04 (5.76) ^{a}	1.48 ± 0.07 (10.38) ^{a}	$p < 0.01$	0.756	0.531
pH	5.76 ± 0.05 (1.93) ^{ab}	5.61 ± 0.02 (0.98) ^{ab}	5.80 ± 0.08 (2.92) ^{a}	5.54 ± 0.04 (1.78) ^{bc}	5.34 ± 0.04 (1.78) ^{c}	5.57 ± 0.08 (3.06) ^{abc}	$p < 0.01$	$p < 0.01$	0.924
DM %	34.8 ± 2.80 (18.00) ^{ab}	37.44 ± 2.94 (17.57) ^{a}	30.66 ± 2.35 (17.12) ^{abc}	26.2 ± 0.45 (3.83) ^{bc}	27.04 ± 0.38 (3.17) ^{bc}	23.59 ± 1.86 (17.62) ^{c}	$p < 0.01$	0.061	0.727

[1] H: humidity; CP: crude protein; IMF: intramuscular fat; DM: dry matter. [2] T1: conventional feed with no corn replacement; T2: corn replacement with 32% of cassava + taro; and T3: corn replacement with 42% of cassava + taro. In addition, $^{a, b, c}$ are for each control; least square means without a common superscript differ significantly ($p < 0.05$) between groups.

3.4. Morphometry Characteristics of Gastrointestinal Tract and Visceral Organs

The pigs reared in Río Chico presented significantly ($p < 0.05$) higher values in almost all of the gastrointestinal tract (GIT) variables, with the exception of the full and empty small intestine weight and the total GIT weight (Table 5). Regarding the diet administered to the animals, it significantly affected ($p < 0.05$) the liver weight, empty stomach weight, full and empty cecum weight, and full GIT total weight, with higher values in pigs fed with 42% cassava and taro.

Table 5. Behaviors of the digestive tracts (mean ± standard error (coefficient of variation)) of backyard pigs fed with different formulations of nutritional alternatives with cassava and taro from two locations in Ecuador (Quinindé and Río Chico).

Traits [1]	Quinindé			Río Chico			*p*		
	Treatments [2]						Location (L)	Treatment (T)	L × T
	T1 (Control)	T2 (32%)	T3 (42%)	T1 (Control)	T2 (32%)	T3 (42%)			
SW (kg)	0.12 ± 0.01 (21.22) ^{b}	0.14 ± 0.02 (32.97) ^{ab}	0.15 ± 0.02 (26.07) ^{ab}	0.22 ± 0.05 (46.47) ^{ab}	0.28 ± 0.06 (49.08) ^{a}	0.23 ± 0.02 (18.95) ^{ab}	$p < 0.01$	0.523	0.608
LW (kg)	1.17 ± 0.03 (5.46) ^{b}	1.29 ± 0.04 (6.3) ^{b}	1.24 ± 0.04 (6.42) ^{b}	1.41 ± 0.09 (14.46) ^{b}	1.41 ± 0.10 (16.51) ^{b}	1.76 ± 0.04 (4.6) ^{a}	$p < 0.01$	$p < 0.05$	$p < 0.05$
PW (kg)	0.13 ± 0.6 (13.55) ^{b}	0.12 ± 0.01 (24.59) ^{b}	0.12 ± 0.01 (14.65) ^{b}	0.13 ± 0.003 (6.54) ^{b}	0.16 ± 0.01 (16.38) ^{a}	0.16 ± 0.01 (12.17) ^{a}	$p < 0.01$	0.420	0.089
FEWL (kg)	1.32 ± 0.34 (56.88) ^{b}	2.08 ± 0.37 (39.99) ^{a}	1.73 ± 0.22 (27.93) ^{ab}	1.73 ± 0.03 (3.26) ^{ab}	1.37 ± 0.16 (25.48) ^{b}	1.65 ± 0.06 (8.51) ^{ab}	$p < 0.05$	0.742	0.168
ESW (kg)	0.48 ± 0.01 (5.89) ^{b}	0.53 ± 0.03 (13.35) ^{b}	0.51 ± 0.03 (12.23) ^{b}	0.79 ± 0.03 (8.61) ^{a}	0.44 ± 0.07 (35.06) ^{b}	0.85 ± 0.03 (6.72) ^{a}	$p < 0.01$	$p < 0.01$	$p < 0.01$
FSIW (kg)	2.17 ± 0.36 (37.26)	2.79 ± 0.43 (34.19)	2.70 ± 0.28 (23.2)	2.01 ± 0.1 (11)	2.58 ± 0.09 (7.83)	2.98 ± 0.21 (15.77)	0.793	$p < 0.05$	0.423

Table 5. *Cont.*

Traits [1]	Quinindé			Río Chico			*p*		
	Treatments [2]								
	T1 (Control)	T2 (32%)	T3 (42%)	T1 (Control)	T2 (32%)	T3 (42%)	Location (L)	Treatment (T)	L × T
ESIW (kg)	1.27 ± 0.06 (11.38)	1.37 ± 0.04 (6.64)	1.40 ± 0.07 (11.94)	1.23 ± 0.04 (7.82)	1.50 ± 0.12 (18.11)	1.56 ± 0.07 (9.82)	0.257	$p < 0.05$	0.433
FCEW kg)	0.46 ± 0.07 (35.18) ^{c}	0.60 ± 0.08 (30.97) ^{bc}	0.76 ± 0.04 (10.97) ^{ab}	0.82 ± 0.05 (12.86) ^{a}	0.62 ± 0.04 (15.34) ^{ab}	0.68 ± 0.03 (8.37) ^{ab}	$p < 0.05$	0.066	$p < 0.05$
ECEW (kg)	0.12 ± 0.01 (15.66) ^{c}	0.13 ± 0.01 (13.28) ^{bc}	0.15 ± 0.01 (12.09) ^{ab}	0.15 ± 0.01 (10.54) ^{ab}	0.17 ± 0.01 (13.15) ^{a}	0.14 ± 0.1 (10.10) ^{ab}	$p < 0.01$	0.11	$p < 0.01$
FPRW (kg)	0.23 ± 0.02 (22.48) ^{c}	0.25 ± 0.01 (10.2) ^{c}	0.25 ± 0.01 (5.98) ^{bc}	0.39 ± 0.02 (12.30) ^{a}	0.34 ± 0.03 (17.52) ^{ab}	0.38 ± 0.03 (15.58) ^{a}	$p < 0.01$	0.719	0.228
EPRW (kg)	0.19 ± 0.01 (15.8) ^{b}	0.22 ± 0.01 (6.48) ^{b}	0.21 ± 0.01 (12.66) ^{b}	0.32 ± 0.02 (16.98) ^{a}	0.25 ± 0.02 (21.43) ^{ab}	0.25 ± 0.02 (13.81) ^{ab}	$p < 0.01$	0.318	$p < 0.05$
FCOWL (kg)	2.28 ± 0.27 (26.84) ^{b}	2.62 ± 0.07 (6.07) ^{ab}	2.00 ± 0.10 (10.68) ^{b}	3.09 ± 0.12 (8.96) ^{a}	3.12 ± 0.2 (14.47) ^{a}	3.30 ± 0.16 (10.77) ^{a}	$p < 0.01$	0.348	0.077
ECOW (kg)	0.87 ± 0.03 (8.35) ^{c}	0.94 ± 0.04 (10.25) ^{bc}	0.91 ± 0.02 (4.76) ^{bc}	1.19 ± 0.12 (23.13) ^{abc}	1.38 ± 0.18 (29.46) ^{a}	1.34 ± 0.09 (15.53) ^{ab}	$p < 0.01$	0.494	0.504
TWTF (kg)	6.43 ± 0.85 (29.67) ^{c}	8.37 ± 0.61 (16.3) ^{bc}	7.51 ± 0.50 (14.92) ^{c}	9.93 ± 0.08 (1.83) ^{b}	10.57 ± 0.15 (3.1) ^{ab}	12.20 ± 0.47 (8.67) ^{a}	$p < 0.01$	$p < 0.01$	0.074
TWTE (kg)	2.93 ± 0.10 (7.38) ^{c}	3.19 ± 0.06 (4.28) ^{bc}	3.16 ± 0.10 (7.34) ^{bc}	3.67± 0.16 (9.66) ^{ab}	3.74 ± 0.21 (12.73) ^{ab}	4.13 ± 0.12 (6.69) ^{a}	0.068	$p < 0.01$	0.095

[1] SW: spleen weight; LW: liver weight; PW: pancreas weight; FEWL: full stomach weight; ESW: empty stomach weight; FSIW: full small intestine weight; ESIW: empty small intestine weight; FCEW: full cecum weight; ECEW: empty cecum weight; FPRW: full pig rectum weight; EPRW: empty pig rectum weight; FCOWL: full colon weight; ECOW: empty colon weight; TWTF: full total gastrointestinal tract weight; TWTE: empty total gastrointestinal tract weight. [2] T1: conventional feed with no corn replacement; T2: corn replacement with 32% of cassava + taro; and T3: corn replacement with 42% of cassava + taro. In addition, $^{a, b, c}$ are for each control; least square means without a common superscript differ significantly ($p < 0.05$) between groups.

3.5. Relationship between Carcass Measurements and Morphometry of Pigs' Gastrointestinal Tracts

Table 6 shows the correlations between the morphometric variables of the GIT and those of the carcass. The significant ($p < 0.05$) correlations found between the total GIT weight and the different parts of GIT weight were expected. The results reveal the negative and significant relationship ($p < 0.01$) between the amount of fat in the different parts and the development of the GIT, especially in the small intestine and colon.

Table 6. Pearson correlation coefficients between carcass characters and those of the gastrointestinal tract.

Traits [1]	HW (kg)	RW (kg)	DBT1 (cm)	LF (cm)	HF (cm)	P (%)	IMF (%)	LW (kg)	ESIW (kg)	ECEW (kg)	ECOW (kg)	TWTE (kg)
HCW (kg)	0.77 **	0.85 **	0.35	−0.01	0.06	0.20	−0.02	0.19	0.03	0.13	−0.01	0.16
HW (kg)		0.48 *	0.16	0.02	0.19	0.24	0.00	0.19	0.03	0.09	−0.05	0.15
RW (kg)			0.29	0.12	0.04	−0.01	−0.04	0.02	−0.09	0.01	−0.17	−0.02
DBT1 (cm)				−0.17	−0.06	−0.07	0.19	0.31	−0.01	0.18	−0.03	0.19
LF (cm)					−0.41 *	−0.13	−0.16	−0.30	−0.09	−0.24	−0.42 *	−0.27
HF (cm)						−0.38 *	−0.19	−0.40 *	−0.5 **	−0.13	−0.39 *	−0.42 *
P (%)							−0.02	0.36	0.58 **	0.15	0.43 *	0.50 **
IMF (%)								0.135	0.01	−0.09	−0.01	0.04
LW (kg)									0.74 **	0.78 **	0.48 **	0.92 **
ESIW (kg)										0.56 **	0.65 **	0.89 **
ECEW (kg)											0.41 *	0.76 **
ECOW (kg)												0.68 **

[1] HCW: hot carcass weight; HW: ham weight; RW: rib weight; DBT1: backfat thickness at first rib level; LF: loin fat; HF: haunch fat; P: protein; IMF: intramuscular fat; LW: live weight; ESIW: empty small intestine weight; ECEW: empty cecum weight; ECOW: empty colon weight; TWTE: empty total gastrointestinal tract weight. * $p < 0.05$; ** $p < 0.01$.

4. Discussion

The present study investigated the effects of the simultaneous addition of cassava and taro to the feed of pigs and their effects on the carcass characteristics of backyard-raised pigs. They are reared under extensive traditional production systems in developed countries characterized by a low number of animals, which are generally fed with feed derived from the farmer's own crops and kitchen waste; Creole or crossbred pigs are often used, and technological advances have been poor [5]. This study follows a previous study examining the effects of these same alternatives on growth and fattening parameters, in which it was possible to verify that the simultaneous use of both feed alternatives yields good productive results, in addition to lowering production costs by considerably reducing the amount of maize in the diet [22].

Differences in the carcass characteristics based on geographic location, as described by Schinckel and De Lange [46], were attributed to both changes in genetic selection and the environment in which the animals are reared. The tests carried out used crossbred pigs purchased from local producers, with only a small selection of animals highly specialized in meat production. Crossbreeding in the Ecuadorian backyard pig is very frequent, expressing very diverse phenotypes that vary from one producer to another [5,47]. Despite the fact that the choice of animals was random when forming the groups, and that the environmental and breeding conditions were similar, there are many differences between the locations, which suggests a genetic heterogeneity in the subjects that make up the sample; this corresponds to the reality of backyard pig farming systems in Ecuador, and the results of the treatments must be interpreted within the context of each of the two experimental locations. These differences were primarily found in weight and performance carcass parameters, fat thickness, and the development of the gastrointestinal tract. The differences in the coefficients of variation of the carcass yield between treatments could reflect variation in the live and carcass weights of each group, as well as the higher development of the gastrointestinal tracts in animals fed with the feed alternative.

Environmental temperature is an aspect to consider in pig farming because animals can suffer from thermal stress when raised in environments with temperatures above 25 °C [48]. A high temperature reduces feed consumption, affecting energy metabolism, increasing the accumulation of subcutaneous fat, which affects the quality of the meat [49]. This could be one of the reasons why the pigs fattened in Quinindé have high fat thickness, in contrast to the pigs from Río Chico that had heavier carcasses and meat pieces with higher yields, as well as greater development of the gastrointestinal tract. Despite the environmental similarities of the two geographical areas, in view of the commercialization of pork produced under backyard farming systems in Ecuador, it should be taken into account that the heterogeneity of crossbred pigs also gives rise to characteristic differences in their carcasses.

The genetic origin of the animals is another aspect to take into account, since backyard pig producers use crossbred pigs [5]. They come from crossing Creole pigs and foreign breed pigs, obtaining an increase in genetic variability, benefiting the pigs' hardiness, immunological efficiency, and productive behavior [50]. In previous research, we studied backyard pigs in Ecuador and found that the breed most commonly used in this system was Creole, following by a crossbreed and Pietrain breed, although some farms reared a white pig breed such as Landrace [5]. However, other studies reveal that among the imported breeds preferentially used in this production system is the Duroc Jersey, since it is a dual-purpose breed, useful for meat and fat [51], followed by the Pietrain breed, whose characteristic is producing lean meat and little fat [52]. The results suggest the need to promote the breeding of standardized genetic models, for which native Creole-based local resources may be a good option; however, studies are needed to characterize the variability of these genetic resources [9,16].

The addition of 10% cassava leaf in the diet during the fattening phase of pigs has been found to improve the carcass characteristics in relation to conventional feed [1]. However, our results have revealed that backyard pigs from Ecuador fed with cassava and taro did not show differences in the morphological characteristics of the carcass. In addition, we did not find that different percentages of the alternative feed affected the size of the carcass cuts or the quality of the meat; thus, we consider that the most optimal formulation is T2, which includes 32% of the alternative feed. This formulation was also previously shown to be the most economically favorable in terms of productivity, without causing negative effects on the health of fattening pigs [22].

Gonzalez et al. [53] observed that cassava flour significantly improved body weight and had an impact on meat quality, with lower fat content being observed following treatments with cassava. However, comparing the results obtained when cassava was administered to the pigs in foliage and flour form at the same time shows that the results were similar to those obtained in our research, with higher carcass weights and yields, as well as the highest backfat thickness [54]. The addition of cassava with rice in the pig feed, replacing corn, produced the lowest carcass yields and lengths, as well as the lowest backfat thickness, although the differences between groups were not significant [55]. In the same way, the use of 40% fermented cassava in pig feed affected the fat content, moisture content, and ash in the carcass, as well as the protein content of the meat [56]. The last was also found in our results, as the meat derived from pigs fed with cassava and taro showed a higher protein content and a lower degree of fatness. Coinciding with our results, the addition of cassava or taro causes the animal to accumulate less fat, which is evidenced by a decrease in the thickness of the subcutaneous and intramuscular fat in pigs [1,57].

Hasan et al. [12] used cassava by-products (foliage, pulp, and peel) in proportions of 20, 40, and 60% in the feeding of weaned pigs. Their results determined that, both at a physical level (pH, color, and water retention capacity) and at a chemical level (protein and fat), the best treatment was the one that contained 20% cassava by-products.

Our results show that the addition of cassava and taro to the diet of pigs leads to an increase in organ weight, which is consistent with the results observed by Caicedo et al. [58] when testing the addition of different percentages of taro as a substitute for corn. Kaensombath and Lindberg [57] found similar results when soybean meal was replaced with ensiled taro leaves. One reason for the increase in organ weight in pigs fed with taro could be related to the ingestion of oxalate [57], but we cooked the alternative feed to avoid the presence of anti-nutritional factors such as oxalates. However, Taysayavong et al. [59] did not find differences in visceral organ weight and length in Moo Lath and Large White breeds, although they affirmed that their results could be explained by the short experimental period, which was only twelve days.

A greater consumption of feed motivates a greater development of the GIT, which justifies its greater development when providing moist feeds, since a greater volume of feed is given per moisture calculation [60]. The increase in the feed allowance leads to important increases in the weight of the total viscera, liver, kidneys, etc. [43]. For their part, Fitzsimons et al. [61] point out that, in general, the amount of energy provided in the food could influence the weight of the liver and the gastrointestinal tract. Coinciding with Ortega et al. [62], our results show that this leads to a decrease in the degree of fattening of the carcass in all parts. The addition of a large amount of fiber in the diet contributes to development of the GIT [57,63], which explains the results obtained in our study. Since the cassava and taro were administered whole, with the peel included, this provided extra fiber content to the diet of the animals [22].

5. Conclusions

The addition of cassava and taro residues as an alternative in the diet of pigs raised in the traditional backyard production system of Ecuador can be considered an alternative to reduce the use of corn without greatly affecting the morphological characteristics of the carcass. However, this affects their performance, as there is an increase in the weight of the gastrointestinal tracts of pigs during fattening. However, the environmental conditions and genetic origin could determine the geographical differences in these aspects.

The proportions of cassava and taro alternatives used in the formulations must be optimized since they directly increase the amount of protein in the meat and decrease the degree of fatness in the carcass.

In the production of backyard pigs in Ecuador based on the use of by-products and agricultural wastes, it is necessary to promote the standardization of the type of pig that is raised, promoting the use of local genetic resources.

Author Contributions: Conceptualization, A.V.L., A.G.-M. and E.R.S.; data curation, A.V.L. and J.G.O.; formal analysis, A.G.-M., A.V.L. and E.R.S.; funding acquisition, A.V.L. and E.R.S.; investigation, A.V.L., A.G.-M. and E.R.S.; methodology, A.V.L., A.G.-M. and E.R.S.; project administration, A.V.L. and E.R.S.; software, A.V.L., A.G.-M. and E.R.S.; supervision, E.R.S.; validation, A.G.-M. and E.R.S. All authors were involved in developing, writing, commenting, editing, and reviewing the manuscript. All authors have read and agreed to the published version of the manuscript.

Funding: This study was supported by the State University of the South of Manabí (Ecuador) in the framework of the livestock development program (PROG-014-PROY-003-2019), as well as by the University of Córdoba (Spain) and the Andalusian Government (Spain) for research by the PAI Group AGR-134 (CORADES) in the framework of cooperation programs for the training of researchers.

Institutional Review Board Statement: The research was approved by the Responsible Researcher of the Scientific Commission of the Agricultural Career of the Southern State University of Manabí, Ecuador (PROG-014-PROY-003-2019).

Informed Consent Statement: Not applicable.

Data Availability Statement: This is not applicable, as the data are not in any data repository with public access. However, if an editorial committee needs access; we will happily provide them with it. Please use this email: erodero@uco.es.

Acknowledgments: We thank the University of the South of Manabí (UNESUM) and the Department of Animal Production of the University of Córdoba (Spain) for the support in developing this research.

Conflicts of Interest: The authors declare no conflict of interest.

References

1. Ekpo, J.; Idorenyin, M.S.; Metiabasi, U.; Grace Idiong, C. Meat Quality and Sensory Evaluation of Pork from Pig Fed Pro-Vitamin A Cassava Leaf Meal, Pumpkin Stem and Moringa Leaf Meal as Dietary Supplements. *J. Agric. Food Sci.* **2022**, *6*, 10–23.
2. FAO. Food and Agriculture Organization of the United Nations. *Revisión del Mercado de la Carne; Índices de Precios de La Carne de La FAO*; FAO: Rome, Italy, 2023. Available online: https://www.fao.org/worldfoodsituation/foodpricesindex/es/ (accessed on 13 August 2023).
3. Asociación de Porcicultores del Ecuador-ASPE. Importancia Economica de la Porcicultura Valor Bruto de la Produccion. Ecuador. 2022. Available online: https://aspe.org.ec/wp-content/uploads/2022/09/DATOS_PORCICULTURA.pdf (accessed on 13 August 2023).
4. Sagaró-Zamora, F.; Ferrer-Hernández, E. Alimentación Para Cerdos de Ceba En Condiciones Locales de Agricultura Familiar. *Ciencia en su PC* **2021**, *1*, 22–35.
5. Valverde Lucio, A.; Gonzalez-Martínez, M.; Alcívar Cobeña, J.L.; Rodero Serrano, E. Characterization and Typology of Backyard Small Pig Farms in Jipijapa, Ecuador. *Animals* **2021**, *11*, 1728. [CrossRef]
6. Sanz, J.G.; Peris, C.; Torres, A. *Productividad de las Explotaciones Porcinas en Sistemas Intensivos*; Generalitat Valenciana: Madrid, Spain, 1994.
7. Herrera, K.; Monar, G. Proyecto de Inversión para la Construcción de una Granja en Vinces, Provincia de Los Ríos que se Dedique al Cuidado, Crianza y Comercialización de Ganado Porcino. Bachelor's Thesis, Higher Polytechnic School of the Litoral, Guayaquil, Ecuador, 2006. Available online: https://www.dspace.espol.edu.ec/bitstream/123456789/56540/1/D-CD35652%20HERRERA%20-%20MONAR.pdf (accessed on 13 August 2023).
8. Segarra Zenteno, E.; Salinas Cueva, L.; López Crespo, G. Calidad de la canal de cerdos en la industria porcina de ecuador (Artículo de Revisión). *Rev. Ecu. Cien. Anim.* **2018**, *2*, 118–132.
9. Mader, C.J.; Montanholi, Y.R.; Wang, Y.J.; Miller, S.P.; Mandell, I.B.; McBride, B.W.; Swanson, K.C. Relationships among measures of growth performance and efficiency with carcass traits, visceral organ mass, and pancreatic digestive enzymes in feedlot cattle. *J. Anim. Sci.* **2009**, *87*, 1548–1557. [CrossRef] [PubMed]
10. Tous, N.; Esteve, E.; Lizardo, R.; 3tres3.com. Nutrición y Grasa Intramuscular: Efecto del Ácido Linoleico Conjugado y la Vitamina A. 2012. Available online: https://www.3tres3.com/latam/articulos/nutricion-y-grasa-intramuscular-efecto-del-cla-y-la-vitamina-a_11189/ (accessed on 13 August 2023).
11. Abu, O.A.; Olaleru, I.F.; Omojola, A.B. Carcass characteristics and meat quality of broilers fed cassava peel and leaf meals as replacements for maize and soyabean meal. *J. Agric. Vet. Sci.* **2015**, *8*, 41–46.

12. Hasan, F.; Zendrato, D.P.; Hanafi, N.D.; Sadeli, A.; Daulay, A.H. The Utilization of Cassava By-Products as Complete Feed on Physical and Chemical Meat Quality of Weaning Male Crossbred Landrace Pigs. *J. Phys. Conf. Ser.* **2020**, *1542*, 012002. [CrossRef]
13. Sánchez, E.; Navarro, C.; Sayas, M.; Sendra, E.; Fernández, J. Efecto de las condiciones ante mortem y post mortem sobre los factores que determinan la calidad de la carne. In *Bienestar Animal y Calidad de Carne*, 1st ed.; Mota, D., Guerrero, I., Trujillo, M.E., Eds.; BM Eds: Mexico City, Mexico, 2010; pp. 329–349.
14. Cheng, Q.; Sun, D.W.; Scannell, A.G. Feasibility of water cooking for pork ham processing as compared with traditional dry and wet air cooking methods. *J. Food Eng.* **2005**, *67*, 427–433. [CrossRef]
15. Tomažin, U.; Batorek-Lukač, N.; Škrlep, M.; Prevolnik-Povše, M.; Čandek-Potokar, M. Meat and Fat Quality of Krškopolje Pigs Reared in Conventional and Organic Production Systems. *Animal* **2019**, *13*, 1103–1110. [CrossRef]
16. Castrillón, H.; Wilson, E.; Fernández, J.A.; Restrepo, L. Determination of PSE (pale, soft and exudative) meat in pork carcasses. *Vitae* **2005**, *12*, 23–28.
17. Campion, D. Sistemas de Producción Porcina y su Incidencia en la Calidad. 2013. Available online: https://repositorio.uca.edu.ar/handle/123456789/455 (accessed on 13 August 2023).
18. Rybarczyk, A.; Bogusławska-Was, E.; Dłubała, A. Effect of Bioplus Yc Probiotic Supplementation on Gut Microbiota, Production Performance, Carcass and Meat Quality of Pigs. *Animals* **2021**, *11*, 1581. [CrossRef] [PubMed]
19. Jonsäll, A.; Johansson, L.; Lundström, K. Sensory quality and cooking loss of ham muscle (M. biceps femoris) from pigs reared indoors and outdoors. *Meat Sci.* **2001**, *57*, 245–250. [CrossRef] [PubMed]
20. Sundrum, A.; Henning y Butfering, L. Efectos de las dietas en la granja para la producción porcina orgánica sobre el rendimiento y la calidad de la canal. *3tres3* **2012**, *78*, 1199–1205.
21. Agudelo Quintero, J.; Mesa-Granda, M. Eficiencia Productiva En Cerdos de Levante Alimentados Con Materias Primas Alternativas de Países Tropicales: Meta-Análisis. *Intropica* **2022**, *17*, 114–132. [CrossRef]
22. Valverde Lucio, A.; Gonzalez-Martínez, A.; Rodero Serrano, E. Utilization of Cooked Cassava and Taro as Alternative Feed in Enhancing Pig Production in Ecuadorian Backyard System. *Animals* **2023**, *13*, 356. [CrossRef] [PubMed]
23. Negreyeva, A.N.; Babushkin, V.A.; Gagloev, A.C. The influence of nontraditional feed in the fattening pig's diet on meat quality. *Int. J. Pharm. Res.* **2018**, *10*, 706–714.
24. Agdeppa-Namoco, R.P.; Chene, K.; Gican, P. Alternative Feed Rations for Grower Stage Backyard Swine. *Mindano. J. Sci. Technol.* **2012**, *10*, 93–106.
25. Cerisuelo, A.; Castellóa, L.; Moseta, V.; Martíneza, M.; Hernández, P.; Piquera, O.; Gómeza, E.; Gasac, J.; Lainez, M. The Inclusion of Ensiled Citrus Pulp in Diets for Growing Pigs: Effects on Voluntary Intake, Growth Performance, Gut Microbiology and Meat Quality. *Livest. Sci.* **2010**, *134*, 180–182. [CrossRef]
26. Brandalise, F.; Martín Grillo, R.; Pinto, L.A.; Serrano, E.A.; Sánchez, M.I. *Conceptualización, Caracterización y Registro de la Agricultura Familiar. La Experiencia de Panamá*; FAO: Panama City, Panama, 2017; Available online: https://www.fao.org/documents/card/en/c/c90b1aa7-f532-4fed-a2eb-a004461bcd71 (accessed on 13 August 2023).
27. AGROCALIDAD. Programa Nacional Sanitario Porcino. Control y Erradicación de la Peste Porcina Clásica por Zonificación en el Ecuador. 2012. Available online: https://www.agrocalidad.gob.ec/wp-content/uploads/2020/05/d1.pdf (accessed on 13 August 2023).
28. Muñoz-Ron Paola, I.; Elizabeth, S.; Ii, S.-C.; Fabricio, A.; Iii, L.-P.; Poma Iv, J. Diagnosis of the production, comercialization, and consumption of pig products in Sachas, Orellana Diagnóstico da produção, comercialização e consumo de produtos suínos no cantão de Sacha, Orellana. *Polo del Conocimiento* **2020**, *4*, 1–10.
29. Hernández Yeltsin, R.J. Caracterización del manejo zootécnico del cerdo criollo (*Sus scrofa domesticus*) en el área rural del municipio de Río Blanco, Matagalpa, 2013. Ph.D. Thesis, Universidad Nacional Agraria Facultad de Ciencia Animal, Sistemas Integrales de Producción Animal, Rivas, Nicaragua, 2013.
30. UNICEF, Desnutrición Crónica Infantil. Uno de los Mayores Problemas de Salud Pública en Ecuador. 2018. Available online: https://www.unicef.org/ecuador/sites/unicef.org.ecuador/files/2021-03/Desnutricion-Cronica-Infantil.pdf (accessed on 13 August 2023).
31. Yapura, S. Importancia del Maíz en la Producción Animal. Veterinaria Digital. 2021. Available online: https://www.veterinariadigital.com/articulos/importancia-del-maiz-en-la-produccion-animal/ (accessed on 13 August 2023).
32. Suárez Guerra, L.; Mederos Vega, V. Revisión Bibliográfica apuntes sobre el cultivo de la yuca (Manihot Esculenta Crantz). Tendencias actuales. *Cultivos Tropicales* **2011**, *32*, 27–35.
33. Aragadvay-Yungán, R.G.; Núñez-Torres, O.P.; Velástegui-Espín, G.P.; Villacís-Aldaz, L.A.; Guerrero-López, J.R. Uso de harina de colocasia esculeta l., en la alimentación de cerdos y su efecto sobre parámetros productivos. *J. Selva Andin. Anim. Sci.* **2016**, *3*, 98–104. [CrossRef]
34. Sánchez, J.; Caicedo, W.; Aragón, E.; Andino, M.; Bosques, F.; Viamonte, M.I.; Ramírez de la Ribera, J.L. La inclusión de la *Colocasia esculenta* (papa china) en la alimentación de cerdos en ceba. *Rev. Electron. Vet.* **2018**, *19*, 1–6.
35. Kaensombath, L. Taro leaf and stylo forage as protein sources for pigs in Laos: Biomass yield, ensiling and nutritive value. Bachelor's Thesis, Swedish University of Agricultural Sciences, Uppsala, Sweden, 2012.
36. Lan, L.T.T.; Hung, L.T.; Thu, N.T.A.; Loc, H.T.; Liang, J.B.; Thiet, N.; Ngu, N.T. Effects of substituting taro (*Colocasia esculenta*) wastes silage in diets on growth and nutrient digestibility in pigs. *J. Anim. Health Prod.* **2021**, *9*, 112–118.

37. Quiñónez, R.; González, C.; Polanco, D.; Perdomo, B.; Araque, H. Evaluación de diferentes tipos de deshidratación de raíz y follaje de yuca amarga (*Manihot esculenta*) sobre su composición química. *Zootec. Trop.* **2007**, *25*, 43–49.
38. Ricaurte, F.A. *La Yuca Como Alternativa en la Alimentacion de Cerdos en la Etapa de ceba Granja los Laureles Vereda Tacarimena Municipio el Yopal Casanare*; Universidad Nacional Abierta y a Distancia: Yopal, Colombia, 2014.
39. Carbajal Basilo, D.S.; Otarola Gamara, A. Efecto del Tiempo de Cocción en la Reducción de Oxalatos en Harina de dos Variedades Depituca (Colocasia Esculenta). Universidad Nacional Daniel Alcides Carrión Facultad de Ciencias Agropecuarias Escuela de Formación Profesional de Industrias Alimentarias. 2019. Available online: http://repositorio.undac.edu.pe/bitstream/undac/2555/1/T026_70137643_T.pdf (accessed on 13 August 2023).
40. Phoneyaphon, V.; Preston, T.R. Protein-enriched cassava (*Manihot esculenta* Crantz) root as replacement for ensiled taro (*Colocasia esculenta*) foliage as source of protein for growing Moo Lat pigs fed ensiled cassava root as basal diet. *Livest. Res. Rural Dev.* **2016**, 28.
41. AGROCALIDAD. *Instructivo para el Sacrificio y Disposición Final de Animales. Agencia Ecuatoriana de Aseguramiento de la Calidad del Agro. Ministerio de Agricultura, Ganadería, Acuacultura y Pesca*; 2013. Available online: https://www.agrocalidad.gob.ec/wp-content/uploads/2020/05/bi3.pdf (accessed on 13 August 2023).
42. Galián Jiménez, M. Características de la Canal y Calidad de la Carne, Composición Mineral y Lipídica del Cerdo Chato Murciano y su Cruce con Ibérico. Efecto del Sistema de Manejo. Bachelor's Thesis, University of Murcia, Murcia, Spain, 2007.
43. Nieto, R.; Lara, L.; García, M.A.; Vílchez, M.A.; Aguilera, J.F. Effects of dietary protein content and food intake on carcass characteristics and organ weights of growing iberian pigs. *Anim. Sci.* **2003**, *77*, 47–56. [CrossRef]
44. de Souza, T.C.R.; Landín, G.M.; García, K.E.; Barreyro, A.A.; Barrón, A.M. Cambios nutrimentales en el lechón y desarrollo morfofisiológico de su aparato digestivo. *Vet. Mex.* **2012**, *43*, 155–170.
45. Maclean, W.; Harnly, J.; Chen, J.; Chevassus-Agnes, S.; Gilani, G.; Livesey, G.; Warwick, P. Food energy–Methods of analysis and conversion factors. In *Food and Agriculture Organization of the United Nations Technical Workshop Report*; The Food and Agriculture Organization: Rome, Italy, 2003; Volume 77, pp. 8–9. Available online: https://www.fao.org/3/Y5022E/y5022e00.htm (accessed on 13 August 2023).
46. Schinckel, A.P.; De Lange, C.F.M. Characterization of growth parameters needed as inputs for pig growth models. *J. Anim. Sci.* **1996**, *74*, 2021–2036. [CrossRef]
47. Conde-Aguilera, J.A.; Aguinaga, M.A.; Lara, L.; Aguilera, J.F.; Nieto, R. Carcass traits and organ weights of 10-25-Kg body weight iberian pigs fed diets with different protein-to-energy ratio. *Anim. Feed Sci. Technol.* **2011**, *164*, 116–124. [CrossRef]
48. Pardo Domínguez, Z. Efectos Fisiológicos y Productivos del Estrés por Calor en el Cerdo Ibérico y Posibles Estrategias Nutricionales para Mitigarlo. Bachelor's Thesis, University of Granada, Granada, Spain, 2022.
49. Romo-Valdez, J.; Silva-Hidalgo, G.; Gaxiola, H.R.G.; Valdez, A.R.; Romo-Rubio, J. Estrés por calor: Influencia sobre la fisiología, comportamiento productivo y reproductivo del cerdo. *Abanico Veterinario* **2022**, *12*, 29.
50. Linares, V.; Linares, L.; Mendoza, G. Caracterización etnozootécnica y potencial carnicero de *Sus scrofa* "cerdo criollo" en Latinoamérica. *Sci. Agropecu.* **2011**, *2*, 97–110. [CrossRef]
51. Castillo, L. Principales Razas Porcinas y Cruzamientos. Instituto Nacional de Investigación Agropecuaria. 1984. Available online: https://repositorio.iniap.gob.ec/bitstream/41000/1317/1/INIAP%20%20Bolet%c3%adn%20Divulgativo%20139.pdf (accessed on 13 August 2023).
52. Chávez Narváez, M.M. Evaluación del Crecimiento y Grasa Dorsal del Cerdo Criollo del Cantón Guamote Provincia de Chimborazo. Bachelor's Thesis, Polytechnic Higher School of Chimborazo, Riobamba, Ecuador, 2022. Available online: http://dspace.espoch.edu.ec/bitstream/123456789/17515/1/17T01754.pdf (accessed on 13 August 2023).
53. González, C.; Vecchionacce, H.; Díaz, I.; Ortiz, V. Utilización de harina cruda de raíz de yuca (Manihot Esculenta C.) y harina cruda de cormos de ocumo chino (*Colocasia esculenta*) en la alimentación de cerdos). *Arch. Latinoam. Prod. Anim.* **1997**, *5* (Supl. 1), 277–279.
54. Aguilar Martínez, R.J. Inclusión de Harina de Follaje y Raíz de Yuca (*Manihot esculenta* Crantz) en Cerdos en Desarrollo y su Efecto Sobre el Comportamiento Productivo y Morfometria del Tracto Gastointestinal. Bachelor's Thesis, Universidad Nacional Agraria, Managua, Nicaragua, 2017. Available online: https://repositorio.una.edu.ni/id/eprint/3533 (accessed on 23 September 2023).
55. Manjarrez, B.; Arteaga, C.; Robles, A.; Avila, E.; Shimada, A. Valor nutritivo de una combinación de harina de yuca (*Manlhot eaculenta*) con puliduras de arroz, como sustituto de maiz en la alimentación de pollos y cerdos. *Tlle. Pec. Uex.* **1974**, *25*, 58–63.
56. Aro, S.O.; Akinjokun, O.M. Meat and carcass characteristics of growing pigs fed microbially enhanced cassava peel diets. *Arch. Zootec.* **2012**, *61*, 23–25. [CrossRef]
57. Kaensombath, L.; Lindberg, J.E. Effect of replacing soybean protein by taro leaf (*Colocasia esculenta (l.) schott*) protein on growth performance of exotic (Landrace × Yorkshire) and native (Moo Lath) Lao pigs. *Trop. Anim. Health Prod.* **2012**, *45*, 45–51. [CrossRef] [PubMed]
58. Caicedo, W.; Rodríguez, R.; Lezcano, P.; Vargas, J.C.; Uvidia, H.; Valle, S.; Flores, L. Effect of taro tubers silage (*Colocasia Esculenta (L.) Schott*) on the morphometry of the gastrointestinal tract of growing pigs. *Cuban J. Agric. Sci.* **2017**, *51*, 183–190.
59. Taysayavong, L.; Lindberg, J.E.; Ivarsson, E. Digestibility, nitrogen retention, gut environment and visceral organ size in Moo Lath and Large White growing pigs fed un-fermented and fermented cassava root pulp and soybean pulp. *Livest. Res. Rural. Dev.* **2018**, *30*.

60. Ly, J.; Almaguel, R.; Lezcano, P.; Delgado, E. Miel rica o maíz como fuente de energía para cerdos en crecimiento. Interdependencia entre rasgos de comportamiento, digestibilidad rectal y órganos digestivos. *Rev. Com. Prod. Porc.* **2014**, *21*, 66–69.
61. Fitzsimons, C.; Kenny, D.A.; McGee, M. Visceral organ weights, digestion and carcass characteristics of beef bulls differing in residual feed intake offered a high concentrate diet. *Animal* **2014**, *8*, 949–959. [CrossRef]
62. Gabriel Ortega, J.; Valverde, L.A.; Indacochea, G.B.; Castro, P.C.; Vera, T.M.; Alcívar, C.J.; Vera, V.R. *Diseños Experimentales: Teoría y Práctica para Experimentos Agropecuarios*, 2nd ed.; Ortega, J.G., Lucio, A.L., Eds.; Grupo Compás: Jipijapa, Ecuador, 2021; Available online: http://142.93.18.15:8080/jspui/handle/123456789/625 (accessed on 13 August 2023).
63. Asmus, M.D.; DeRouchey, J.M.; Tokach, M.D.; Dritz, S.S.; Houser, T.A.; Nelssen, J.L.; Goodband, R.D. Effects of lowering dietary fiber before marketing on finishing pig growth perfor-mance, carcass characteristics, carcass fat quality, and intestinal weights. *J. Anim. Sci.* **2014**, *92*, 119–128. [CrossRef]

animals

MDPI

Review

Potential of Organic Acids, Essential Oils and Their Blends in Pig Diets as Alternatives to Antibiotic Growth Promoters

Rumbidzai Blessing Nhara [1,2,3,*], Upenyu Marume [1,3,*] and Carlos Wyson Tawanda Nantapo [1,3]

1 Department of Animal Sciences, School of Agriculture Science, Faculty of Natural and Agricultural Science, North-West University, P Bag X 2046, Mmabatho 2735, South Africa; nantapocarlos@gmail.com

2 Department of Livestock, Wildlife and Fisheries, Great Zimbabwe University, Masvingo P.O. Box 1235, Zimbabwe

3 Food Security and Safety Niche Area, Faculty of Natural and Agricultural Science, North-West University, P Bag X 2046, Mmabatho 2735, South Africa

* Correspondence: rbnhara@gmail.com (R.B.N.); upenyu.marume@nwu.ac.za (U.M.); Tel.: +27-81-0390670 (U.M.)

Simple Summary: Organic acids and essential oils have been shown to be effective alternatives to antibiotic growth promoters in pig production. Organic acids and essential oils have antibacterial, antiviral, and antioxidant properties. The article will focus on the effectiveness of organic acids, essential oils, and their blends in pig diets as alternative antibiotic growth promoters. Furthermore, the effects of organic acids, essential oils, and their blends on growth performance, oxidative stress, and meat quality are examined. Organic acids and essential oils, which have antimicrobial properties, can be used in place of antibiotic growth promoters. The use of organic acids and essential oils as growth promoters enhances pig welfare and aids in the fight against antimicrobial resistance.

Abstract: Over the years, the use of management and feeding strategies to enhance pig productivity while minimizing the use of antibiotic growth promoters has grown. Antibiotic growth promoters have been widely used as feed additives to reduce diet-related stress and improve pig performance. However, increasing concern about the consequences of long-term and increased use of antibiotic growth promoters in animal production has led to a paradigm shift towards the use of natural organic alternatives such as plant essential oils and organic acids in pig nutrition to enhance growth. Antibiotic growth promoters endanger human health by allowing multidrug-resistant genes to be transferred horizontally from non-pathogenic to pathogenic bacteria, as well as directly between animals and humans. Scientific research shows that alternative growth promoters such as essential oils and organic acids appear to improve pigs' ability to prevent pathogenic bacteria from colonizing the intestinal system, stabilizing the gut microflora and promoting eubiosis, as well as improving immunity and antioxidant stability. The purpose of this review was to provide an in-depth review of organic acids and essential oils as growth promoters in pig production, as well as their effects on productivity and meat quality. Organic acids and essential oils in pig diets are a safe way to improve pig performance and welfare while producing antibiotic-free pork.

Keywords: antibiotic growth promoters; essential oils; organic acids; pig nutrition; welfare; meat quality

1. Introduction

Over the years, there has been an increasing global interest in the development of management and feeding strategies that maximize pig productivity while minimizing the use of antibiotic growth promoters [1]. Antibiotic growth promoters were first used as feed additives to prevent the stress resulting from changing feed in the diet [2]. However, growing concern about the long-term and increased use of antibiotic growth promoters in animal production has resulted in a shift toward using natural organic alternatives to boost pig growth, such as plant essential oils and organic acids [3]. Antibiotic growth promoters are harmful to human health because multidrug-resistant genes can be transferred horizontally from non-pathogenic to pathogenic bacteria, resulting in the direct transfer of antibacterial-resistant bacteria from animals to humans [4]. Moreover, antibiotic growth promoters are not eco-friendly as their residues are found in soils and water, negatively impacting the ecosystem and its functions [5]. Alternative growth promoters such as essential oils and organic acids have been reported to improve a pig's ability to prevent pathogenic bacteria from colonizing the intestinal system [2], stabilize the gut microflora, and promote eubiosis [6]. They also improve mineral utilization, act as an energy source, promote endogenous enzyme secretion, and improve immunity and antioxidant stability [2]. As a result, improved pig performance is comparable to that of antibiotic growth promoters [6–8]. The focus of this review is to provide a comprehensive account of organic acids and essential oils as growth promoters in pig production and how they impact productivity and meat quality in pigs. It will provide an overview of the mode of action, performance responses, and potential of essential oils and organic acids in the pig industry.

2. Material and Methods

A systematic literature review was conducted using Web of Science, Google Scholar, Scopus, and PubMed to gather peer-reviewed papers from 1990 to 2023. The search criteria focused on antibiotic growth promoters, alternative growth promoters, essential oils, organic acids, pig nutrition, immune system, pig welfare, and pork quality. The data were analyzed to determine if essential oils and organic acids can be used as alternative antibiotic growth promoters in pig diets and to learn how their inclusion affects pig performance and pork quality. Data were analyzed, synthesized, and presented based on the key questions raised in the development of a review.

3. Mode of Action of Antibiotic Growth Promoters

A large proportion of the pigs produced in the world received antimicrobials in their feeds to counter post-weaning challenges. This equates to 70–80% of all pig starters, 70–80% of grower diets, and 50–60% of finisher feeds in Europe for the last few years [9]. Weaning piglets causes altered stomach pH, post-weaning diarrhea, and performance issues due to a lack of hydrochloric acid, which activates digestive enzymes. Insufficient hydrochloric acid and other environmental stressors disturb intestinal flora balance leading to a proliferation of pathogenic coliforms [10]. Antibiotics enhance feed conversion but do not impact carcass quality [11]. Antibiotics used in swine production can suppress or inhibit the growth of certain microorganisms [9]; however, their chemical composition and bacterial spectrum of antimicrobials vary widely. Antibiotics induce bacterial cell death by inhibiting essential cellular functions. Antibiotics can be classified based on the cellular component or system effect that may induce cell death, or merely inhibit cell growth [11]. Antibiotics can suppress the growth of pathogenic microbes by reducing competition for nutrients, hence reducing microbial metabolites that affect growth rate [11].

However, administering antibiotics to livestock has resulted in the problem of antimicrobial resistance. Antimicrobial resistance compromises the efficacy of preventing and treating a growing number of microbial infections. It arises as a result of natural selection and mutations resulting in antibiotics being ineffective, giving a survival advantage to the mutated strain [12]. Additionally, antibiotics used in livestock production are not fully absorbed and metabolized in animals, resulting in a large dose being highly active when excreted, causing the enrichment of antibiotic-resistant genes and a huge risk to the environment [13].

4. Organic Acids

4.1. Characteristics of Organics Acids

Organic acids are classified as any organic carboxylic acid, with or without keto, hydroxyl, or others from the non-amino functional group, including some short-chain fatty acids, but not all amino acids with the general R-COOH structure [14]. Organic acids are categorized into three main functional classes: short-chain fatty acids, medium-chain fatty acids, and tricarboxylic fatty acids. Short-chain fatty acids (SCFA), or simple mono-carboxylic acids (maximum 5 carbon atoms) such as acetic, formic, propionic, and butyric acids, are organic acids that are synthesized in the lower intestine by the microbial fermentation of indigestible sugars and amino acids. Medium-chain fatty acids (MCFA), or carboxylic acids containing a hydroxyl group (6–12C) such as malic, citric, tartaric, and lactic acids, represent an important energy source with higher antimicrobial activity due to their higher pKa. Lastly, tricarboxylic acids (TCA), or simply carboxylic acids with double bonds such as sorbic and fumaric acids, are intermediates in the Krebs cycle, and are involved in energy metabolism [14–16]. Other organic acids, such as sorbic, benzoic, and lactic acids, follow different structures. Organic acids are widely distributed in nature as normal constituents of plants and animal tissues, produced either by chemical synthesis or microbial fermentation of carbohydrates in the large intestine [17]. The dissociation constant (pKa) and carbon chain length (C1–C7) of common organic acids determine their antimicrobial efficacy. The sodium, potassium, and calcium salts of these acids, such as sodium benzoate, calcium formate, and calcium propionate, also have antimicrobial properties. Organic acids can be applied directly to feeds in solid or sprayed form, and are classified as feed preservatives or acidifiers. The efficacy of dietary organic acids depends largely on animal species, chemical composition (acid, salt), molecular weight, MIC value of the acid, targeted microbe species, gastrointestinal tract site, and buffering capacity of the feed [17,18]. Safety, odor, taste, and solubility in water are aspects to consider when applying organic acids to animal nutrition. Organic acids are weak acids that partly dissociate, and upon entering the bacterial cell membrane, they detach themselves in the inner, more alkaline part. The undissociated part then reduces the pH in the cytoplasmic area, disrupting the normal metabolic processes of certain types of bacteria, including *E. coli.*, *Listeria* spp., *Salmonella* spp., *Clostridia* spp., and some coliforms, thereby killing the cell [19]. In European markets, the demand for feed acidifiers grew by 6.6% from 2006–2012. The global market for feed acidifiers is projected to increase, with high demand in developing economies as well as demand for safe meat products from developed economies and an increasing world population [6]. The global market size for animal nutrition organic acids in 2020 was estimated at US$113.3 million, and is expected to expand by a 6.5% compound annual growth rate (CAGR) in 2021 to 187.1 million by 2028 [20]. As stated in this report, growth is driven by demand for low-cost renewable energy sources, as well as an increased need to replace traditional growth promoters. Furthermore, pigs and poultry were in high demand, with lactic acid accounting for 49.0% of total revenue. Table 1 shows the common organic acids used as dietary acidifiers for pigs and poultry.

Table 1. Chemical Properties of common organic acids used in animal nutrition.

OA	Chemical Name	Dissociation Constant, kP_a	Physical Form
Tartaric	2,3-Dihydroxybutanedioic acid	2.93/4.23	Liquid
Formic	Methanoic Acid	3.75	Colorless liquid
Acetic	Ethanoic Acid	4.76	Colorless liquid
Propionic	2-Propanoic Acid	4.88	Colorless oily liquid
Caprylic acid	1-Octanoic acid	4.89	Colorless to light-yellow oily liquid
Butyric	Butanoic acid	4.82	Colorless oily liquid
Lactic	2-Hydroxypropanoic Acid	3.08	Colorless to yellow viscous liquid
Sorbic	2,4-Hexandienoic Acid	4.76	White crystalline powder or granules
Fumaric	2-Butenedioic Acid	3.02	White crystalline powder
Benzoic	Benzenecarboxylic acid	4.20	Colorless crystalline powder
Malic	Hydroxybutanedioic Acid	3.40/5.1	Liquid
Citric	2-Hydroxy-1,2,3-Propanetricarboxylic Acid	3.13/5.95/6.39	White or crystalline powder

Source [15–17,21].

4.2. Mode of Action of Organic Acids

The mode of action of common OAs is not yet fully understood. However, their mode of action may be partially due to factors such as, (a) inhibition of the development of pathogenic microbes in the gastrointestinal tract by reducing gut pH, (b) reduction of gastric emptying rates and maintenance of endogenous enzyme secretion, (c) mineral chelation and stimulation on intermediary metabolism, and (d) facilitation of proper digestion due to lower gastric pH and enhanced pepsin secretion [16].

4.3. Bactericidal Properties

Organic acids are weak acids, and in their undissociated state they can easily diffuse across cell membranes. In the cytoplasm, they dissociate and release hydrogen (H^+) ions, which increases intracellular acidity of the cell, influencing cell metabolism and disrupting the normal microbial cell functioning [14]. Bacterial cells are forced to expend energy to expel the protons, leading to intracellular accumulation of $RCOO^-$ acid anion. Accumulation of anions will interfere with RNA and DNA synthesis, resulting in impaired cell growth and multiplication as well as osmotic cell pressure, inducing both bactericidal and bacteriostatic effects [15]. The acid regulation properties of OAs allow them to reduce activity of harmful bacteria by altering the ambient pH value in the bacterial cell [6]. Organic acids have a stronger effect on the inhibition of gram-positive bacteria than gram negative bacteria due to structural differences in the cell membrane [19]. The efficiency of organic acids in reducing the microbial count is affected by the type of acid, temperature, buffering capacity, and water activity. Table 2 shows some common organic acids used in pig production.

Table 2. Common organic acids used in pig production.

Organic Acid	Dietary Dose	Observations	References
Formic	1.4 g/kg	Positive auxinic effects, improved ADG, ADFI, and FCR during initial post-weaning 3 week period	[22]
Formic	6.4 g/kg	Higher microbiota diversity	[22]
Acid blend 1 Acid blend 2	231 FO, 124 AA, 127 LA, 133 PA and, 76 g kg^{-1} of CA. 50% acid (290 FO, 170 AA; 160 PA; 85 g kg^{-1} CA) + 50% LA on silica (517 LA; 7 FO and 20 g kg^{-1} AA)	Lower fecal coliform counts, inhibit early ileal microbiota development for all acids Acid 2 improved ADG in week 2	[23]
Malic	1.5% in weanling pigs	No improvement on growth performance	[24]
Benzoic	0.5% in nursery pigs	Inhibit pathogenic microbes maintain intestinal microecological balance, improve growth performance, and protein digestibility	[25]
Provenic	BA 50%, CF 3%, and FA 1%	Improved apparent total tract digestibility, fecal score, intestinal microbiota, and volatile fatty acid	[26]
Carbadox (PA and LA)	50 mg/kg	Reduced diarrhea scores	[27]
Orgacids^{tm} (FO, PA, LA, MA, TA, and CA)	2 kg/tonne	Low fecal pH and Enterobacteriaceae counts, higher *Lactobacillus* spp. counts, low meat cholesterol	[28]
Benzoic	5 g/kg	Improved BWG, ADFI, FCE	[29]
Citric	4 mmol/L	Improved immune function, reduced enterotoxigenic *E. coli* induced damage to the intestinal barrier of weaned piglets	[13]
Citric	5, 10, or 15 g/kg, during gestation and lactation	Increased total tract apparent digestibility of Cp and P Enhanced plasma and colostrum IgG and IgA Improved total protein of milk and colostrum	[30]
Matrix coated OA blend	0.2% in growing pigs	Enhanced growth performance and improved gut microbial population with no adverse effect on nutrient digestibility	[31]

PA—phosphoric acid; LA—lactic acid; FA—fumaric acid; BA—benzoic acid; CF—calcium formate; FO—formic acid; CA—citric acid; AA—acetic acid. Matrix coated OA blend—17% fumaric acid, 13% citric acid, 10% malic acid, and 1.2% MCFA (capric and caprylic acid) and carrier. ADG—average daily gain; BWG—body weight gain; ADFI—average daily feed intake; FCE—feed convection efficiency.

4.4. Lowering Stomach pH and Endogenous Enzyme Secretion

Short-chain fatty acids have a stimulating effect on both the endocrine and exocrine pancreatic secretions. Organic acids, when ingested, can create an acidic environment [2]. Low stomach pH alters gut microflora by reducing the non-acid tolerant bacterial species such as *E. coli* and *Salmonella* [14]. The acidic environment in the stomach activates the conversion of enzyme precursor pepsinogen to pepsin, which is responsible for protein digestion [1,32]. Organic acids elevate serum secretin content, stimulating pancreatic exocrine secretions and resulting in improved nutrient digestibility in the duodenum [33].

4.5. Energy Source and Mineral Utilization

Organic acids act as an energy source in the gut, as they are intermediary products of the tricarboxylic acid (TCA) cycle [34]. Their inclusion in diets helps in preventing tissue breakdown from gluconeogenesis and lipolysis [6,19]. Organic acid anions can form complexes with minerals like calcium, phosphorous, magnesium, and zinc, enhancing mineral digestion and reducing the excretion of supplemental minerals and nitrogen [2]. Organic acids can improve P solubility and phytate P utilization by competitively chelating Ca^{2+}, reducing the formation of insoluble Ca phytate complexes [14]. Figure 1 illustrates the mode of action for organic acids when they are included in pig diets.

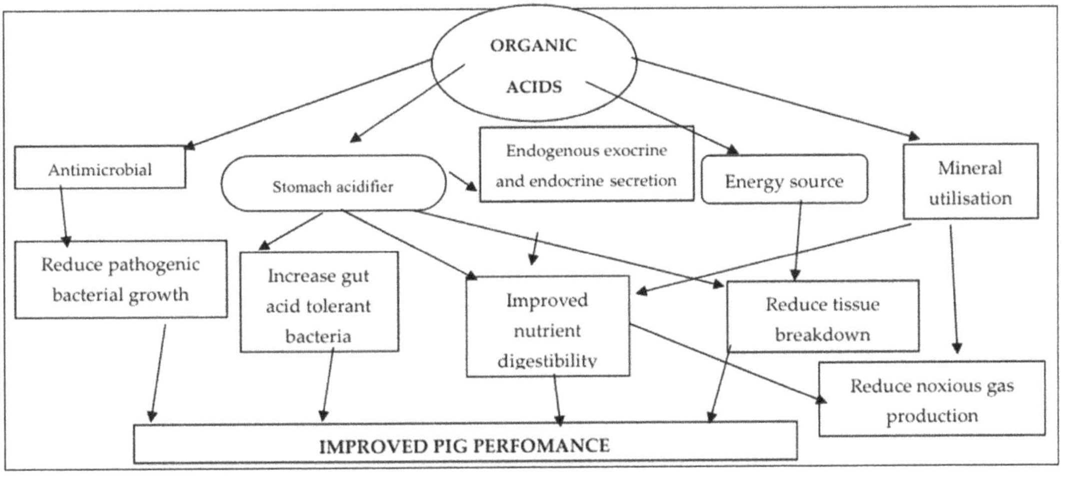

Figure 1. Summary of organic acidifier mode of action in pigs.

5. Essential Oils

5.1. Characteristics of Essential Oils

Essential oils are a mixture of various compounds, mainly terpenes and terpene derivatives. They are concentrated hydrophobic liquids containing volatile aromatic compounds produced by plants, stored in cavities, secretory cells, and epidermal cells. They are produced as secondary metabolites and they have antibacterial, antifungal, and antiviral properties [35]. These provide essential oils with the ability to replace antibiotic growth promoters and improve animal performance and health [26]. Ecological factors, species, climatic conditions, harvest time, the part of the plant used, and the method of isolation affect the chemical composition of essential oils and their efficacy [36]. Table 3 shows the commercial and non-commercial application of essential oils in pig nutrition and health.

Table 3. Application of commercial and non-commercial essential oils in pig nutrition and health.

Name	Components	Dietary Dose and Duration	Main Findings	References
Delacon blend	40% Common Fenugreek seed, 12.5% Subterranean Clove, 7.5% Cinnamomum Cassia Presl, and 40% Kaolin ($2SiO_2.Al_2O_3.2H_2O$)	0.04%. 42–60 days	Improved growth performance, apparent ileal digestibility	[37]
ORSENTIAL	1.1% Thymol + 2.2% Carvacrol	300–1000 g/tonne. 54 days	Higher ADG, lower incidences of diarrhea, reduced fecal ammonia emissions and blood urea nitrogen, increased serum IgG	[38]
ColiFit Icaps C	Trans-cinnamal Ehyde, eugenol, carvacrol, thymol, and diallyl disulfure at 101,218; 12,400; 6514; 4359 and 1123 mg/kg, respectively.	1 kg/tonne, 7 days	Higher fecal lactobacilli, increased lactobacilli/coliform ratio against enterotoxigenic *Escherichia coli* (ETEC) F4 strain COLI30/14-3	[39]
PEP1000-1, Biomin Inc.	Anis oil, citrus oil, oregano oil	0.1%	Improved diarrhea score	[27]
Next enhance 150, NE150	thymol 25% and carvacol 25%		Improved nutrient digestibility, antioxidant ability, intestinal morphology, and digestive enzymes in weaned pigs	[26]
Essential oil blend	*Cinnamomum zeylanicum* and *Trachyspermum capticum*	0.3 and 0.4 g/kg, resp duration: 63 days	Increased HDL concentration at day 28 Increase ImmunoglobulinM from day 28–56 Serum pro-inflamattory cytokines (IL-6) decreased from day 28–56, higher lactobacilli and lower fecal enterobacterial populations	[40]

ADG—average daily weight gain; HDL—high density lipoprotein.

5.2. Antibacterial

Essential oils exhibit a wide spectrum of in vitro antibacterial activities against gram-negative and gram-positive bacteria including *E. coli*, *Salmonella*, *Staphylococcus*, *Klebsiella*, *Proteus*, *Bacillus*, and *Clostridium* species. Plant extracts kill pathogens due to their hydrophobicity and a high percentage of phenolic compounds. Bioactive compounds in essential oils prevent the development of virulent structures in bacteria, and active compounds disturb the enzyme system of bacteria blocking the virulence of the microbe [2]. Hydrophobicity properties of essential oils enable them to separate lipids present in the cell membrane of bacteria and mitochondria, making it more permeable and disturbing the cell structure [34]. This leads to cell death, due to the leakage of critical molecules and ions from the bacteria [41]. Essential oil containing phenolic groups exhibit antimicrobial properties through their delocalized electrons and the presence of a hydroxyl group on the phenolic ring. The oils initiate damage to bacterial cell membranes by compromising the pH homeostasis of the bacterial cell membranes [41]. Essential oils have a certain degree of selectivity towards gram-negative bacteria than gram-positive bacteria [1].

5.3. Antioxidant and Anti-Inflammatory Ability

The presence of phenolic OH and other pKa in essential oils contributes to their antioxidant properties. They act as hydrogen donors to the peroxy radical produced during lipid oxidation, inhibiting hydroxyl peroxide formation [26]. Essential oils improve redox balance in various organs and protect against oxidative damage caused by psychological stressors [42]. They also improve the oxidative capacity of meat, which influences its quality. Essential oils inhibit the production of pro-inflammatory cytokines and chemokines by endotoxin-stimulated immune and epithelial cells. Anti-inflammatory properties are partially mediated by inhibiting the NF-kB activation pathway [26], which prevents gut morphological changes, mucosa damage, increased mucosal permeability, impaired gut development, and poor nutrient absorption capacity [42].

5.4. Immune Stimulation

Essential oils have immune-stimulating effects on the gastrointestinal tract commonly referred to as the gut-associated lymphoid tissue (GALT) (Figure 2). The gastrointestinal tract possesses the largest mass of lymphoid tissue and plays an important role in antigen defense in the body [32]. Supplementing essential oils improves the immune status of animals by increasing lymphocyte proliferation rate, phagocytosis rate, IgG, IgA, and IgM concentrations, as well as changes in lymphocyte distribution in the gut [9]. Table 4 indicates secondary metabolites found in essential oils.

Table 4. Secondary metabolites in essential oils.

Compound Name	Classification	References
α-amyrin	Pentacyclic triterpene	[43]
1α,4α-dihydroxybishopsolicepolide	Guaianolide sesquiterpene lactone	[44]
12α,4α-dihydroxybishopsolicepolide	Sequiterpene	[43]
3,5-dicaffeoyl quinic acid	Phenylpropanoid	[45]
Acacetin	Flavone	[43]
Betulinic acid	Pentacyclic triterpenoid	[43]
Caffeic acid	Phenylpropanoid	[45]
Chlorogenic acid	Phenylpropanoid	[45]
Isoalantolactone	Sesquiterpene lactone	[46]
Phytol	Diterpene	[43]
Scopoletin	Coumarin	[43]
Yomogiartemin	Guaianolide sesquiterpene lactone	[44]

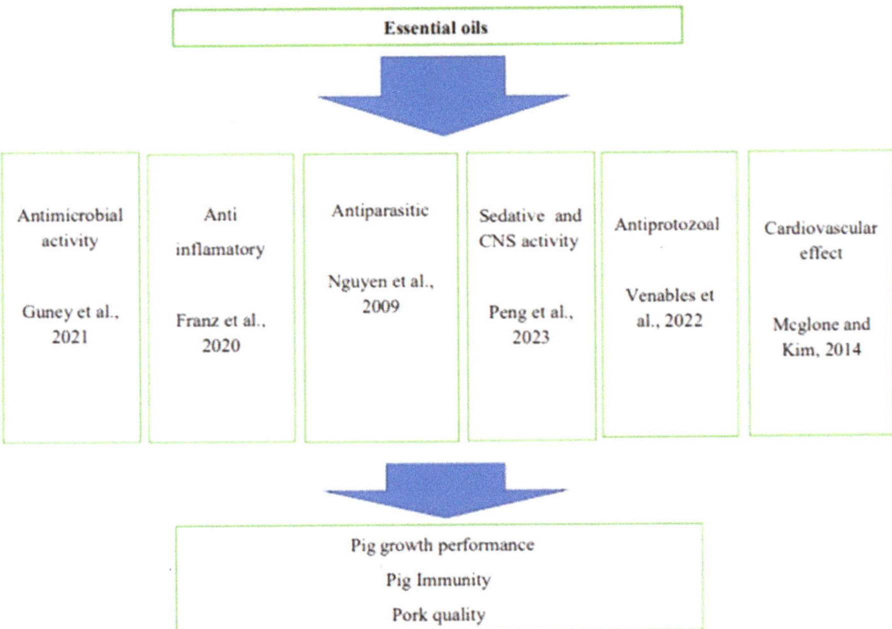

Figure 2. Summary of biological activities of essential oils [45–50].

6. Effect of Organic Acids and Essential Oils on Pig Performance

6.1. Influence on Voluntary Feed Intake (VFI)

Voluntary feed intake in pigs can be influenced by many factors, including dietary characteristics. Essential oils possess an intense smell which makes feed appealing, and pigs tend to consume feed more frequently and/or a larger amount at each meal before gut fill [4]. However, some essential oils result in reduced feed intake which can be attributed to an irritating smell that renders palatability displeasing [45]. Feed intake increased relative to control diets due to supplementation of essential oils ranged from 9% to 12% [49]. Lan, et al. [36] reported a range of 3–19% increase in feed intake. In weaned pigs' supplementation of organic acids, essential oils and their blend showed to improve average daily feed intake [26]. Essential oils can increase feed palatability and intake due to their flavor-enhancing properties and odor. An increase in palatability associated with the supplementation of essential oils can be attributed to their antioxidative properties that help preserve feed quality and prevent the formation of unpleasant odors [42]. Some organic acids show no effect on feed intake as their performance effect in pigs is mainly on the gastrointestinal tract and nutrient metabolism [6]. Table 5 shows the effect of organic acids and essential oils on voluntary feed intake.

Table 5. Effects of organic acids and essential oils on voluntary feed intake.

Feed Additive	Pig Group	Effect on VFI	References
EO	Piglets	Increase	[26]
EO	Piglets	Decrease	[46]
EO	Piglets	Decrease	[47]
EO	Piglets	Increase	[6]
EO blend	Piglets	NS	[48]
OA	Piglets	NS	[49]
OA	Finishing pigs	Increase	[49]
OA blend	Weaned piglets	NS	[50]
EO + OA	Nursary piglets	Increase	[26]
EO + OA	Finishing pigs	NS	[51]

VFI—voluntary feed intake; EO—essential oils; OA—organic acids; NS—not significant.

6.2. Influence on Nutrient Digestibility and Growth Efficiency

Studies on organic acids and essential oils have shown that they are beneficial and improve nutrient digestion. Formic acid and its salts lower the pH of the GIT, which increases the activity of digestive enzymes [34]. Organic acids have been shown to improve protein digestion by as much as 4%. Formic acid and its salts increased protein apparent tract digestibility, but did not improve ileal amino acid digestion (this could be due to diet acidification [46]). Citric acid improved the apparent total tract digestibility of protein, calcium, and phosphorous in sows [30]. Dietary benzoic acid improved the apparent digestibility of calcium and phosphorus in growing pigs, as well as crude protein in weanlings. Sows fed benzoic acid diets also had a high digestibility coefficient for organic matter, ether extract, crude protein, and crude fiber [52]. Dietary supplementation with protected acid blends increased the digestibility of dry matter, nitrogen, and energy in lactating sows [53]. Essential oils improved the apparent digestibility of crude protein and dry matter in swine. Studies have shown that phytogenic compounds can regulate ileal mucus gene expression and stimulate digestive secretions, thereby improving nutrient digestibility [42]. Essential oils act as a digestive stimulant by activating three (3) peripheral sensing mechanisms, known as oronasal. Oronasal sensing prepares the GI tract for food reception while also stimulating digestive secretion and gut motility [54].

6.3. Effect on Fecal Characteristics and Noxious Gas Production

Organic acids and essential oils can lower diarrheal incidence due to their ability to alter gut pH and microflora [2,6]. *Escherichia coli* is a major factor in causing infection and diarrhea in weaned pigs, and it affects growth performance [55]. Organic acids have been shown to increase the number of beneficial bacteria in the GIT in pigs and reduce the concentration of *E. coli* in feces. OAs penetrate bacterial cells in a non-dissociated form and disrupt the normal physiology of certain bacteria [14]. Organic acids and EO blends work in synergy to suppress growth of pathogenic microbes and promote the growth of beneficial microbes [55]. Formic acid, fumaric acid, and citric acid reduced the incidence and severity of diarrhea, increased microbial diversity in the GIT, and reduced *E. coli* counts while increasing lactobacilli counts [22,56,57]. Herb and plant extracts reduced the *E. coli* count and improved energy digestion [58,59]. Improved nutrient digestibility in pigs due to essential oils and organic acid supplementation had an impact on fecal noxious gas production. Kiarie et al. [60] state that protective acids reduced fecal emission of ammonia and hydrogen sulfide in lactating sows. The inclusion of humic substances in pig diets reduced ammonia emission by 3 to 18% in pig manure. Reduced aerial ammonia concentrations have beneficial effects on human health [52].

6.4. Effects on Gut Morphology and Gut Microflora

Low gastric pH due to the addition of acidifiers in diets maximizes the growth of beneficial bacteria in the GIT [61]. Organic acids and medium-chain fatty acids have been demonstrated to reduce pathogenic activity in pigs when fed in combinations rather than individually. They reduce the expression of pro-inflammatory cytokines and increase the proliferation of *Lactobacillus* bacteria. Formic acid added to diets of weaned pigs showed an increase in intestinal microbial diversity and a change in the concentration of certain microbes. A blend of organic acids also increased fecal *Lactobacillus* species and decreased *E. coli* fecal counts [62]. Plant extracts fed to pigs indicated improved gut health by modulating gut microbiota. Supplementation of essential oils decreased ileal total microbial mass and increased the lactobacilli to enterobacterial ratio. In vivo studies showed that essential oils increased the lactobacilli group and decreased *E. coli* and total coliform in piglets [42].

Organic acid and essential oil blends can increase the villous height of the duodenum. Essential oil supplements increased the villous height of the jejunum and the villous height to crypt ratio [26]. Supplements of essential oils reduced the number of intra epithelial and increased the villus height to crypt depth in the distal small intestines [2]. Essential oils decrease the number of pathogenic bacteria in the gut, favoring an increase in villus length, gut surface area, and crypt depth in the jejunum and colon [32]. Blends of medium-chain fatty acids and short-chain organic acids can be utilized by enterocytes as energy sources and attenuate the negative effect of weaning on villus length and crypt depth in pigs [34]. In weaner pigs, benzoic acid showed an increase in the villus height to crypt depth ratio [63]. However, in a study by Kong et al. [64], butyric acid did not affect the histology of grower-finishing pigs; only their mucosal depth was larger, and this can be attributed to better gut integrity in older animals.

6.5. Effect on Immune Status and Oxidative Stress

Supplementing pig diets with organic acid and essential oil has an impact on the immune system and the regulation of oxidative stress. Studies showed that essential oils reduced the numbers of intra epithelium lymphocytes in the mesenteric lymph nodes. Essential oils improve immunity and reduce the need for immune defense activity in the gut [32]. A mixture of carvacrol, cinnamaldehyde, and capsicum oleoresin decreased the population of intra epithelium lymphocytes in the jejunum and ileum of pigs [26]. Supplementation with butyric acid and essential oil reduced the white blood cell counts in growing pigs [62]. Li et al. [65] also stated that organic acid and essential oil blends can reduce the total white blood cell and neutrophil counts during the post-weaning period. An increase in WBC counts indicates systematic inflammation and the risk of bacterial infection [66].

Oxidative stress is commenced when the amount of ROS produced exceeds the neutralization ability of antioxidants. Excessive ROS leads to oxidative damage of proteins, lipids, and DNA, thereby destroying cell function [67]. Oxidative stress and inflammation are correlated physiological processes [68]. To prevent the accumulation of free radicals, cells develop defense mechanisms including the antioxidant enzymes and non-enzymatic antioxidants. Superoxide dismutase (SOD), glutathione peroxidase (GSH P_x), and catalases (CAT) constitute antioxidant enzymes, whereas ascorbic acid, α tocopherol, Glutathione (GSH), carotenoids, and flavonoids are part of non-enzymatic antioxidants [69]. Oxidative stress represents an important chemical mechanism that leads to biological damage. Exposure of pigs to varied stressors leads to the increased production of ROS and the overwhelming of the antioxidant system. Oxidative stress is associated with reduced performance, decreased feed intake, diarrhea, and destruction of liver tissues [70]. In vivo experiments in pigs have shown that the antioxidant effects of essential oils reduce oxidative stress. Frankic et al. [69] state that carvacrol added in drinking water reduced the level of DNA lesions induced in freshly isolated hepatocytes and testicular cells in pigs. Mounir et al. [71] demonstrated that supplementation of plant extracts in pigs reduced DNA damage in lymphocytes which can be a potential benefit to the immune system. Organic acid Na-butyrate supplementation in gestating sow diets and pre-weanling diets had a positive effect on muscle and adipose tissue oxidative genes [70,71]. Improved antioxidant indices can prevent villi from radical-induced damage, which is correlated with better intestinal morphology and nutrient digestibility [72].

6.6. Effect on Growth Performance and Carcass Characteristics

Organic acids and essential oils improve the productivity of pigs to levels comparable to antibiotic growth promoters. According to Lückstädt et al. [6], organic acids improved daily weight gain, feed conversion rate, birth weight, weaning weight, and back fat thickness in pigs. The addition of fulvic acid in pig diets improved average daily gain and growth. Fulvic acid can improve the metabolism of proteins and carbohydrates [73]. A blend of organic acids showed an improvement in growth performance in older pigs and newly weaned pigs [74,75]. In the grower-finisher period, application of different levels and different sources of plant extracts showed positive effects on the growth performance of pigs [76]. [77] reported a higher average daily gain and feed conversion ratio in pigs fed garlic-treated diets. Bedford and Gong [78] observed a significant improvement in average daily gain and feed conversion ratio with the use of an herb mixture in pig diets from 25 to 105 kg.

6.7. Effect on Physicochemical Meat Properties

Meat quality characteristics are generally not affected by changes in diet composition, but can be influenced when diet alters carcass composition. According to Peng et al. [79], supplementing organic acids had no effect on marbling, meat color, cooking losses, drip losses, and water holding capacity in finisher pigs. Organic acids are considered growth promoters. However, there are no scientific studies confirming that they do not alter carcass composition. The addition of an organic acid and essential oil blend in grower-finisher pigs did not affect the pork's ultimate pH, cooking loss, and shear force. Essential oils showed an effect on meat color by improving the oxidative stability of meat [12]. Rosemary essential oils reduced indicators of lipid oxidation and protein oxidation in pork [8]. Oregano essential oils when supplemented in pig diets prevented lipid oxidation but did not affect the cooking loss, drip loss, shear force, and chemical composition of pork [80]. Supplementing pigs with butyrate showed an effect on boar taint impacting pork sensory attributes. Butyrate has a regulatory effect on cell apoptosis and accumulation of androsterone in pigs, which causes boar taint [81]. However, there is a need for more research on the sensory effects of organic acids and essential oils in pork. Table 6 shows the effects of organic acids and essential oils on pig performance in relation to age.

Table 6. Effects of organic acid and essential oil blends on pig performance in relation to age.

OA and EO Blends	Target	Dose	Results	References
* BA + EO	Weanling pigs	3.0 and 0.1%	No effect on growth performance, metabolites, cytokines, intestinal microbiota	[81]
OA (FA, AA, CA, PA, and Ca) + EO (thyme, nettle, oak, and balm)	Weaned piglets	CR + 0.5% EA + 0.3% OA	Improved daily gains at later growth stage, higher protein quality, 6.9% cholesterol reduction	[82]
BA + EO (thymol, 2-methoxyphenol, eugenol, piperine, and curcumin)	Weanling pigs	2/3/4 g/kg at 1.8 BA + 0.072 EO, 1.8 BA + 0.072 EO and 1.8 BA + 0.072 EO levels	Increased net revenue when BA + EO at 3 or 4 g/kg	[83]
FA + FormaXOL™	Finisher pigs	4 kg/tonne	Reduces Salmonella shedding and seroprevalence at longer supplementation duration but increased feed cost per live weight gain	[84]
BA 50%, Calcium formate 3% and FA 1% + Thymol 25% and carvacrol 25%	Weaned piglets	1.5 g/kg OA + 30 mg/kg EO	Complementary effect on growth performance, little interactive effects on intestinal health between EO and OA	[26]
Cinnamaldehyde 15%, thymol 5%, CA 10%, SA 10%, MA 6.5% and FA 13.5%	Weaned piglets	1 kg/tonne	Improved growth performance and fecal microbes, modulate serum immune parameters, increased isovaleric acid	[85]
PEP1000-1®, (anis, citrus, oregano oils, and natural Flavors) + Biotronic®, (PA and LA)	Nursery piglets	0.4% and 0.2%, resp.	Matched growth performance of antibiotic supplement	[40]
Bamboo Vinegar + Acidifier I Bamboo Vinegar + Acidifier II	Weaned piglets	0.4% BV + 0.25% Acidifier	Wider species richness and bacterial community diversities in feces	[81]

CR—complete ration; PA—phosphoric acid, LA—lactic acid; FA—fumaric acid; BA—benzoic acid; CF—calcium formate; FO—formic acid; CA—Citric acid; AA—acetic acid; MA—malic acid; SA—sorbic acid. Acidifier I—LA, CA, MA, TA, and PA mixed at 20:20:10:15:35; Acidifier II—LA, CA, MA, TA, and PA mixed at 40:20:20:20:0; * FormaXOL™—encapsulated blend of formic acid, citric acid, and essential oils from citrus fruit extract, cinnamon, oregano, thyme, and capsicum, Kemin Industries, Inc., Southport, UK, UK*EO—essential oils (CRENA; DSM Nutritional Products, LLC, Belvidere, IL, USA); BA—benzoic acid ((Vevovitall®, DSM Nutritional Products Inc., Parsippany, NJ, USA).

6.8. Potential of Organic Acids and Essential Oils as Feed Additives in the Pig Industry

Figure 3 summarizes the potential benefits of organic acids, essential oils, and their blend in pig diets. The application of organic acidifiers and essential oils in pig diets has great potential in improving pig performance, pork quality, and reducing environmental pollution. This will, in turn, assist in meeting the ever-increasing demand for animal protein, positively affecting food and nutritional security. The use of organic acid and essential oils has the potential to reduce noxious gas emissions from pig manure, impacting climate change mitigation. The overall adoption of organic acids and essential oils in pig nutrition will lead to a drastic shift in the provision of safe pork that is antibiotic-free [2].

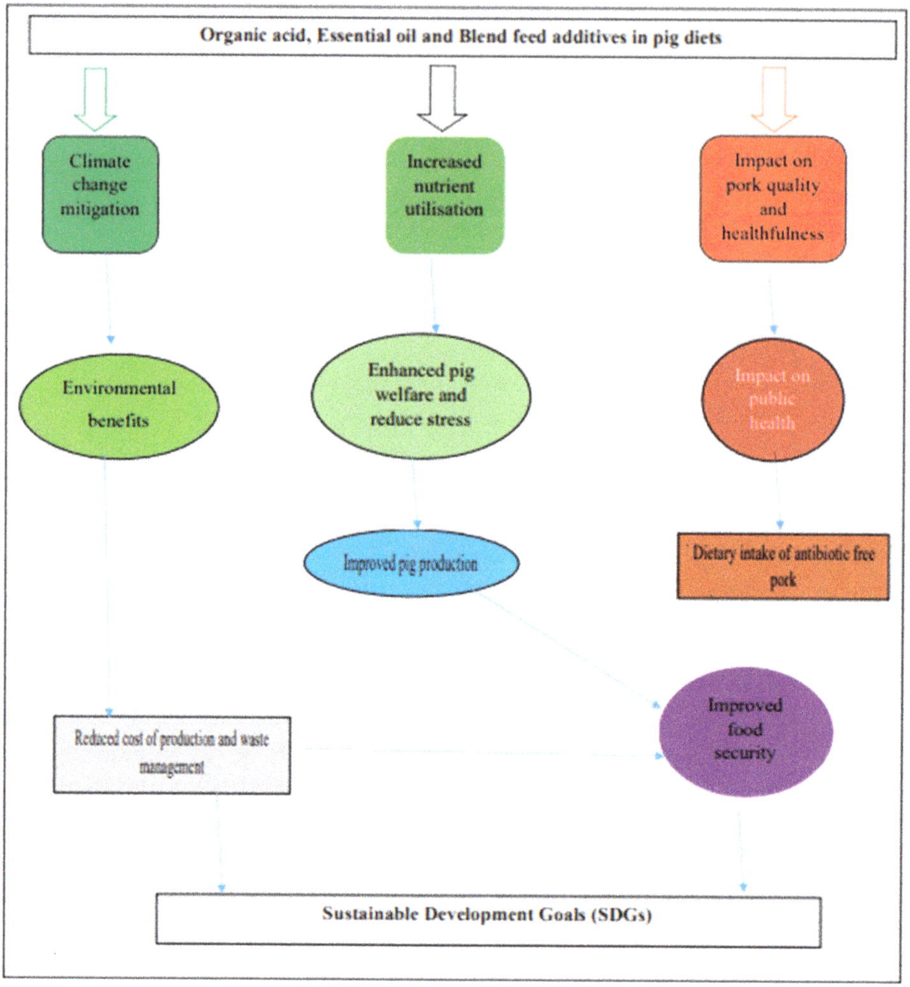

Figure 3. Summary on potential of organic acids and essential oil feed additives.

7. Conclusions

Organic acid and essential oils can improve nutrient digestibility, growth performance, carcass traits, gut morphology, microflora, meat quality, and chemical composition in pigs. The potential of organic acids and essential oils to improve pig performance and pork quality is comparable to that of antibiotic growth promoters and can be an alternative in smart pig production practices and the production of safe meat. However, information on their specific mode of action in growing pigs is still lacking, and there is a need for further research. Future studies are recommended on the effects of organic acid and essential oils on fermentation indices, immune and enzyme gene expression, fatty acid profile, and lipid quality indices.

Author Contributions: Conceptualization, R.B.N., U.M. and C.W.T.N.; review and editing, R.B.N., U.M. and C.W.T.N.; supervision; U.M. All authors have read and agreed to the published version of the manuscript.

Funding: This research received no external funding.

Conflicts of Interest: The authors declare no conflict of interest.

References

1. De Lange, C.F.M.; Pluske, J.; Gong, J.; Nyachoti, C.M. Strategic use of feed ingrédients and feed additives to stimulate gut health and development in young pigs. *Livest. Sci.* **2010**, *134*, 124–134. [CrossRef]
2. Liu, Y.; Espinosa, C.D.; Abelilla, J.J.; Casas, G.A.; Lagos, L.V.; Lee, S.A.; Stein, H.H. Non-antibiotic feed additives in diets for pigs: A review. *Anim. Nutr.* **2018**, *4*, 113–125. [CrossRef]
3. Long, S.F.; Xu, Y.T.; Pan, L.; Wang, Q.Q.; Wang, C.L.; Wu, J.Y.; Wu, Y.Y.; Han, Y.M. Mixed organic acids as antibiotic substitutes improve performance, serum immunity, intestinal morphology and microbiota for weaned piglets. *J. Anim. Feed Sci.* **2018**, *235*, 23–32. [CrossRef]
4. Han, Y.S.; Tang, C.H.; Zhao, Q.Y.; Zhan, T.F.; Zhang, K.; Han, Y.M.; Zhang, J.M. Effects of dietary supplementation with combinations of organic and medium chain fatty acids as replacements for chlortetracycline on growth performance, serum immunity, and faecal microbiota of weaned piglets. *Livest. Sci.* **2018**, *216*, 210–218. [CrossRef]
5. Kuppusamy, S.; Kakarlab, D.; Venkateswarluc, K.; Megharajd, M.; Yoone, Y.E.; Lee, Y.B. Agriculture, Ecosystems and Environment Veterinary antibiotics (VAs) contamination as a global agro-ecological issue: A critical view. *Agric. Ecosyst. Environ.* **2018**, *257*, 47–59. [CrossRef]
6. Lückstädt, C.; Wylie, L.; Remmer, R.; De Kok, R.; Costa, H.R.; Brebels, M.; van Heusden, C. *Organic Acids in Animal Nutrition*; Anitox: Lawrenceville, GA, USA, 2014.
7. Pateiro, M.; Francisco, B.; Domínguez, J.; Sant, A. Essential oils as natural additives to prevent oxidation reactions in meat and meat products: A review. *Food Res. Int.* **2018**, *113*, 156–166. [CrossRef]
8. Zhai, H.; Liu, H.; Wang, S.; Wu, J.; Kluenter, A.M. Potential of essential oils for poultry and pigs. *Anim. Nutr.* **2018**, *4*, 179–186. [CrossRef]
9. Cromwell, G.L. Why and how antibiotics are used in swine production. *Anim. Biotechnol.* **2002**, *13*, 7–27. [CrossRef] [PubMed]
10. Kenworthy, R.; Crabb, W.E. The intestinal flora of young pigs, with reference to early weaning, *Escherichia coli* and scours. *J. Comp. Pathol. Ther.* **1963**, *73*, 215–228. [CrossRef] [PubMed]
11. Li, J. Current status and prospects for in-feed antibiotics in the different stages of pork production—A review. *Asian-Australas. J. Anim. Sci.* **2017**, *30*, 1667–1673. [CrossRef] [PubMed]
12. Kohanski, M.A.; Dwyer, D.J.; Collins, J.J. How antibiotics kill bacteria: From targets to networks. *Nat. Rev. Microbiol.* **2010**, *8*, 423–435. [CrossRef]
13. Liu, X.; Qian, M.; Shen, Y.; Qin, X.; Huang, H.; Yang, H.; He, Y.; Bai, W. An high-through put sequencing approach to the preliminary analysis of bacterial communities associated with changes in amino acid nitrogen, organic acid and reducing sugar contents during soy sauce fermentation. *Food Chem.* **2021**, *349*, 129131. [CrossRef]
14. Khan, S.H.; Iqbal, J. Recent advances in the role of organic acids in poultry nutrition. *J. Appl. Anim. Res.* **2016**, *44*, 359–369. [CrossRef]
15. Tugnoli, B.; Giovagnoni, G.; Piva, A.; Grilli, E. From acidifiers to intestinal health enhancer: How organic acids can improve growth efficiency of pigs. *Animals* **2020**, *10*, 134. [CrossRef]
16. Rathnayake, D.; Mun, H.S.; Dilawar, M.A.; Baek, K.S.; Yang, C.J. Time for a paradigm shift in animal nutrition metabolic pathway: Dietary inclusion of organic acids on the production parameters, nutrient digestibility, and meat quality traits of swine and broilers. *Life* **2021**, *11*, 476. [CrossRef]
17. Hajati, H. Application of organic acids in poultry nutrition. *Int. J. Avian Wildl. Biol.* **2018**, *3*, 324–329. [CrossRef]
18. Kumar, A.; Toghyani, M.; Kheravii, S.K.; Pineda, L.; Han, Y.; Swick, R.A.; Wu, S.-B. Potential of blended organic acids to improve performance and health of broilers infected with necrotic enteritis. *Anim. Nutr.* **2021**, *7*, 440–449. [CrossRef]
19. Suiryanrayna, M.V.A.N.; Ramana, J.V. A review of the effects of dietary organic acids fed to swine. *J. Anim. Sci. Biotechnol.* **2015**, *6*, 45. [CrossRef]
20. Partanen, K.H.; Mroz, Z. Organic acids for performance enhancement in pig diets. *Nutr. Res. Rev.* **1999**, *12*, 117–145. [CrossRef]
21. Luise, D.; Motta, V.; Salvarani, C.; Chiappelli, M.; Fusco, L.; Bertocchi, M.; Mazzoni, M.; Maiorano, G.; Costa, L.N.; Van Milgen, J.; et al. Long-term administration of formic acid to weaners: Influence on intestinal microbiota, immunity parameters and growth performance. *Anim. Feed Sci. Technol.* **2017**, *232*, 160–168. [CrossRef]
22. Duarte, M.E.; Kim, S.W. Phytobiotics from Oregano Extracts Enhance the Intestinal Health and Growth Performance of Pigs. *Antioxidants* **2022**, *11*, 2066. [CrossRef]
23. Gottlob, R.O.; Neill, C.R.; Groesbeck, C.N.; Schneider, J.D.; Frantz, N.Z.; Tokach, M.D.; Goodband, R.D.; DeRouchey, J.M.; Nelssen, J.L.; Dritz, S.S. Effects of water-soluble and in-feed organic acids on the growth performance of weanling pigs. *Kans. Agric. Exp. Stn. Res. Rep.* **2006**, *10*, 60–66. [CrossRef]
24. Kiarie, E.; Voth, C.; Wey, D.; Zhu, C.; Vingerhoeds, P.; Borucki, S.; Squires, E.J. Comparative efficacy of antibiotic growth promoter and benzoic acid on growth performance, nutrient utilization, and indices of gut health in nursery pigs fed corn–soybean meal diet. *Can. J. Anim. Sci.* **2018**, *98*, 868–874. [CrossRef]
25. Xu, Y.T.; Liu, L.I.; Long, S.F.; Pan, L.; Piao, X.S. Effect of organic acids and essential oils on performance, intestinal health and digestive enzyme activities of weaned pigs. *Anim. Feed Sci. Technol.* **2018**, *235*, 110–119. [CrossRef]

26. Kommera, S.K.; Mateo, R.D.; Neher, F.J.; Kim, S.W. Phytobiotics and organic acids as potential alternatives to the use of antibiotics in nursery pig diets. *Asian-Australas. J. Anim. Sci.* **2006**, *19*, 1784–1789. [CrossRef]
27. Loh, T.C.; Low, Y.S.; Chang, Y.K. Effects of Orgacidstm on Growth Performance, Gut Microflora and pH and Meat Cholesterol in Growing Pigs. In Proceedings of the 11th International Conference of the Association of Institutions for Tropical Veterinary Medicine, Petaling Jaya, Malaysia, 23–27 August 2004; p. 392.
28. Rodrigues, L.M.; Neto, T.O.D.A.L.; Garbossa, C.A.P.; Martins, C.C.D.S.; Garcez, D.; Alves, L.K.S.; de Abreu, M.L.T.; Ferreira, R.A.; Cantarelli, V.D.S. Benzoic acid combined with essential oils can be an alternative to the use of antibiotic growth promoters for piglets challenged with *E. coli* F4. *Animals* **2020**, *10*, 1978. [CrossRef]
29. Liu, F.J.; Ding, G.S.; Tang, A.N. Simultaneous separation and determination of five organic acids in beverages and fruits by capillary electrophoresis using diamino moiety functionalized silica nanoparticles as pseudostationary phase. *Food Chem.* **2014**, *145*, 109–114. [CrossRef]
30. Choi, J.; Wang, L.; Liu, S.; Lu, P.; Zhao, X.; Liu, H.; Lahaye, L.; Santin, E.; Liu, S.; Nyachoti, M.; et al. Effects of a Microencapsulated Formula of Organic Acids and Essential Oils on Nutrient Absorption, Immunity, Gut Barrier Function, and Abundance of Enterotoxigenic *Escherichia coli* F4 in Weaned Piglets Challenged with *E. coli* F4. *J. Anim. Sci.* **2020**, *98*, skaa259. [CrossRef] [PubMed]
31. Ayalew, H.; Zhang, H.; Wang, J.; Wu, S.; Qiu, K.; Qi, G.; Tekeste, A.; Wassie, T.; Chanie, D. Potential Feed Additives as Antibiotic Alternatives in Broiler Production. *Front. Vet. Sci.* **2022**, *9*, 916473. [CrossRef] [PubMed]
32. Papatsiros, V.G.; Christodoulopoulos, G.; Filippopoulos, L.C. The use of organic acids in monogastric animals (swine and rabbits). *J. Cell Anim. Biol.* **2012**, *6*, 154–159. [CrossRef]
33. Li, S.; Zheng, J.; Deng, K.; Chen, L.; Zhao, X.L.; Jiang, X.; Fang, Z.; Che, L.; Xu, S.; Feng, B.; et al. Supplementation with organic acids showing different effects on growth performance, gut morphology, and microbiota of weaned pigs fed with highly or less digestible diets. *J. Anim. Sci.* **2018**, *96*, 3302–3318. [CrossRef]
34. Tariq, S.; Wani, S.; Rasool, W.; Bhat, M.A.; Prabhakar, A.; Shalla, A.H.; Rather, M.A. A comprehensive review of the antibacterial, antifungal and antiviral potential of essential oils and their chemical constituents against drug-resistant microbial pathogens. *Microb. Pathog.* **2019**, *134*, 103580. [CrossRef]
35. Zeng, Z.; Zhang, S.; Wang, H.; Piao, X. Essential oil and aromatic plants as feed additives in non-ruminant nutrition: A review. *J. Anim. Sci. Biotechnol.* **2015**, *6*, 7. [CrossRef] [PubMed]
36. Lan, R.X.; Li, T.S.; Kim, I.H. Effects of essential oils supplementation in different nutrient densities on growth performance, nutrient digestibility, blood characteristics and fecal microbial shedding in weaning pigs. *Anim. Feed Sci. Technol.* **2016**, *214*, 77–85. [CrossRef]
37. Tan, B.F.; Lim, T.; Boontiam, W. Effect of dietary supplementation with essential oils and a Bacillus probiotic on growth performance, diarrhoea and blood metabolites in weaned pigs. *Anim. Prod. Sci.* **2020**, *61*, 64–71. [CrossRef]
38. Montoya, D.; D'angelo, M.; Martín-Orúe, S.M.; Rodríguez-Sorrento, A.; Saladrigas-García, M.; Araujo, C.; Chabrillat, T.; Kerros, S.; Castillejos, L. Effectiveness of two plant-based in-feed additives against an *Escherichia coli* F4 oral challenge in weaned piglets. *Animals* **2021**, *11*, 2024. [CrossRef]
39. Samanta, A.K.; Gali, J.M.; Dutta, T.K.; Rajkhowa, T.K.; Mandal, G.P.; Patra, A.K. Effect of dietary phytobiotic mixture on growth performance, nutrient utilization, and immunity in weaned piglets. *Trop. Anim. Health Prod.* **2021**, *53*, 459. [CrossRef]
40. Omonijo, F.A.; Ni, L.; Gong, J.; Wang, Q.; Lahaye, L.; Yang, C. Essential oils as alternatives to antibiotics in swine production. *Anim. Nutr.* **2018**, *4*, 126–136. [CrossRef] [PubMed]
41. More, G.; Lall, N.; Hussein, A.; Tshikalange, T.E. Antimicrobial constituents of *Artemisia afra* Jacq. ex Willd. against periodontal pathogens. *Evid.-Based Complement. Altern. Med.* **2012**, *2012*, 252758. [CrossRef] [PubMed]
42. Moyo, P.; Kunyane, P.; Selepe, M.A.; Eloff, J.N.; Niemand, J.; Louw, A.I.; Maharaj, V.J.; Birkholtz, L.-M. Bioassay-Guided Isolation and Identification of Gametocytocidal Compounds from *Artemisia afra* (Asteraceae). *Malar. J.* **2019**, *18*, 65. [CrossRef]
43. Liu, N.Q.; Cao, M.; Frédérich, M.; Choi, Y.H.; Verpoorte, R.; van der Kooy, F. Metabolomic investigation of the ethnopharmacological use of *Artemisia afra* with NMR spectroscopy and multivariate data analysis. *J. Ethnopharmacol.* **2010**, *128*, 230–235. [CrossRef]
44. Venables, L.; Koekemoer, T.C.; Van de Venter, M.; Goosen, E.D. Isoalantolactone, a sesquiterpene lactone from *Artemisia afra* Jacq. ex Willd and its in vitro mechanism of induced cell death in HeLa cells. *S. Afr. J. Bot.* **2016**, *103*, 216–221. [CrossRef]
45. Güney, M.; Karaca, S.; Erdogan, S.; Kor, A.; Kale, C.; Onalan, S.; Demirel, M.; Bingol, N.T. Effects of Dietary Supplementation with Rosemary Oil on Methanogenic Bacteria Density, Blood and Rumen Parameters and Meat Quality of Fattening Lambs. *Ital. J. Anim. Sci.* **2021**, *20*, 794–805. [CrossRef]
46. Franz, C.; Baser, K.H.C.; Windisch, W. Essential oils and aromatic plants in animal feeding—A European perspective. A review. *Flavour Fragr. J.* **2010**, *25*, 327–340. [CrossRef]
47. Michiels, J.; Missottenb, J.A.M.; Fremauta, D.; DeSmetb, S.; Diericket, N.A. Animal Feed Science and Technology In vitro characterisation of the antimicrobial activity of selected essential oil components and binary combinations against the pig gut flora. *J. Anim. Feed Sci.* **2009**, *151*, 111–127. [CrossRef]
48. Nguyen, N.P.K.; Tran, K.N.; Nguyen, L.T.H.; Shin, H.M.; Yang, I.J. Effects of Essential Oils and Fragrant Compounds on Appetite: A Systematic Review. *Int. J. Mol. Sci.* **2023**, *24*, 7962. [CrossRef] [PubMed]

49. Peng, X.; Zhou, Q.; Wu, C.; Zhao, J.; Tan, Q.; He, Y.; Hu, L.; Fang, Z.; Lin, Y.; Xu, S.; et al. Effects of dietary supplementation with essential oils and protease on growth performance, antioxidation, in fl ammation and intestinal function of weaned pigs. *Anim. Nutr.* **2022**, *9*, 39–48. [CrossRef]
50. Mcglone, J.J.; Kim, S.W. Effects of dietary humic substances on pig growth performance, carcass characteristics, and ammonia emission 1. *J. Anim. Sci.* **2014**, *84*, 2482–2490. [CrossRef]
51. Ma, J.; Piao, X.; Shang, Q.; Long, S.; Liu, S.; Mahfuz, S. Mixed organic acids as an alternative to antibiotics improve serum biochemical parameters and intestinal health of weaned piglets. *Anim. Nutr.* **2021**, *7*, 737–749. [CrossRef]
52. Resende, M.; Chaves, R.F.; Garcia, R.M.; Barbosa, J.A.; Marques, A.S.; Rezende, L.R.; Peconick, A.P.; Garbossa, C.A.P.; Mesa, D.; Silva, C.C.; et al. Benzoic Acid and Essential Oils Modify the Cecum Microbiota Composition in Weaned Piglets and Improve Growth Performance in Finishing Pigs. *Livest. Sci.* **2020**, *242*, 104311. [CrossRef]
53. Pluske, J.R.; Turpin, D.L.; Sahibzada, S.; Pineda, L.; Han, Y.; Collins, A. Impacts of Feeding Organic Acid-Based Feed Additives on Diarrhea, Performance, and Fecal Microbiome Characteristics of Pigs after Weaning Challenged with an Enterotoxigenic Strain of *Escherichia coli*. *Transl. Anim. Sci.* **2021**, *5*, txab212. [CrossRef] [PubMed]
54. Kluge, H.; Broz, J.; Eder, K. Effects of dietary benzoic acid on urinary pH and nutrient digestibility in lactating sows. *Livest. Sci.* **2010**, *134*, 119–121. [CrossRef]
55. Devi, S.M.; Lee, K.Y.; Kim, I.H. Analysis of the effect of dietary protected organic acid blend on lactating sows and their piglets. *Rev. Bras. Zootec.* **2016**, *45*, 39–47. [CrossRef]
56. Brenes, A.; Roura, E. Essential oils in poultry nutrition: Main effects and modes of action. *Anim. Feed Sci. Technol.* **2010**, *158*, 1–14. [CrossRef]
57. Tsiloyiannis, V.K.; Kyriakis, S.C.; Vlemmas, J.; Sarris, K. The effect of organic acids on CTR of porcine postweaning diarrhea. *Res. Vet. Sci.* **2001**, *70*, 287–293. [CrossRef]
58. Upadhaya, S.D.; Lee, K.Y.; Kim, I.H. Influence of protected organic acid blends and diets with different nutrient densities on growth performance nutrient digestibility and faecal noxious gas emission in growing pigs. *Vet. Med.* **2014**, *59*, 491–497. [CrossRef]
59. Bontempo, V.; Jiang, X.R.; Cheli, F.; Lo Verso, L.; Mantovani, G.; Vitari, F.; Domeneghini, C.; Agazzi, A. Administration of a novel plant extract product via drinking water to post-weaning piglets: Effects on performance and gut health. *Animal* **2014**, *8*, 721–730. [CrossRef] [PubMed]
60. Radzikowski, D.; Milczarek, A. Efficiency of Herbs and Botanicals in Pig Feeding. *Anim. Sci. Genet.* **2022**, *18*, 73–87. [CrossRef]
61. Kiarie, E.; Walsh, M.C.; Nyachoti, C.M. Performance, digestive function, and mucosal responses to selected feed additives for pigs. *J. Anim. Sci.* **2016**, *94*, 169–180. [CrossRef]
62. Huang, Y.; Yoo, J.S.; Kim, H.J.; Wang, Y.; Chen, Y.J.; Cho, J.H.; Kim, I.H. Effects of Dietary Supplementation with Blended Essential Oils on Growth Performance, Nutrient Digestibility, Blood Profiles and Fecal Characteristics in Weanling Pigs. *Asian-Australas. J. Anim. Sci.* **2010**, *23*, 607–613. [CrossRef]
63. Lee, J.; Kim, J.W.; Hall, H.; Nyachoti, C.M. Effect of dietary organic acids supplementation on growth performance, nutrient digestibility, and gut morphology in weaned pigs. *Can. J. Anim. Sci.* **2021**, *102*, 255–265. [CrossRef]
64. Morel, P.C.H.; Chidgey, K.L.; Jenkinson, C.M.C.; Lizarraga, I.; Schreurs, N.M. Effect of benzoic acid, sodium butyrate and sodium butyrate coated with benzoic acid on growth performance, digestibility, intestinal morphology and meat quality in grower-finisher pigs. *Livest. Sci.* **2019**, *226*, 107–113. [CrossRef]
65. Wang, J.P.; Lee, J.H.; Yoo, J.S.; Cho, J.H.; Kim, H.J.; Kim, I.H. Effects of phenyllactic acid on growth performance, intestinal microbiota, relative organ weight, blood characteristics, and meat quality of broiler chicks. *Poult. Sci.* **2010**, *89*, 1549–1555. [CrossRef] [PubMed]
66. Kong, N.; Chen, G.; Wang, H.; Li, J.; Yin, S.; Cao, X.; Wang, T.; Li, X.; Li, Y.; Zhang, H.; et al. Blood leukocyte count as a systemic inflammatory biomarker associated with a more rapid spirometric decline in a large cohort of iron and steel industry workers. *Respir. Res.* **2021**, *22*, 254. [CrossRef] [PubMed]
67. Li, Q.; Yang, S.; Chen, F.; Guan, W.; Zhang, S. Nutritional strategies to alleviate oxidative stress in sows. *Anim. Nutr.* **2022**, *9*, 60–73. [CrossRef]
68. Ahasan, A.S.M.L.; Invernizzi, G.; Farina, G.; Pilotto, A.; Barbé, F.; Bontempo, V.; Rossi, R.; Bellagamba, F.; Lecchi, C.; Savoini, G.; et al. The effects of superoxide dismutase-rich melon pulp concentrate on inflammation, antioxidant status and growth performance of challenged post-weaning piglets. *Animal* **2019**, *13*, 136–143. [CrossRef]
69. Slamenova, D.; Horvathova, E.; Marsalkova, L.; Wsolova, L. Carvacrol given to rats in drinking water reduces the level of DNA lesions induced in freshly isolated hepatocytes and testicular cells by H_2O_2. *Neoplasma* **2008**, *55*, 394–399.
70. Frankic, T.; Levart, A.; Salobir, J. The effect of vitamin E and plant extract mixture composed of carvacrol, cinnamaldehyde and capsaicin on oxidative stress induced by high PUFA load in young pigs. *Animal* **2010**, *4*, 572–578. [CrossRef]
71. Le Gall, M.; Gallois, M.; Seve, B.; Louveau, I.; Holst, J.J.; Oswald, I.P.; Lallès, J.-P.; Guilloteau, P. Comparative effect of orally administered sodium butyrate before or after weaning on growth and several indices of gastrointestinal biology of piglets. *Br. J. Nutr.* **2009**, *102*, 1285–1296. [CrossRef] [PubMed]
72. Zhang, H.; Ren, E.; Xu, R.; Su, Y. Transcriptomic Responses Induced Inmuscle and Adipose Tissues of Growing Pigs by Intravenous Infusion of Sodium Butyrate. *Biology* **2021**, *10*, 559. [CrossRef]

73. Mounir, M.; Ibijbijen, A.; Farih, K.; Rabetafika, H.N.; Razafindralambo, H.L. Synbioticsand Their Antioxidant Properties, Mechanisms, and Benefits on Human and Animal Health: A Narrative Review. *Biomolecules* **2022**, *12*, 1443. [CrossRef]
74. Grilli, E.; Messina, M.R.; Tedeschi, M.; Piva, A. Feeding a microencapsulated blend of organic acids and nature identical compounds to weaning pigs improved growth performance and intestinal metabolism. *Livest. Sci.* **2010**, *133*, 173–175. [CrossRef]
75. Ahmed, S.T.; Hwang, J.A.; Hoon, J.; Mun, H.S.; Yang, C.J. Comparison of Single and Blend Acidifiers as Alternative to Antibiotics on Growth Performance, Faecal Microflora, and Humoral Immunity in Weaned Piglets. *Asian-Australas. J. Anim. Sci.* **2014**, *27*, 93–100. [CrossRef]
76. Cho, J.H.; Liu, S.D.; Kim, I.H. Effects of Dietary Korean Garlic Extract Aged by Leuconostoc Mesenteroides KCCM35046 on Growth Performance, Digestibility, Blood Profiles, Gas Emissions, and Microbiota in Weanling Pigs. *Can. J. Anim. Sci.* **2020**, *100*, 462–469. [CrossRef]
77. Cui, Y.; Lu, H.; Tian, Z.; Deng, D.; Ma, X. Current Trends of Chinese Herbal Medicines on Meat Quality of Pigs. A Review. *J. Anim. Feed Sci.* **2021**, *30*, 187–205. [CrossRef]
78. Gumus, R.; Gelen, S.U. Effects of dietary thyme and rosemary essential oils on performance parameters with lipid oxidation, water activity, pH, colour and microbial quality of breast and drumstick meats in broiler chickens. *Arch. Anim. Breed.* **2023**, *66*, 17–29. [CrossRef]
79. Ranucci, D.; Miraglia, D.; Trabalza-Marinucci, M.; Acuti, G.; Codini, M.; Ceccarini, M.R.; Forte, C.; Branciari1, R. Dietary effects of oregano (*Origanum vulgaris* L.) plant or sweet chestnut (*Castanea sativa* Mill.) wood extracts on microbiological, characteristics and lipid oxidation of cooked ham during storage. *Int. J. Food Sci.* **2015**, *4*, 5497. [CrossRef]
80. Bedford, A.; Gong, J. Implications of butyrate and its derivatives for gut health and animal production. *Anim. Nutr.* **2018**, *4*, 151–159. [CrossRef]
81. Peng, S.; Zhu, J.; Liu, Z.; Hu, B.; Wang, M.; Pu, S. Prediction of Ammonia Concentration in a Pig House Based on Machine Learning Models and Environmental Parameters. *Animals* **2022**, *13*, 165. [CrossRef]
82. Sístková, M.; Bartoš, P. Effects of phytogenic feed additives on growth performance and on ammonia and greenhouse gases emissions in growing-finishing pigs. *J. Anim. Feed Sci.* **2016**, *212*, 143–148. [CrossRef]
83. Kutlu, T. The Effects of Feed Additive, Selenium Source, Polyphenol, Lysine Level, and Sow Lactation Feeder Type on Pigs. Ph.D. Thesis, Kansas State University, Manhattan, KS, USA, 2023.
84. Walia, K.; Argüello, H.; Lynch, H.; Leonard, F.C.; Grant, J.; Yearsley, D.; Kelly, S.; Duffy, G.; Gardiner, G.E.; Lawlor, P.G. Effect of strategic administration of an encapsulated blend of formic acid, citric acid, and essential oils on Salmonella carriage, seroprevalence, and growth of finishing pigs. *Prev. Vet. Med.* **2017**, *137*, 28–35. [CrossRef] [PubMed]
85. Yang, C.; Zhang, L.; Cao, G.; Feng, J.; Yue, M.; Xu, Y.; Dai, B.; Han, Q.; Guo, X. Effects of dietary supplementation with essential oils and organic acids on the growth performance, immune system, fecal volatile fatty acids, and microflora community in weaned piglets. *J. Anim. Sci.* **2019**, *97*, 133–143. [CrossRef] [PubMed]

animals

MDPI

Review

Importance of Selected Nutrients and Additives in the Feed of Pregnant Sows for the Survival of Newborn Piglets

Paloma Islas-Fabila [1], Patricia Roldán-Santiago [2,*], Luis Alberto de la Cruz-Cruz [3,4], Ofelia Limón-Morales [5], Anna Dutro-Aceves [3], Héctor Orozco-Gregorio [4] and Herlinda Bonilla-Jaime [5,*]

1 Programa de Doctorado en Ciencias Biológicas y de la Salud, Universidad Autónoma Metropolitana, Unidad Iztapalapa, Mexico City 09340, Mexico; paloma_islas@hotmail.com
2 Departamento de Reproducción, Facultad de Medicina Veterinaria y Zootecnia, Universidad Nacional Autónoma de México, Avenida Universidad, Mexico City 04510, Mexico
3 Escuela de Medicina Veterinaria y Zootecnia, Universidad del Valle de México-Coyoacán, Calzada de Tlalpan, Mexico City 04910, Mexico; ladelacruzcc@gmail.com (L.A.d.l.C.-C.); annadutro@gmail.com (A.D.-A.)
4 Departamento de Producción Agrícola y Animal, Universidad Autónoma Metropolitana, Unidad Xochimilco, Calzada del Hueso 1100, Coapa, Villa Quietud, Coyoacán, Mexico City 04960, Mexico; gohector72@yahoo.com.mx
5 Departamento de Biología de la Reproducción, Universidad Autónoma Metropolitana, Unidad Iztapalapa, Mexico City 09340, Mexico; ofelia.limon@yahoo.com
* Correspondence: patriciaroldan@fmvz.unam.mx (P.R.-S.); bjh@xanum.uam.mx (H.B.-J.); Tel.: +52-01555622-5676 (ext. 45281) (P.R.-S.)

Simple Summary: According to the National Research Council (NRC), during gestation, sows have higher nutritional requirements to meet their needs and those of their fetuses. Therefore, an optimal feeding strategy is essential. Despite the importance of nutrition during gestation, the impact of supplementing the diets of gestating sows with foods rich in fatty acids, protein, amino acids, and dietary fiber on their offspring has not been thoroughly explored, so empirical evidence is scarce. The objective of this review is to evaluate the effect of gestating sows' nutrition on the survival and postnatal growth of neonate piglets. Sixty percent of the publications reviewed discussed the effect of supplementing diets with one or two of these nutrients, indicating the importance of the topic. Better overall postnatal survival and growth was found to be associated with supplementation with these nutrients during gestation. The studies mainly evaluated the effect of amino acids and fiber, likely because the former are the primary source of protein for the fetus, while the latter exerts an effect on the immune system. Additional research is needed to support these findings.

Abstract: This systematic review analyzed the effect of selected nutrients and additives in the feed of pregnant sows on the survival of newborn piglets. We analyzed 720 peer-reviewed publications in English in PubMed® and Web of Science®, dated July 2023 to January 2024, related to the effect of dietary supplementation with fatty acids and various percentages of protein, amino acids, and/or sources of dietary fiber on the offspring of gestating sows. While several papers evaluated the effect of nutrition on gestating sows, only a few delved into the distinct feeding strategies required at each stage of gestation to meet the NRC's nutritional requirements for maternal tissue gain and postnatal neonatal survival and growth. This body of research suggests that as gestation progresses the sow's nutritional requirements increase, as the NRC established, to satisfy their own metabolic needs and those of their fetuses. Additional research is needed to determine an optimal feeding strategy.

Keywords: nutrition; pregnant sows; fatty acid; protein; amino acids; fiber

1. Introduction

According to the NRC [1], sows have higher nutritional requirements during gestation to meet their metabolic needs and those of their fetuses [2]. The demand for nutrients increases throughout gestation because sows undergo several significant changes, including fetal growth, mammary growth, and colostrum production [3,4]. Therefore, inadequate maternal nutrition in relation to the increased requirements established by the NRC [1] to maintain the highest number of fetuses in utero can result in delayed fetal growth, reduced litter uniformity, low birthweights, and a higher number of stillborn piglets [3]. According to Kim et al. [4], during the first 70 days of gestation, fetuses present limited growth, so sows need an increase of only 0.25 g of protein/day. After 70 days, however, they require a significant increase (19-fold) of 4.63 g protein/day due to the growth of the placenta and the heart, liver, and intestines of the fetuses [4–6]. The feeding of gestating sows is generally classified into 3 stages: (1) early gestation (days 1–28), when they normally receive 2.0 kg/day of feed (depending on body condition); (2) mid-gestation (days 29–84), when feed intake should be increased by 0.15–0.20 kg/day to meet the energy required to maintain the sow and ensure adequate body weight gain [7]; and (3) late gestation (days 85–115), when the focus shifts to fetal and mammary growth, and feed intake is usually increased by 0.3–0.5 kg/day [3,7]. As gestation progresses, optimizing nutrition becomes a key factor that can lead to greater total litter weight at birth (15.06 vs. 14.36 kg), increased weight at weaning (5.37 vs. 5.20 kg), and higher individual birthweights (1.48 vs. 1.44 kg). Likewise, it can help sows produce more piglets per litter (+0.35) and more live piglets per litter (+0.34) [3,8]. A possible explanation of why optimal alimentation improves reproductive performance is that the dam's nutritional status affects circulating progesterone that can modify endometrial development and secretory activity, and impact the composition of the allantoic fluids that carry nutrients to the fetuses [3]. Since alimentation during gestation plays an extremely important role in fetal growth and development, and in the survival and postnatal growth of neonates, several studies have evaluated feeding strategies during gestation to determine their consequences for fetal growth and development. The aim of this review is to analyze the effect of selected nutrients and additives in the feed for pregnant sows on the survival of newborn piglets.

2. Materials and Methods

This systematic review was written following the Preferred Reporting Items for Systematic Reviews and Meta-Analyses (PRISMA) [9] (Figure 1).

Exclusion Criteria

Duplicate records and records marked as ineligible by automation tools were eliminated. Studies over a time period from 2013 to 2023 were prioritized; studies that had titles and subjective information and studies that did not provide sufficient statistical information were eliminated.

Information Sources, Search, and Selection

We analyzed 720 peer-reviewed publications in English in PubMed® and Web of Science® dated July 2023 to January 2024, related to the effect of dietary supplementation with fatty acids and various percentages of protein, amino acids, and/or sources of dietary fiber on the offspring of gestating sows. The search terms were (gestating sows* OR primiparous sows* OR multiparous sows*) AND (Newborn piglets* OR neonate porcine*) AND (dietary* OR supplementation* OR additives* OR feed* OR nutritional strategies*) AND (fatty acid*) AND (protein* OR percentages of protein*) AND (amino acids* OR basic amino acids* OR several neutral amino acids*) AND (dietary fiber*). In addition, the use of * in terms allows for broadening the search results. The search terms were used in PubMed® and Web of Science®: title, abstract and keyword (TITLE-ABS-KEY) parts of documents. EndNoteTM 20 software was used to analyze the results found in the databases.

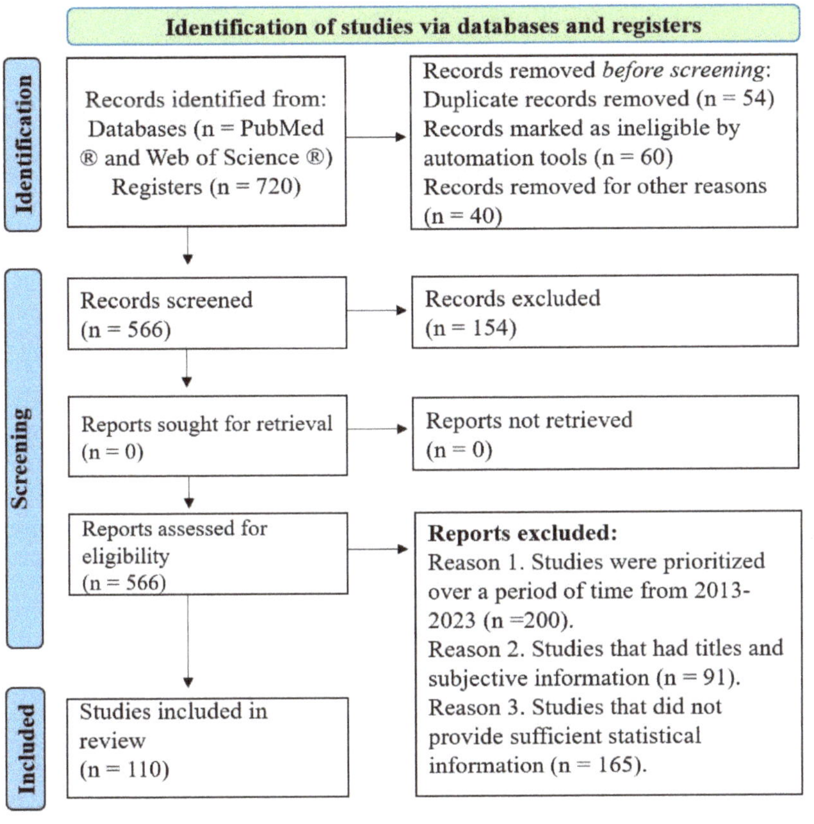

Figure 1. The search protocol and the resulting inclusions and exclusion. Adapted from Page et al. [9].

3. Results and Discussion

3.1. Effect of Feeding throughout Gestational Periods on Offspring

During early gestation (days 1–28) the goals of providing sows with adequate nutrition are to ensure the maximum number of quality embryos and replenish the body reserves lost during previous lactations, weaning, and services [7]. In cases where sows lose considerable body reserves and exhibit poor body condition, it may be beneficial to increase the amount of feed provided during early gestation to maintain the correct metabolic and endocrine status that is vital for the development and survival of embryos and fetuses [10]. Observations show that increasing the amount of feed from 2.5 to 3.25 kg/day during early gestation in sows that present low body weight can increase litter size from 13.2 to 15.2 piglets [11].

According to Blavi et al. [12], the recommended values of standardized ileal digestible (SID) Lys and total Lys/kg of feed with a feed energy content of 12.12 MJ ME/kg are as follows: a SID Lys (g)/ME (Mcal) ratio of 1.6 is enough to satisfy the recommendations of hyperprolific sows (12–14 total piglets born), except the young animals at the end of the gestation period (85–114 days) and the multiparous highly hyperprolific (>14 total piglets born; leaner animals) ones for the period 0–85 days. The 1.9 ratio satisfies the needs of the hyperprolific gilts at the end of gestation and the highly hyperprolific throughout the first two thirds of gestation (0–85 days). The ratio should be increased to 2.3 to satisfy the requirements of highly hyperprolific sows during the last third of gestation. The recommendations for the other AA should be considered using the "Ideal Protein" concept reported in most nutrient requirement systems for swine, applied according to the recommendations of SID Lys [12].

Mid-gestation (days 29–84): during this period, sows need to increase energy inputs by 2–3 MJ/day for body maintenance and ensure adequate weight gain. This means increasing feed intake by 0.15–0.20 kg/day [10,13]. It should be noted that maternal nutrition is especially important during this stage because the formation of primary muscle fibers occurs (days 20–50 of gestation), which then serves as a template for the myogenesis of secondary muscle fibers from days 54 to 90 of gestation [10,14]. The distribution and number of muscle fibers can significantly impact the birthweight, growth, and performance of neonates, with especially large effects on daily weight gain and lean mass composition [14].

Late gestation (days 85–115): this is the period of greatest growth of the fetuses and mammary tissue, so the sow's nutritional needs to increase substantially [10]. McPherson et al. [5] determined that fetuses require 0.25 g/d of protein up to day 69 of gestation, but that this figure increases to 4.63 g/d in late gestation. As a result, it is estimated that in the last 10 days of gestation, each fetus may gain up to one-third of its final birthweight. Consequently, meeting nutrient demands during late gestation is important to maximize fetal growth [10]. On the other hand, according to Feyera and Theil [15], from d 105 to 115 of gestation, sows require approximately 39 MJ/d of metabolizable energy and, by far, the highest proportion (79%) is lost as heat (30.5 MJ/d) [16]. The remaining 21% is retained in reproductive tissues or products, such as colostrum (3.6 MJ/d), fetal growth (2.6 MJ/d), mammary growth (1.6 MJ/d), and uterus, placenta, fluids, and membranes (0.3 MJ/d). Heat loss is required for maintenance purposes and colostrum production, fetal growth, mammary growth and growth of uterine tissues [17]. Studies show that incorporating fat into the diet in the last 10–14 days of gestation can increase the survival of swine neonates by raising birthweights from 1.36 to 1.45 kg [13]. A study by Chen et al. [18] found that feeding sows at this stage of gestation diets that do not meet the recommended energy requirements can cause piglets to exhibit lighter body weight at birth and weaning. This can reduce the weight of the small intestine and affect the height–depth relation of the crypts of the ileum and jejunum villi. Thus, it is clear that when the maternal energy requirements stipulated by the NRC [1] are not met during gestation, nutrient utilization in the growing fetus becomes selective and the development of the gastrointestinal tract may be compromised [10].

However, it is important to understand, as well, that overfeeding during late gestation can cause birth problems, such as prolonged parturition [19], likely due to a lower uterine muscle tone, especially in older sows [20]. This condition can increase the number of stillborn piglets. Moreover, even though the fetus is fully formed in late gestation, the functionality of its organ systems may be limited until a few weeks or days before birth [21], so this final maturation period is potentially an ideal time for nutrition to influence piglet quality. Other studies stress (for example, Gonçalves et al. [20] or Mallmann et al. [22]) that inadequate nutrition during gestation results in loss of body condition that may be more pronounced in this stage because, after maintenance, fetal growth is the main reason for using available nutrients. If the supply of nutrients is inadequate, the sow will mobilize body tissues to provide the nutrients needed to maintain fetal growth [23], but the manifestations of maternal tissue mobilization include reduced maternal body weight (BW) and backfat, the latter an important factor that affects the amount of colostrum, an essential element for piglet growth. Amdi et al. [24] determined that when sows have high backfat (19 mm) during gestation their piglets have higher birthweight (1.49 ± 0.02 kg; $p < 0.05$), while sows that lose backfat in late gestation tend to have low colostrum production ($R^2 = 0.12$, $p = 0.032$) [25] and 25% less milk fat on day 21 of lactation [24]. Clearly, as gestation progresses, the nutritional NRC's nutritional requirements [1] for both the dam and her fetuses change. Undoubtedly, nutrition during gestation is a main factor associated with the welfare of sows, and one that exerts a significant effect on fetal and postnatal survival, since the dam nourishes her fetuses through the placenta, and neonates through the mammary transfer system. Both delivery systems depend on appropriate nutritional intake by the dam [26]. Researchers have developed and evaluated several nutritional plans to determine the effect of supplementing the diet of gestating sows on their progeny.

3.2. Diets Focused on Fatty Acid Supplementation

Administering diets rich in fatty acids (fish oil and flax seed oil, among others) has been assessed, reporting that long-chain polyunsaturated fatty acids (LC-PUFA) like 20: 5n-3 (EPA), 22: 6n-3 (DHA), 22: 5n-3 (DPA), and 20: 4n-6 (ARA) participate in regulating the immune system, blood coagulation, neurotransmitters, cholesterol metabolism, and the structure of membrane phospholipids in the brain and retina [27], thus exerting important effects on fetal growth and development [28]. In contrast, a deficit of fatty acids during gestation can lead to an irreversible impairment of cognitive and/or physiological functions [28–30]. Similarly, administering diets rich in polyunsaturated fatty acids ensures sufficient energy intake for swine neonates [31], as these feeding regimens during late gestation and lactation increase the fat content of milk and, depending on the source of fat, modulate fatty acid profiles [32], thus favoring the development of the immune system in the early life stages of piglets. This suggests that piglets can benefit from polyunsaturated fatty acid supplementation in the sow's diet during gestation in two ways: (1) prenatally, when developing embryos have access to docosahexaenoic acid (DHA); and (2) postpartum, when litters consume colostrum and milk with high concentrations of eicosapentaenoic acid (EPA) and DHA [33]. The work by Liu et al. [34], mentioned earlier, found that supplementing the sow's diet with 2.5% conjugated linoleic acid (CLA) from day 85 of gestation causes a significant increase ($p < 0.05$) in colostral immunoglobulin G (IgG) concentrations, and can increase litter weight linearly ($p < 0.05$) and litter size at day 21 of lactation, while causing a linear ($p = 0.01$) decrease in pre-weaning mortality. One mechanism through which LC-PUFA may influence the growth and survival of neonates is by enhancing the immune system. Immunoglobulin G (IgG) in the colostrum is the main source of antibodies that stimulate the passive immune system of newborn piglets [27]. Another study in this field administered salmon oil at 1.79% to pregnant multiparous sows from day 105 of gestation to day 14 of lactation. The results showed that this rate of supplementation increased the total proportion of omega-3 fatty acids in the colostrum ($p < 0.001$), milk ($p < 0.01$), piglet plasma ($p < 0.01$), and adipose ($p < 0.001$), liver ($p < 0.001$), and muscle tissues ($p < 0.001$) [35]. This is important because omega-3 fatty acids play a key role in fetal brain and cognitive development, since the phospholipids that make up the cell membranes of the nervous system contain large amounts of this type of fatty acid [36].

In a separate study, 5% of hemp seeds (*Cannabis sativa*) were added to a diet from day 108 of gestation to weaning (4 weeks post-farrowing). These researchers observed that piglet body weight was influenced by this dietary treatment of the sows during the first week of lactation (2.66 vs. 3.18 kg; $p = 0.03$) [27]. Similarly, a study in which pregnant sows were supplemented with fish oil (16.5–100 g/kg) found that this type of diet reduced pre-weaning mortality rates and increased postnatal piglet growth ($p < 0.05$), mainly due to a lower number of crushed piglets and an increase in suckling behavior by the neonates [37,38]. Studies by Laws et al. [39,40] showed that supplementation with monounsaturated fatty acids (MUFA) (18:1 n-9) (100 g/kg extra) during the first semester of gestation can reduce the incidence of low-birthweight piglets (<1 kg), perhaps due to enhanced placental growth [7].

3.3. Diets Focused on Protein Supplementation

Proteins play roles in the body structure, nutrition, enzymatic catalysts, and molecular transport and defense of organisms, among other aspects [3]. Adding protein to the diet of gestating sows alters their metabolic characteristics [41] and impacts postnatal development and the performance of their offspring [42]. Therefore, the availability, quantity, and quality of dietary protein participate significantly in the developing embryos and fetuses [41]. For example, a 50% lower supply of dietary protein (compared to the required amount of 121 g/kg) during gestation can reduce birthweight, impair myogenesis, and restrict muscle growth potential and postnatal lean growth in neonates [43,44]. Studies also show that excessive or inadequate protein intake by the gestating sow results in a higher percentage of neonates with intrauterine growth retardation (IUGR), characterized by low birthweight (1.1 kg or less) [43]. Although newborn piglets with IUGR may experience catch-up growth after birth, they show increased adipose tissue deposition, hypercholesterolemia, reduced locomotor activity, and high mortality [45,46].

In this field, Campos et al. [3] pointed out that protein deficiency in the maternal diet (only 0.5% protein) decreases concentrations of basic amino acids (arginine, lysine, ornithine) and several neutral amino acids (alanine, glutamine, glycine, branched chain amino acids, proline, serine, taurine, threonine) in the placenta and endometrium by 16–30%, with possible negative impacts on birthweight and litter uniformity [3,47]. Similarly, a study that explored the effect of administering diets supplemented with low percentages of protein (9%) during pregnancy and lactation showed that those feeding regimens during gestation cause a significant decrease in the body weight of weaned piglets and in the daily weight gain of weaning piglets ($p < 0.05$) (Table 1) [47–52].

Table 1. Effect of diets supplemented with different percentages of protein on the reproductive performance of gestating sows.

Animals	Experimental Design	Results	Conclusion	References
59 multiparous sows (Yorkshire × Landrace) with bodyweights (BW) around 241.67 ± 8.86 kg	(1) two levels of dietary metabolizable energy (ME) density were provided (13.40 or 13.82 MJ/kg); 2) three dietary protein levels were provided from day 35 of gestation (crude protein = CP: 10.5, 12, 13.5%).	Backfat thickness in lactating sows decreased and the % of CP increased ($p = 0.03$). CP level in the diet had a negative effect on colostrum quality: % casein: $p = 0.03$; % protein: $p = 0.04$; % lactose: $p = 0.06$; total solids: $p = 0.03$; lean solids: $p = 0.03$, all decreased.	Backfat thickness and colostrum quality decreased as the CP level in the diet increased (10.5–13.5%). A diet for gestating sows containing 13.82 MJ/kg ME and 10.5% CP may improve reproductive and litter performance, and colostrum quality.	[48]
47 Landrace × Yorkshire gilts; 190 kg at insemination	Gilts were fed one of two iso-energetic compound feeds in which dietary protein differed by 12%.	Milk yield peaked at 12.9 kg/d around day 20. Sows fed the low protein compound feed had a lower milk yield from day 20 to day 40 than controls (8.0 vs. 10.3 kg/d; $p < 0.05$).	Sows on a low-protein diet had decreased milk production at the end of lactation, so it seems problematic to reduce the protein content of the lactation diet in winter, especially in gilts with limited gastric capacity.	[49]
32 Landrace × Yorkshire sows at parity two, with a similar mean bodyweight of 164.2 kg	One diet had normal crude protein (CP = 13.3%), the other had a low CP of 10.1%.	Sows receiving low levels of CP had higher serum levels of Lys and Thr and lower levels of Try, Ile, and Val ($p < 0.05$), but no effect on the serum levels of other AAs were found ($p > 0.05$).	Maternal protein deposition was decreased by a low CP.	[50]
72 F1 multiparous sows (Yorkshire × Landrace) with an average BW of 218.69 kg	Experimental diets with different CP levels, as follows: (i) CP11 containing 11% CP; (ii) CP12, 12% CP; (iii) CP13, 13% CP; (iv) CP14, 14% CP; (v) CP15, 15% CP; and (vi) CP16, 16% CP.	Increasing CP levels in the gestation diet caused a significant increase in creatinine at days 35 and 110 of gestation (linear, $p = 0.01$; linear, $p = 0.01$).	Reducing dietary CP levels from 16 to 11% in a gestation diet did not have detrimental effects on the sows' body condition or piglet performance.	[51]

Studies by Jia et al. [52], meanwhile, observed that neonates from sows that ingested low protein levels (6%) exhibited low body and liver weight ($p < 0.05$). This finding is consistent with earlier reports which observed that maternal protein deprivation during gestation reduces the birthweight of piglets and decreases liver, brain, heart, and kidney weights [52–54]. Finally, stunted growth of piglets from gestating sows supplemented with low protein diets has been associated with low serum glucose levels and high liver glycogen at birth. Increased hepatic glycogen content suggests an adaptive mechanism of energy conservation through reduced glycolysis, or increased gluconeogenesis, in response to fetal nutritional deficiency [52]. However, it is also important to clarify that not only diets with low percentages of protein have negative effects on the sow and her progeny, but that regimens with high percentages (14–18%) can also have adverse effects on fetuses and dams, since a secondary consequence of high levels of ammonia and possibly other metabolites in plasma from a high-protein diet can create a toxic environment for both [55,56] and may reduce the size and number of skeletal muscle fibers in newborns. Regarding gestating females, an unbalanced protein intake has consequences on body weight and fat gain [44,46]. Studies emphasize that these changes can affect mammary gland development, lactation, and the interval between weaning and estrus [44]. Rehfeldt et al. [43] found that diets for gestating sows with high protein concentrations (30%) produce piglets with intrauterine growth restriction and low thymus and bone weights. As an organ of the immune system, a reduced thymus gland may be related to decreased immune function [56]. For all these reasons, the results of several studies indicate the importance of providing adequate protein levels in the diets of gestating sows.

3.4. *Diets Focused on Amino Acid Supplementation*

The amino acid (AA) family is important in gestating sows because it regulates metabolic pathways that play fundamental roles in improving the health, survival, growth, development, lactation, and reproduction of organisms, while also participating in placental angiogenesis and placental, embryonic, and fetal development in most mammals [3,57]. According to Wu et al. [57], AAs are classified as essential or non-essential [58]. Essential AAs are defined as those of which the carbon skeletons cannot be synthesized, or are inadequately synthesized by the body relative to its needs and, hence, must be provided through the diet to meet NRC requirements [1,59,60]. Non-essential AAs are ones that the body can synthesize in adequate amounts. There is also a category of conditionally essential AAs, which the body can normally synthesize in adequate amounts, but may have to be added to the diet to meet NRC requirements [1] under conditions where utilization rates exceed synthesis rates [59]. Because mammary and fetal tissue growth is rapid during late gestation, AA needs are greater, especially in primiparous sows (Figure 2A–D). Muscle tissue growth must be taken into account among the reproductive needs of younger sows since fetal and mammary gland growth in these females occurs mainly during this stage [6], when the fetus is estimated to gain 17.5 g of protein in body tissues from day 0 to 70 (0.25 g protein/day) and 203.7 g from day 70 to 114 (4.63 g protein/day). If a sow has 14 fetuses, protein gain is 3.5 g/d and 64.8 g/d for early and late gestation, a difference of 61.3 g/d, or an 18.5-fold increase in the rate of tissue protein gain between early and late gestation [5,6]. Thus, as gestation progresses, the composition of AA varies as a consequence of changes in the rate and composition of tissue gain for fetal growth [6]. For example, observations of tryptophan (Trp) show that the supplementation of this AA during gestation reduces fetal mortality while promoting viability [61,62], perhaps because this AA serves as a precursor of several molecules (serotonin, melatonin kynurenic acid, etc. [63]) and scavenging free radicals, reactive nitrogen species, and chlorine, so it limits cellular damage [62]. During gestation, glutamine also plays a role in the immune response, and in fetal growth, survival, and metabolic regulation [64], while leucine is a key element for the development of blastocysts that can proceed to embryonic implantation [65,66].

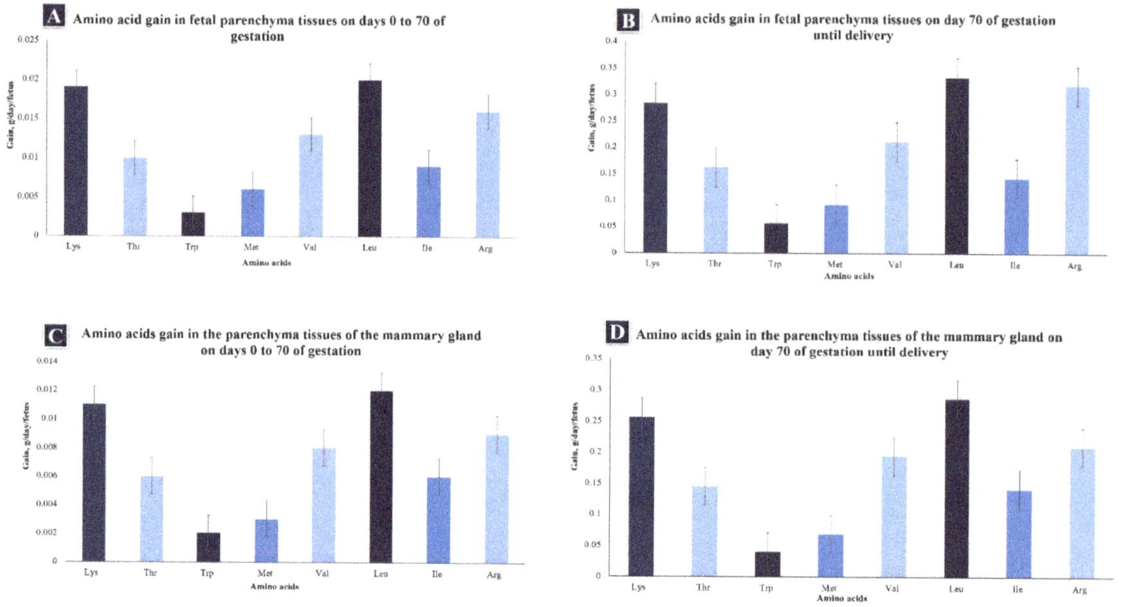

Figure 2. (**A**–**D**) Amino acid gain in fetal parenchymal and mammary gland tissues of gilts from day 0 of gestation to parturition. Lys = lysine, Thr = threonine, Trp = tryptophan, Met = methionine, Val = valine, Leu = leucine, Ile = isoleucine, Arg = arginine (data from Wu [59]).

One AA widely used in dietary supplementation of pregnant sows is arginine (Arg), an essential element for fetal growth. Arginine exists in especially high levels during early gestation in porcine allantoic fluid (4–6 mM) and can be metabolized to nitric oxide (NO) in animal cells. Nitric oxide functions as an endothelium-derived relaxing factor, neurotransmitter, and modulator of immune responses [2,44], indicating its significant metabolic role in fetal development, as a decrease in this AA during gestation can reduce NO synthesis and may alter angiogenesis and placental and endometrial tissue growth. A low Arg concentration in the placenta can reduce placental–fetal blood flow and the supply of nutrients from the dam to the fetus, ultimately delaying fetal growth [2]. Studies by Che et al. [67] demonstrated that sows fed a diet supplemented with Arg (1% L-arginine HCl up to day 114 of gestation) produced more live piglets (+1.6 piglets, $p < 0.05$) and higher total litter weight (+1.6–2.1 kg, $p < 0.05$), indicating that Arg has an important effect on fetal growth during late gestation [2,67]. Another study showed that Arg may be physiologically necessary during late gestation by playing a critical role in increasing placental angiogenesis, since extreme vascular growth and proliferation in the placenta and increased placental angiogenesis in that period allow for sufficient placental (or umbilical) blood flow and nutrient transfer for rapid fetal growth [67,68].

A recent study by Nuntapaitoon et al. [69] showed that supplementation with 0.5% L-arginine HCl reduced the proportion of piglets with restricted growth and increased the proportion of neonates with birthweights > 1.35 kg ($p < 0.05$). It is likely that these dietary effects are due to an increase in placental blood flow that allowed for more nutrients and oxygen to be transferred across the placenta. High birthweight in piglets is advantageous for survival rates during lactation [69]. In this regard, Mateo et al. [70] found that on day 7 of lactation, milk yield and the concentrations of most AAs in mother's milk were higher in response to Arg supplementation during lactation compared to a control group ($p < 0.05$). This increase could be due to the positive effect of L-arginine on vascularization, which improves blood flow and makes nutrient absorption by the lactating mammary gland more efficient [44]. Moreover, supplementation with 0.4% of Arg from day 30 to day 114 of gestation has been shown to cause a variation of 24% in the birthweights of liveborn piglets and 22% in the proportion of live-born piglets with birthweights of 1.29 kg ($p < 0.05$) [56,59].

Lysine is considered the primary limiting AA in diets for lactating sows based on cereals and soy [71]. According to Hojgaard et al. [72], estimates of the dietary requirements of digestible standardized ileal Lys for lactating sows vary widely, from 27 to 70 g/d, or from 4.9 to 10.5 g/kg, because factors like genetics, age, litter size, appetite, and feed ingredients can all affect the dietary requirement for Lys [71,72]. Liu et al. [73] affirmed that primiparous sows eat 10–15% less than multiparous ones, so the percentage of SID Lys consumed during lactation must be increased in the former compared to the latter. Administering adequate supplies of Lys during lactation allows those sows to maximize milk production and their reproductive yield [73]. It is also important to emphasize that a low ingestion of lysine during lactation can have a negative effect on the sow's metabolic balance, secretion of reproductive hormones, and the interval between weaning and estrus while, in contrast, a high Lys consumption can improve metabolic states in sows and increase total litter weight at birth and the weight of piglets at weaning [74]. Given these findings, diverse studies have evaluated the effect of dietary supplementation with various percentages of lysine on milk production and reproductive performance in primiparous and multiparous sows (Table 2).

Table 2. Effect of diets supplemented with different percentages of lysine on the reproductive performance of primiparous and multiparous sows.

Animals	Experimental Design	Results	Conclusion	References
48 gilts (Yorkshire × Landrace), with an initial bodyweight of 168.1 ± 9.71 kg at day 35 of gestation	The first factor was metabolizable energy levels in the diet (3.265 or 3.365 kcal of ME/kg); the second was dietary lysine levels: gestation—0.55, 0.65, 0.75, and 0.85%. (total methionine 0.23%; threonine, 0.48%; tryptophan, 0.13%); Lactation—0.70, 0.85, 1, 1.15% (total methionine 0.25%; threonine 0.62%; tryptophan 0.18%).	The sows fed 3.365 kcal of EM/kg showed a tendency to present greater weight gain ($p = 0.07$). Their piglets had a higher tendency to exhibit greater weight at day 21 of lactation ($p = 0.08$). Plasma urine nitrogen levels increased as the level of lysine in the diet was raised on day 110 of gestation ($p = 0.03$).	Supplementation with lysine at 0.75% during gestation, and at 1% for lactation, with 3.365 kcal of EM/kg in primiparous sows can improve their performance and the growth of their offspring.	[74]
33 Yorkshire × Landrace multiparous sows (parities 2 and 3)	From day 90 to 110 of gestation, the sows were divided into 2 groups: control ($n = 17$) (2.6 kg/d that provided 14.8 g/d of SID Lys), and digestible ileal Lys (SID) at 40% ($n = 16$) (20.8 g/d of SID Lys, administered in soy flour).	The diets did not cause changes in the body fat or body weight of the sows in the late gestation period ($p > 0.10$), or changes in mammary tissue ($p > 0.10$).	Ingesting Lys above levels currently recommended by the NRC did not improve mammary development, so it is not necessary to use two phases to provide additional Lys protein to sows during this period.	[75]
On day 42 of gestation, 200 multiparous sows (parity = 5.1 ± 2.0) were randomly allocated to five dietary treatment groups	Experimental diets: (1) SID Lys for the mid-gestation period (days 42 to 76-indispensable amino acids). (2) SID Lys for the late gestation period (days 77 to 103-indispensable amino acids.	Total liveborn piglets per litter increased lineally and quadratically ($p < 0.001$) as the level of SID Lys in the diet increased.	Supplementation with SID Lys at 11.1 and 16.1 g/d (1.36 and 1.79 g/Mcal of metabolizable energy; 0.4% and 0.58%) for the middle and final periods of gestation, can increase the number of liveborn piglets per litter.	[76]
105 sows in their initial reproductive cycle (1.4 ± 0.5) were assigned randomly to either a precision program (PF; $n = 50$) or a control group (CON; $n = 55$)	The PF sows received two isocaloric diets (2518 kcal/kg NE; 0.80% and 0.20% standardized ileal digestible Lys [SID], respectively), while the CON sows received a diet with 0.56% SID Lys.	The sows that received the PF program had greater weight gain from day 38 to 72 (614 vs. 518 g/d; $p < 0.05$) and from day 73 to 108 (719 vs. 618 g/d; $p = 0.063$) of gestation, with greater gain in back thickness between days 63 and 110 (0.7 vs. −1.1 ± 1.6 mm; $p < 0.05$).	Using programs that include daily requirements of energy and Lys in sows during gestation helped reduce the use of feed during lactation without affecting their reproductive performance.	[77]

For example, the work by Liu et al. [73] that evaluated the effect of dietary supplementation with 0.84, 0.94, 1.04, and 1.14% of standardized ileal digestibility (SID) Lys, balanced with Met, Thr, Trp, and Val in primiparous Yorkshire sows demonstrated that lactation increased lineally with higher levels of Lys in the diet ($p = 0.04$). These authors further showed that survival rates improve when primiparous sows are fed diets that contain 1.14% of Lys during lactation ($p = 0.04$), accompanied by higher weight ($p = 0.04$) and greater weight gain in piglets at day 21 ($p = 0.03$) [73]. Another study in this area evaluated increases in the ingestion of SID Lys (11.0, 13.5, 16.0, 18.5 g/d) in primiparous and multiparous sows (22.2 and 24.3 MJ of net energy per day, respectively), showing that the percentage of liveborn piglets increased ($p = 0.01$) with a greater ingestion of SID Lys by the multiparous sows, though not the primiparous ones, due to a treatment–group interaction ($p = 0.04$) related to the percentage of stillborn piglets. These results suggest that 11 g/day of SID Lys is an adequate level for both primiparous and multiparous gestating sows, as it provided 18.5 g/day and reduced ($p = 0.01$) the rate of fetal death by 2.3 percent [78]. All these findings highlight the importance of optimal maternal nutrition during gestation and providing the correct amount of nutrients to meet the metabolic needs of sows and their fetuses.

3.5. Diets Focused on Dietary Fiber Supplementation

Dietary fiber, generally defined as the non-digestible portion of plant-derived feeds, is a key component of many swine diets. Though not fully digested, dietary fiber can impact a wide range of physiological processes, either directly (e.g., by intestinal filling) or indirectly, by producing physiologically active gases and by-products after fermentation in the colon [79]. In addition, because dietary fibers are not hydrolyzed by endogenous enzymes in the small intestine, they are available for bacterial fermentation in the large intestine, where they can significantly modify the microbial balance with positive or negative impacts on animal health, depending on the source of the dietary fiber and the physiological state of the pig [79,80]. Adding cellulose to a standard swine diet, for example, can increase ileal populations of bifidobacteria and enterobacteria in growing pigs [80], while a selective inclusion of fiber can alter the gut microbiome and promote gut health [81]. This occurs primarily because intestinal bacteria hydrolyze dietary fibers and metabolize their constituent sugars, leading to the production of ATP [79,82]. The main end-products of microbial fermentation of dietary fiber are short-chain fatty acids (acetate, propionate, N-butyrate) and gases (carbon dioxide, hydrogen sulfide, methane) [82] (Figure 3: Part 1 and 2). Short-chain fatty acids released by anaerobic bacteria after fiber fermentation contribute to the animal's energy supply and regulate both the growth of intestinal epithelial cells and the composition of the intestinal flora [79].

Due to the foregoing, dietary fiber (DF) supplementation in the diet of gestating sows has beneficial effects on their gut microbiota, immunity, welfare, colostrum production, physiology, and overall performance [83]. This measure can also improve farrowing and increase colostrum production [84], as the amount of feed allowable is often reduced just before farrowing, and glucose is only net-absorbed during the first 4/6 h post-feeding [84,85]. Therefore, adding dietary fiber may prove beneficial in stabilizing the post-absorption energy status in sows [86]. Another study that examined supplementation with high dietary fiber during late gestation (2 weeks before the probable date of parturition) found that this reduced the proportion of stillborn piglets from 8.8 to 6.6% ($p < 0.001$), lowered the proportion of deaths due to low vitality ($p < 0.001$; 2.8 vs. 1.5% in the control and treatment groups, respectively), and decreased the prevalence of piglet diarrhea ($p = 0.004$; 0.7 vs. 0.3% in the control group) [84]. A study by Zhuo et al. [87] that compared multiparous sows throughout gestation (30, 60, 90 days and at birth) in relation to the supplementation of a control diet and two diets with different sources of dietary fiber, e.g., guar gum and cellulose, showed that the total number of piglets born tended to be affected by the type of diet ($p = 0.071$), as this value increased linearly in the treatments that provided sources of DF ($p < 0.01$) [84]. In addition, colostral lipid content was linearly affected by DF levels

($p < 0.05$), as the sows fed DF exhibited higher colostral lipid concentrations. Despite these benefits, however, excessive dietary fiber supplementation can decrease the birth and weaning weights of neonates. A study that evaluated four diets with different proportions of soluble fiber (diet 1: 89%; diet 2: 5.19%; diet 3: 9.12%; diet 4: 12.8%) found that litter weight at birth and average piglet weight at weaning were significantly higher in the litters of the sows that were supplemented with 3.89 and 5.19% of soluble fiber ($p = 0.010$), as both average litter weight (diet 1: 1.40 ± 0.05 kg; diet 2: 1.32 ± 0.05 kg; diet 3: 1.33 ± 0.04 kg; diet 4: 1.28 ± 0.12 kg) and piglet weight at weaning (diet 1: 7.88 ± 0.12 kg; diet 2: 7.46 ± 0.15 kg; diet 3: 6.80 ± 0.18 kg; diet 4: 6.95 ± 0.18 kg) decreased linearly as the proportion of soluble fiber increased ($p < 0.05$) [88].

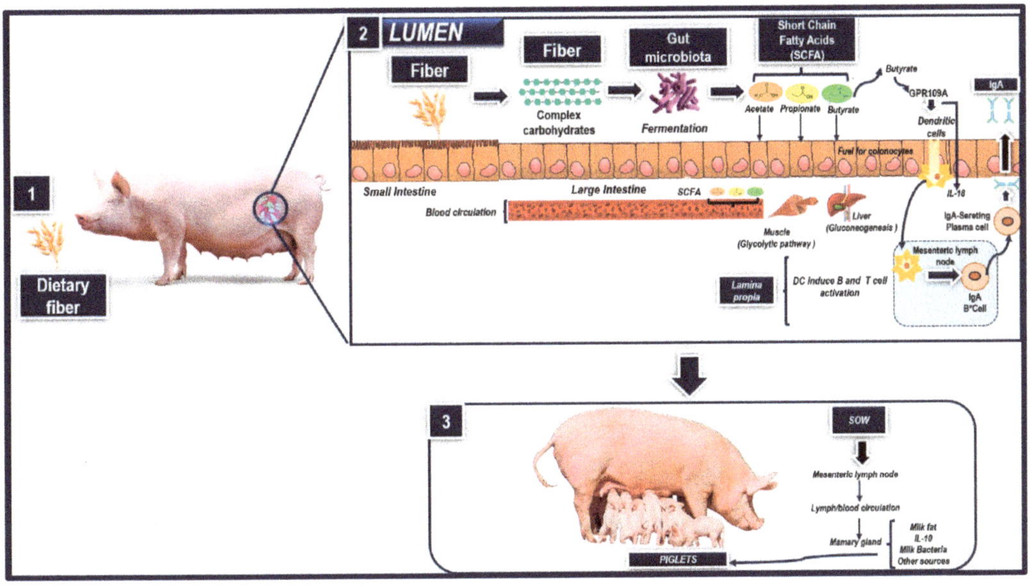

Figure 3. Effect of fiber on the composition of milk and colostrum. (1) Administering dietary fiber during gestation impacts colostrum quality) (2) When DF is fermented by intestinal microorganisms, it produces short-chain fatty acids (SCFAs) and (3) The sow's mammary glands use SCFAs as precursors of milk fat synthesis which piglets consume.

It is important to note that DF supplementation impacts the composition of colostrum and milk since during gestation, a large amount of nutrients absorbed by the intestine are transported to the mammary glands through the bloodstream, so the level of nutrients in the diet affects milk and colostrum synthesis and composition in sows [89], because when DF is fermented by intestinal microorganisms, it produces short-chain fatty acids (SCFAs). The sow's mammary glands use SCFAs as precursors of milk fat synthesis, so a high percentage of DF in their diet during gestation increases milk fat content in their colostrum (Figure 3: Part 3) [84,90]. Other observations show that administering dietary fiber during gestation affects the secretion of immunoglobulins (Ig) and interleukins (IL). In this case, a study by Shang et al. [83] compared diets with two distinct fiber sources, i.e., sugar beet pulp (SBP) and wheat bran (WB), and a control diet (corn and soybean meal), in multiparous sows at day 85 of gestation. They found that the sows fed diets supplemented with SBP had higher ($p < 0.05$) levels of immunoglobulin A (IgA) and interleukin-10 (IL-10) in their colostrum compared to the sows that received the control diet. Regarding milk composition, higher levels of IgA ($p < 0.05$) and IL-10 ($p < 0.05$) were found in the sows fed diets rich in dietary fiber (SBP and WB) compared to controls. Therefore, including dietary fiber is essential for promoting the intestinal health of piglets [83,91]. Both the third trimester of pregnancy and the lactation period are characterized by an important outflow of intestinal immune cells toward the mammary glands, since intestinal microbes can be transferred to the lymph nodes [92,93]. As a result, studies have found that certain bacteria in the intestine coexist in maternal peripheral blood and milk [94]. The dominant bacteria in sow milk are *Ruminococcaceae*, *Streptococcus*, *Lactobacillus*, and *Clostridiales*, which exist mainly in the intestine of animals [89,95]. *Ruminococcaceae* and *Lactobacillus* are especially important bacterial genera for the fermentation of dietary fiber in the intestine, so the composition of DF in the diet of pregnant sows can alter the microbial composition of her milk and increase the intestinal health of neonates [89].

Likewise, it is important to understand that dietary fiber is fermented and used by intestinal microbes to produce various metabolites, including short-chain fatty acids (SCFAs) (Figure 3: Part 1 and 2), especially acetate, propionate, and butyrate [96], which form an important substrate of gluconeogenesis and participate in regulating metabolism, immunity, and cell proliferation in sows [89,97]. Mainly short-chain fatty acids are transported to peripheral circulation through the portal vein, where they act on the liver and peripheral tissues. One proposal holds that they act as signal molecules that regulate various physiological activities of the host, such as immunity and the expression of antioxidant enzymes and inflammatory and proinflammatory factors [89]. In the mammary gland, SCFAs are transferred through the bloodstream and used as substrates for synthesizing milk fat (Figure 3: Part 3). In addition, some immune factors (e.g., IL-10) from the intestine are transported to the mammary gland through the intestinal lymphatic circulation system [89,96].

Another possibility is that intestinal microbes enter the lymph nodes through dendritic cells (DCs) in the intestinal lamina propria because DCs can phagocyte some bacterial antigens that penetrate the mucous layer, and then present them in the mesenteric lymph nodes. DCs induce B cells to differentiate into plasma cells that secrete large amounts of immunoglobulin A (IgA) in the intestinal cavity [96,97]. In addition, it has been observed that butyrate, a product of bacterial fermentation of dietary fiber, induces the expression of IL-18 in intestinal epithelial cells (IECs) through signaling via the 109 A receptor coupled to protein G (GPR109A). Likewise, butyric acid can promote the anti-inflammatory properties of colonic dendritic cells through GPR109A signaling, allowing them to induce the differentiation of Treg cells and IL-10-producing CD4+ T cells [96]. Finally, the beneficial effects of the interaction between dietary fiber and gut microbes are transmitted from sow to piglet through lactation [89].

3.6. Nutritional Strategies for Primiparous and Multiparous Sows

The value of post-insemination alimentary strategies in primiparous and multiparous sows has long been debated, mainly due to their potential impact on reproductive performance [22,98]. The main observations of researchers are that sows with lower parity are more sensitive to changes in body weight during lactation, and more prone to suffering later reproductive alterations. Therefore, sows use the early and mid-gestation periods to recover their body reserves [1,22]. A study that assessed the effect of increasing feed levels (1.8, 2.5, 3.2 kg/d) on early gestation in primiparous (PO1) and second-time sows (PO2) showed that those with a lower parity (PO1, PO2) and adequate body condition exhibited increases in body weight, body condition scores, and backfat ($p < 0.001$) as feed consumption increased from 1.8 to 3.2 kg/d during the first month of gestation [22,99]. However, this increase can have a negative effect on the total number of piglets born (PO1: 13.4, PO2: 15.1) likely by reducing systemic progesterone and, as a result, embryo survival [22]. Moreover, greater feed consumption (3.2 kg/d) during early gestation did not increase the number of piglets born, above all in the primiparous sows, perhaps because they required more growth to reach their target weight during their first pregnancy, since if food intake is insufficient, their bodies may prioritize growth instead of reproduction [22,100]. In this vein, a study that increased feed consumption (1.8, 2.3, 2.8, and 3.3 kg/d) in the final stage of gestation in primiparous sows found that increasing feed from day 120 of gestation to parturition increased maternal bodyweight (200.7–213.1 kg; $p < 0.001$) and the number of stillborn piglets (3.4–5.5%), but reduced feed consumption (4.2–3.9 kg; lineal: $p = 0.001$) and colostrum yield (3.6–3.2 kg) during lactation [101]. These findings concur with the work of Pedersen et al. [102], who pointed out that primiparous sows have lower colostrum yields than multiparous dams (5.2 vs. 7.1 kg; $p < 0.01$). In fact, observations at 24 h postpartum found differences in colostrum composition, as the primiparous sows had greater body fat with lower protein and casein levels than their multiparous counterparts. This suggests that the former utilize dietary nutrients differently than the latter [102,103]. Koketsu et al. [104] affirmed that primiparous sows have lower reproductive performance, likely because their endocrine system is still immature, and they have a lower capacity to consume feed. These results contrast those from Gianluppi et al.'s [105] work, which did not find greater reproductive performance or follicular size in weaned primiparous and multiparous sows that were fed 4.3 kg/day of gestation (58.78 MJ of EM and 26.66 g SID Lys) or a lactation diet (61.66 MJ of EM and 51.60 g SID Lys). These authors recommended feeding weaned sows 2.7 kg/day of a gestation diet (36.91 MJ of EM and 16.74 g SID Lys).

It Is important to emphasize, as well, that in general practice, all sows receive the same standard gestation diet and only the level of alimentation can be adjusted [106]. In most cases, the nutritional contribution of the AAs and minerals is limited, principally at the end of gestation in sows with lower parity, while excesses were observed in earlier stages and with greater frequency in higher-parity sows [107]. As a result, the development of precision feeding (PF) is providing new opportunities to identify, in real time, the factors that affect the nutritional needs of sows [106]. In this regard, models and decision support systems (DSSs) have been developed based on nutritional models that predict individual daily requirements, considering the characteristics of the animals, phases of physiological development, and housing conditions [108]. According to Gaillard et al. [106], the PF strategy makes it possible to reduce the cost of feeding by 3.6% per sow during gestation, and reduces the ingestion of nitrogen and phosphorus by 11.0 and 13.8%, respectively, and excretions by 16.7 and 15.4%, respectively, compared to sows fed under conventional alimentary systems. This suggests that the PF of gestating sows plays an important role in satisfying their requirements for amino acids while, at the same time, lowering feed costs and supplies and excretions of nitrogen and phosphorus [106]. That study, however, analyzed only one gestation cycle per sow, so it would be interesting and valuable to move beyond that to perform follow-up on the effects of PF on the performance of sows and feeding costs over various consecutive cycles, combined with the use of PF during lactation [109]. Indeed, one recent study demonstrated that applying PF during lactation also reduced feeding costs and lysine ingestion [110].

4. Conclusions

Given that according to the NRC, as gestation progresses, sows require greater nutritional requirements to satisfy their own metabolic needs and those of their fetuses, maternal nutrition that is inadequate for maintaining the maximal number of fetuses in the uterus can delay fetal growth, reduce litter uniformity and birthweights, and increase the number of stillborn piglets. For these reasons, numerous studies have evaluated the effect of dietary supplementation with rich amounts of fatty acids and various percentages of proteins, amino acids, and/or dietary fiber on pregnant sows and their progeny. Results show that providing dietary protein is essential for key functions such as structural roles, nutrition, enzymatic catalysts, molecular transport, and the organism's defense system, among others. It is clear, then, that supplying protein in the diet of pregnant sows alters their metabolic characteristics. During pregnancy, amino acids regulate essential metabolic pathways for improving the health, survival, growth, development, lactation, and reproduction of organisms, while dietary fiber is crucial for the development of the microbiota and immune system of newborn piglets. Therefore, optimal feeding strategies designed for each stage of gestation must be sufficiently flexible to meet the NRC's nutritional requirements and support both maternal tissue gain and fetal development.

Author Contributions: Conceptualization, P.I.-F., L.A.d.l.C.-C., H.O.-G., P.R.-S. and H.B.-J.; methodology: P.I.-F.; data curation: P.I.-F., L.A.d.l.C.-C., H.O.-G., P.R.-S., A.D.-A. and H.B.-J.; writing—original draft preparation: P.I.-F., P.R.-S. and H.B.-J.; writing, reviewing, editing: L.A.d.l.C.-C., O.L.-M., A.D.-A., H.O.-G., P.R.-S. and H.B.-J.; visualization: P.I.-F., L.A.d.l.C.-C., O.L.-M., H.O.-G., P.R.-S. and H.B.-J.; supervision: L.A.d.l.C.-C., O.L.-M., H.O.-G., P.R.-S. and H.B.-J. All authors have read and agreed to the published version of the manuscript.

Funding: This research received no external funding.

Institutional Review Board Statement: Not applicable.

Informed Consent Statement: Not applicable.

Data Availability Statement: The data presented in this study are available in this paper.

Acknowledgments: Paloma Islas Fabila is enrolled in the Doctoral Program in Biological Sciences and Health at the Universidad Autónoma Metropolitana (Mexico), supported by Consejo Nacional de Humanidades, Ciencias y Tecnologías (Mexico, CONAHCYT) Fellowship 795465. De la Cruz-Cruz LA., Limón-Morales O., Orozco-Gregorio H., Roldán-Santiago P., and Bonilla-Jaime H. are members of the SNI (Mexico).

Conflicts of Interest: The authors declare no conflicts of interest.

References

1. National Research Council (NRC). *Nutrient Requirements of Swine*, 11th ed.; The National Academies Press: Washington, DC, USA, 2012; 400p. Available online: https://nap.nationalacademies.org/catalog/13298/nutrient-requirements-of-swine-eleventh-revised-edition (accessed on 4 January 2024).
2. Kim, S.W.; Wu, G.; Baker, D.H. Amino acid nutrition of breeding sows during gestation and lactation. *Pig News Inf.* **2005**, *26*, 89N–99N. Available online: https://pubmed.ncbi.nlm.nih.gov/19098235/ (accessed on 20 November 2023).
3. Campos, P.; Silva, B.; Donzele, J.; Oliveira, R.; Knol, E. Effects of sow nutrition during gestation on within-litter birthweight variation: A review. *Animal* **2012**, *6*, 797–806. [CrossRef] [PubMed]
4. Kim, S.W.; Weaver, A.C.; Shen, Y.B.; Zhao, Y. Improving efficiency of sow productivity: Nutrition and health. *J. Anim. Sci. Biotechno.* **2013**, *4*, 26. Available online: http://www.jasbsci.com/content/4/1/26 (accessed on 20 November 2023). [CrossRef]
5. McPherson, R.L.; Ji, F.; Wu, G.; Blanton, J.R.; Kim, S.W. Growth and compositional changes of fetal tissues in pigs. *J. Anim. Sci.* **2004**, *82*, 2534–2540. [CrossRef]
6. Kim, S.W.; Hurley, W.L.; Wu, G.; Ji, F. Ideal amino acid balance for sows during gestation and lactation. *J. Anim. Sci.* **2009**, *87*, E123–E132. [CrossRef]
7. De Vos, M.; Che, L.; Huygelen, V.; Willemen, S.; Michiels, J.; Van Cruchten, S.; Van Ginneken, C. Nutritional interventions to prevent and rear low-birthweight piglets. *J. Anim. Physiol. Anim. Nutr.* **2014**, *98*, 609–619. [CrossRef]
8. Ferguson, E.M.; Slevin, J.; Edwards, S.A.; Hunter, M.G.; Ashworth, C.J. Effect of alterations in the quantity and composition of the pre-mating diet on embryo survival and foetal growth in the pig. *Anim. Reprod. Sci.* **2006**, *96*, 89–103. [CrossRef] [PubMed]
9. Page, M.J.; Moher, D.; Bossuyt, P.M.; Boutron, I.; Hoffmann, T.C.; Mulrow, C.D.; McKenzie, J.E. PRISMA 2020 explanation and elaboration: Updated guidance and exemplars for reporting systematic reviews. *BMJ* **2021**, *372*, n160. [CrossRef] [PubMed]
10. Hansen, L. Precision Feeding Gestating Sows: Effects on Offspring Growth, Immune System Robustness, and Carcass and Meat Quality. Ph.D. Dissertation, University of Guelph, Guelph, ON, Canada, January 2021. Available online: https://atrium.lib.uoguelph.ca/server/api/core/bitstreams/94051345-0e96-4d0599ccc16b69c7b9a8/content (accessed on 20 November 2023).
11. Hoving, L.L.; Soede, N.M.; Graat, E.A.; Feitsma, H.; Kemp, B. Reproductive performance of second parity sows: Relations with subsequent reproduction. *Liv. Sci.* **2011**, *140*, 124–130. [CrossRef]
12. Blavi, L.; Solà-Oriol, D.; Llonch, P.; López-Vergé, S.; Martín-Orúe, S.M.; Pérez, J.F. Management and feeding strategies in early life to increase piglet performance and welfare around weaning: A review. *Animals* **2021**, *11*, 302. [CrossRef]
13. Ampaire, A. Effects of Phase Feeding during Gestation on Gilt Performance, Offspring Quality and Robustness. Ph.D. Dissertation, South Dakota State University, Brookings, SD, USA, 2017. Available online: https://openprairie.sdstate.edu/cgi/viewcontent.cgi?article=2143&context=etd (accessed on 20 November 2023).
14. Beaulieu, A.D.; Aalhus, J.L.; Williams, N.H.; Patience, J.F. Impact of piglet birth weight, birth order, and litter size on subsequent growth performance, carcass quality, muscle composition, and eating quality of pork. *J. Anim. Sci.* **2010**, *88*, 2767–2778. [CrossRef]
15. Feyera, T.; Theil, P.K. Energy and lysine requirements and balances of sows during transition and lactation: A factorial approach. *Livest. Sci.* **2017**, *201*, 50–57. [CrossRef]
16. Theil, P.K.; Jørgensen, H.; Jakobsen, K. Energy and protein metabolism in pregnant sows fed two levels of dietary protein. *J. Anim. Physiol. Anim. Nutr.* **2002**, *86*, 399–413. [CrossRef]
17. Theil, P.K.; Krogh, U.; Bruun, T.S.; Feyera, T. Feeding the modern sow to sustain high productivity. *Mol. Reprod. Dev.* **2023**, *90*, 517–532. [CrossRef] [PubMed]

18. Chen, Y.; Mou, D.; Hu, L.; Zhen, J.; Che, L.; Fang, Z.; Wu, D. Effects of maternal low-energy diet during gestation on intestinal morphology, disaccharidase activity, and immune response to lipopolysaccharide challenge in pig offspring. *Nutrients* **2017**, *9*, 1115. [CrossRef]
19. Ji, F.; Hurley, W.L.; Kim, S.W. Characterization of mammary gland development in pregnant gilts. *J. Anim. Sci.* **2006**, *84*, 579–587. [CrossRef]
20. Gonçalves, M.; Gourley, K.; Dritz, S.; Tokach, M.; Bello, N.; DeRouchey, J.; Goodband, R. Effects of amino acids and energy intake during late gestation of high-performing gilts and sows on litter and reproductive performance under commercial conditions. *J. Anim. Sci.* **2016**, *94*, 1993–2003. [CrossRef]
21. Herpin, P.; Damon, M.; Le Dividich, J. Development of thermoregulation and neonatal survival in pigs. *Livest. Prod. Sci.* **2002**, *78*, 25–45. [CrossRef]
22. Mallmann, A.L.; Oliveira, G.S.; Ulguim, R.R.; Mellagi, A.P.G.; Bernardi, M.L.; Orlando, U.A.; Bortolozzo, F.P. Impact of feed intake in early gestation on maternal growth and litter size according to body reserves at weaning of young parity sows. *J. Anim. Sci.* **2020**, *98*, skaa075. [CrossRef] [PubMed]
23. Yang, Y.X.; Heo, S.; Jin, Z.; Yun, J.H.; Choi, J.Y.; Yoon, S.Y.; Chae, B. Effects of lysine intake during late gestation and lactation on blood metabolites, hormones, milk composition and reproductive performance in primiparous and multiparous sows. *Anim. Reprod. Sci.* **2009**, *112*, 199–214. [CrossRef]
24. Amdi, C.; Giblin, L.; Ryan, T.; Stickland, N.C.; Lawlor, P.G. Maternal backfat depth in gestating sows has a greater influence on offspring growth and carcass lean yield than maternal feed allocation during gestation. *Animal* **2014**, *8*, 236–244. [CrossRef]
25. Decaluwé, R.; Maes, D.; Declerck, I.; Cools, A.; Wuyts, B.; De Smet, S.; Janssens, G. Changes in back fat thickness during late gestation predict colostrum yield in sows. *Animal* **2013**, *7*, 1999–2007. [CrossRef] [PubMed]
26. Molina-Montes, M.E.; Martín-Islán, Á.P. Ácidos grasos esenciales. Omega-3 y Omega-6. *Offarm* **2010**, *29*, 66–72. Available online: https://www.elsevier.es/es-revista-offarm-4-pdf-X0212047X10475135 (accessed on 20 November 2023).
27. Vodolazska, D.; Lauridsen, C. Effects of dietary hemp seed oil to sows on fatty acid profiles, nutritional and immune status of piglets. *J. Anim. Sci. Biotechnol.* **2020**, *11*, 28. [CrossRef]
28. Luo, W.; Xu, X.; Luo, Z.; Yao, J.; Zhang, J.; Xu, W.; Xu, J. Effect of fish oil supplementation in sow diet during late gestation and lactation period on litter characteristics, milk composition and fatty acid profile of sows and their offspring. *Ital. J. Anim. Sci.* **2020**, *19*, 8–17. [CrossRef]
29. Catalan, J.; Moriguchi, T.; Slotnick, B.; Murthy, M.; Greiner, R.S.; Salem, N. Cognitive deficits in docosahexaenoic acid-deficient rats. *Behav. Neurosci.* **2002**, *116*, 1022–1031. [CrossRef] [PubMed]
30. Yaktine, A.L.; Nesheim, M.C. (Eds.) *Seafood Choices: Balancing Benefits and Risks*; National Academies Press: Washington, DC, USA, 2007. Available online: https://nap.nationalacademies.org/catalog/11762/seafood-choices-balancing-benefits-and-risks (accessed on 20 November 2023).
31. Sampels, S.; Pickova, J.; Högberg, A.; Neil, M. Fatty acid transfer from sow to piglet differs for different polyunsaturated fatty acids (PUFA). *Physiol. Res.* **2011**, *60*, 113. [CrossRef] [PubMed]
32. Quesnel, H.; Farmer, C. Nutritional and endocrine control of colostrogenesis in swine. *Animal* **2019**, *13*, s26–s34. [CrossRef]
33. Smit, M.N.; Spencer, J.D.; Almeida, F.; Patterson, J.L.; Chiarini-Garcia, H.; Dyck, M.K.; Foxcroft, G.R. Consequences of a low litter birthweight phenotype for postnatal lean growth performance and neonatal testicular morphology in the pig. *Animal* **2013**, *7*, 1681. [CrossRef]
34. Liu, H.; Wu, F.; Bai, L.L.; Chen, Y.F.; Lai, C.H.; Ren, L.Q.; Wang, F.L. Effect of dietary conjugated linoleic acid supplementation during late gestation on colostrum yield, fatty acid composition, and IgG concentrations in primiparous sows. *Can. J. Anim. Sci.* **2018**, *98*, 732–740. [CrossRef]
35. Lavery, A.; Lawlor, P.G.; Miller, H.M.; Magowan, E. The effect of dietary oil type and energy intake in lactating sows on the fatty acid profile of colostrum and milk, and piglet growth to weaning. *Animals* **2019**, *9*, 1092. [CrossRef] [PubMed]
36. Rodríguez-Cruz, M.R.; Tovar, A.R.; del Prado, M.; Torres, N. Molecular mechanisms of action and health benefits of polyunsaturated fatty acids. *Rev. Investig. Clin.* **2005**, *57*, 457–472. Available online: https://pubmed.ncbi.nlm.nih.gov/16187707/ (accessed on 20 November 2023).
37. Rooke, J.A.; Sinclair, A.G.; Edwards, S.A.; Cordoba, R.; Pkiyach, S.; Penny, P.C.; Horgan, G.W. The effect of feeding salmon oil to sows throughout pregnancy on pre-weaning mortality of piglets. *Anim. Sci.* **2001**, *73*, 489–500. [CrossRef]
38. Rooke, J.A.; Sinclair, A.G.; Edwards, S.A. Feeding tuna oil to the sow at different times during pregnancy has different effects on piglet long-chain polyunsaturated fatty acid composition at birth and subsequent growth. *Brit. J. Nutr.* **2001**, *86*, 21–30. [CrossRef]
39. Laws, J.; Litten, J.C.; Laws, A.; Lean, I.J.; Dodds, P.F.; Clarke, L. Effect of type and timing of oil supplements to sows during pregnancy on the growth performance and endocrine profile of low and normal birthweight offspring. *Brit. J. Nutr.* **2009**, *101*, 240–249. [CrossRef]
40. Laws, J.; Amusquivar, E.; Laws, A.; Herrera, E.; Lean, I.J.; Dodds, P.F.; Clarke, L. Supplementation of sow diets with oil during gestation: Sow body condition, milk yield and milk composition. *Livest. Sci.* **2009**, *123*, 88–96. [CrossRef]
41. Metges, C.C.; Lang, I.S.; Hennig, U.; Brüssow, K.P.; Kanitz, E.; Tuchscherer, M.; Otten, W. Intrauterine growth retarded progeny of pregnant sows fed high protein: Low carbohydrate diet is related to metabolic energy deficit. *PLoS ONE* **2012**, *7*, e31390. [CrossRef]

42. Mickiewicz, M.; Zabielski, R.; Grenier, B.; Le Normand, L.; Savary, G.; Holst, J.J.; Guilloteau, P. Structural and functional development of small intestine in intrauterine growth retarded porcine offspring born to gilts fed diets with differing protein ratios throughout pregnancy. *J. Physiol. Pharmacol.* **2012**, *63*, 225–239. Available online: https://www.jpp.krakow.pl/journal/archive/06_12/pdf/225_06_12_article.pdf (accessed on 20 November 2023). [PubMed]
43. Rehfeldt, C.; Lefaucheur, L.; Block, J.; Stabenow, B.; Pfuhl, R.; Otten, W.; Kalbe, C. Limited and excess protein intake of pregnant gilts differently affects body composition and cellularity of skeletal muscle and subcutaneous adipose tissue of newborn and weanling piglets. *Eur. J. Nutr.* **2012**, *51*, 151–165. [CrossRef] [PubMed]
44. Bee, G. Gestational strategies affecting sow reproduction and piglet birth weight. In Proceedings of the 11th International Symposium Modern Trends in Livestock Production, Belgrade, Serbia, 11–13 October 2017; pp. 121–145. Available online: https://scholar.google.com.mx/citations?view_op=view_citation&hl=es&user=aoSSgDQAAAAJ&citation_for_view=aoSSgDQAAAAJ:u-x6o8ySG0sC (accessed on 20 November 2023).
45. Kucia, M.; Langhammer, M.; Görs, S.; Albrecht, E.; Hammon, H.M.; Nürnberg, G.; Metges, C.C. High-protein diet during gestation and lactation affects mammary gland mRNA abundance, milk composition and pre-weaning litter growth in mice. *Animal* **2011**, *5*, 268–277. [CrossRef] [PubMed]
46. Rehfeldt, C.; Lang, I.S.; Görs, S.; Hennig, U.; Kalbe, C.; Stabenow, B.; Otten, W. Limited and excess dietary protein during gestation affects growth and compositional traits in gilts and impairs offspring fetal growth. *J. Anim. Sci.* **2011**, *89*, 329–341. [CrossRef] [PubMed]
47. Jia, Y.; Gao, G.; Song, H.; Cai, D.; Yang, X.; Zhao, R. Low-protein diet fed to crossbred sows during pregnancy and lactation enhances myostatin gene expression through epigenetic regulation in skeletal muscle of weaning piglets. *Eur. J. Nutr.* **2015**, *55*, 1307–1314. [CrossRef] [PubMed]
48. Fang, L.H.; Jin, Y.H.; Jeong, J.H.; Hong, J.S.; Chung, W.L.; Kim, Y.Y. Effects of dietary energy and protein levels on reproductive performance in gestating sows and growth of their progeny. *JAST* **2019**, *61*, 154. [CrossRef]
49. Eskildsen, M.; Krogh, U.; Sørensen, M.T.; Kongsted, A.G.; Theil, P.K. Effect of reduced dietary protein level on energy metabolism, sow body composition and metabolites in plasma, milk and urine from gestating and lactating organic sows during temperate winter conditions. *Livest. Sci.* **2020**, *240*, 104088. [CrossRef]
50. Yang, M.; Hua, L.; Mao, Z.; Lin, Y.; Xu, S.; Li, J.; Huang, J. Effects of dietary fiber, crude protein level, and gestation stage on the nitrogen utilization of multiparous gestating sows. *Animals* **2022**, *12*, 1543. [CrossRef] [PubMed]
51. Kim, H.; Jin, X.; Kim, C.; Pan, N.; Kim, Y.Y. Effects of different levels of dietary crude protein on the physiological response, reproductive performance, blood profiles, milk composition and odor emission in gestating sows. *Anim. Biosci.* **2023**, *36*, 1263–1273. [CrossRef] [PubMed]
52. Jia, Y.; Cong, R.; Li, R.; Yang, X.; Sun, Q.; Parvizi, N.; Zhao, R. Maternal low-protein diet induces gender-dependent changes in epigenetic regulation of the glucose-6-phosphatase gene in newborn piglet liver. *J. Nutr.* **2012**, *142*, 1659–1665. [CrossRef] [PubMed]
53. Pond, W.G.; Maurer, R.R.; Klindt, J. Fetal organ response to maternal protein deprivation during pregnancy in swine. *J. Nutr.* **1991**, *121*, 504–509. [CrossRef]
54. Wu, G.; Ott, T.; Bazer, F.W. Maternal dietary protein deficiency decreases amino acid concentrations in fetal plasma and allantoic fluid of pigs. *J. Nutr.* **1998**, *128*, 894–902. [CrossRef]
55. Wu, G.; Bazer, F.W.; Satterfield, M.C.; Li, X.; Wang, X.; Johnson, G.A.; Wu, Z. Impacts of arginine nutrition on embryonic and fetal development in mammals. *Amino. Acids.* **2013**, *45*, 241–256. [CrossRef]
56. Herring, C.M.; Bazer, F.W.; Johnson, G.A.; Wu, G. Impacts of maternal dietary protein intake on fetal survival, growth, and development. *Exp. Biol. Med.* **2018**, *243*, 525–533. [CrossRef]
57. Wu, G. Amino acids: Metabolism, functions, and nutrition. *Amino Acids.* **2009**, *37*, 1–17. [CrossRef] [PubMed]
58. Elango, R.; Ball, R.O. Protein and amino acid requirements during pregnancy. *Adv. Nutr.* **2016**, *7*, 839S–844S. [CrossRef] [PubMed]
59. Wu, G. Functional amino acids in growth, reproduction, and health. *Adv. Nutr.* **2010**, *1*, 31–37. [CrossRef] [PubMed]
60. Chen, Y.; Azad, M.B.; Gibson, S.B. Superoxide is the major reactive oxygen species regulating autophagy. *Cell Death Diff.* **2009**, *16*, 1040–1052. [CrossRef]
61. Xu, K.; Liu, H.; Bai, M.; Gao, J.; Wu, X.; Yin, Y. Redox properties of tryptophan metabolism and the concept of tryptophan use in pregnancy. *Int. J. Mol. Sci.* **2017**, *18*, 1595. [CrossRef] [PubMed]
62. Hussain, T.; Tan, B.; Murtaza, G.; Metwally, E.; Yang, H.; Kalhoro, M.S.; Yin, Y. Role of dietary amino acids and nutrient sensing system in pregnancy associated disorders. *Front. Pharmacol.* **2020**, *11*, 586979. [CrossRef] [PubMed]
63. Liu, G.; Chen, S.; Zhong, J.; Teng, K.; Yin, Y. Crosstalk between tryptophan metabolism and cardiovascular disease, mechanisms, and therapeutic implications. *Oxid. Med. Cell Longev.* **2017**, *e2017*, 1602074. [CrossRef]
64. Rogero, M.M.; Borelli, P.; Fock, R.A.; Borges, M.C.; Vinolo, M.A.R.; Curi, R.; Tirapegui, J. Effects of glutamine on the nuclear factor-kappaB signaling pathway of murine peritoneal macrophages. *Amino Acids.* **2010**, *39*, 435–441. [CrossRef]
65. Teodoro, G.F.R.; Vianna, D.; Torres-Leal, F.L.; Pantaleao, L.C.; Matos-Neto, E.M.; Donato, J., Jr.; Tirapegui, J. Leucine is essential for attenuating fetal growth restriction caused by a protein-restricted diet in rats. *J. Nutr.* **2012**, *142*, 924–930. [CrossRef]
66. Zhang, S.; Zeng, X.; Ren, M.; Mao, X.; Qiao, S. Novel metabolic and physiological functions of branched chain amino acids: A review. *J. Anim. Sci. Biotechnol.* **2017**, *8*, 10. [CrossRef]

67. Che, L.; Yang, P.; Fang, Z.; Lin, Y.; Wu, D. Effects of dietary arginine supplementation on reproductive performance and immunity of sows. *Czech J. Anim. Sci.* **2013**, *58*, 167–175. [CrossRef]
68. Vallet, J.L.; Miles, J.R.; Freking, B.A. Development of the pig placenta. *Soc. Reprod. Fertil. Suppl.* **2009**, *66*, 265–279. Available online: https://pubmed.ncbi.nlm.nih.gov/19848293/ (accessed on 21 November 2023). [CrossRef] [PubMed]
69. Nuntapaitoon, M.; Muns, R.; Theil, P.K.; Tummaruk, P. L-arginine supplementation in sow diet during late gestation decrease stillborn piglet, increase piglet birthweight and increase immunoglobulin G concentration in colostrum. *Theriogenology* **2018**, *121*, 27–34. [CrossRef]
70. Mateo, R.D.; Wu, G.; Moon, H.K.; Carroll, J.A.; Kim, S.W. Effects of dietary arginine supplementation during gestation and lactation on the performance of lactating primiparous sows and nursing piglets. *J. Anim. Sci.* **2008**, *86*, 827–835. [CrossRef] [PubMed]
71. Aymerich, P.; Soldevila, C.; Bonet, J.; Gasa, J.; Coma, J.; Solà-Oriol, D. Increasing dietary lysine impacts differently growth performance of growing pigs sorted by body weight. *Animals* **2020**, *10*, 1032. [CrossRef] [PubMed]
72. Hojgaard, C.K.; Bruun, T.S.; Theil, P.K. Optimal lysine in diets for high-yielding lactating sows. *J. Anim. Sci.* **2019**, *97*, 4268–4281. [CrossRef]
73. Liu, B.; Zhou, Y.; Xia, X.; Wang, C.; Wei, H.; Peng, J. Effects of dietary lysine levels on production performance and milk composition of high-producing sows during lactation. *Animals* **2020**, *10*, 1947. [CrossRef]
74. Hong, J.; Fang, L.H.; Kim, Y.Y. Effects of dietary energy and lysine levels on physiological responses, reproductive performance, blood profiles, and milk composition in primiparous sows. *JAST* **2020**, *62*, 334. [CrossRef]
75. Farmer, C.; Gillies, C.; Johannsen, J.C.; Hovey, R.C.; Huber, L.A. Dietary supplementation with lysine (protein) in late pregnancy does not enhance mammary development in multiparous sows. *J. Anim. Sci.* **2023**, *101*, skad385. [CrossRef]
76. Jo, H.; Kim, B.G. Standardized ileal digestible lysine requirement of pregnant sows under commercial conditions. *Anim. Biosci.* **2023**, *36*, 1880–1888. [CrossRef]
77. Stewart, V.; Buis, R.Q.; Christensen, B.; Hansen, L.L.; de Lange, C.F.; Mandell, I.B.; Huber, L.A. The effects of precisely meeting estimated daily energy and lysine requirements for gestating sows over three consecutive pregnancies on sow reproductive and lactation performance. *Transl. Anim. Sci.* **2021**, *5*, txab226. [CrossRef]
78. Thomas, L.L.; Herd, L.K.; Goodband, R.D.; Tokach, M.D.; Woodworth, J.C.; DeRouchey, J.M.; Jones, D.B. Effects of increasing standardized ileal digestible lysine during gestation on reproductive performance of gilts and sows. *Animal* **2021**, *15*, 100221. [CrossRef]
79. Jarrett, S.; Ashworth, C.J. The role of dietary fibre in pig production, with a particular emphasis on reproduction. *J. Anim. Sci. Biotechnol.* **2018**, *9*, 59. [CrossRef]
80. Owusu-Asiedu, A.; Patience, J.; Laarveld, B.; Van Kessel, A.; Simmins, P.; Zijlstra, R. Effects of guar gum and cellulose on digesta passage rate, ileal microbial populations, energy and protein digestibility, and performance of grower pigs. *J. Anim. Sci.* **2006**, *84*, 843–852. [CrossRef]
81. Lindberg, J.E. Fiber effects in nutrition and gut health in pigs. *J. Anim. Sci. Biotechnol.* **2014**, *5*, 15. [CrossRef]
82. Bindelle, J.; Leterme, P.; Buldgen, A. Nutritional and environmental consequences of dietary fibre in pig nutrition: A review. *Biotechnol. Agron. Soc. Environ.* **2008**, *12*, 69–80. Available online: https://popups.uliege.be/17804507/index.php?id=17245&file=1&pid=2179 (accessed on 21 November 2023).
83. Shang, Q.; Liu, H.; Liu, S.; He, T.; Piao, X. Effects of dietary fiber sources during late gestation and lactation on sow performance, milk quality, and intestinal health in piglets. *J. Anim. Sci.* **2019**, *97*, 4922–4933. [CrossRef] [PubMed]
84. Feyera, T.; Højgaard, C.K.; Vinther, J.; Bruun, T.S.; Theil, P.K. Dietary supplement rich in fiber fed to late gestating sows during transition reduces rate of stillborn piglets. *J. Anim. Sci.* **2017**, *95*, 5430–5438. [CrossRef] [PubMed]
85. Serena, A.; Jørgensen, H.; Bach Knudsen, K.E. Absorption of carbohydrate-derived nutrients in sows as influenced by types and contents of dietary fiber. *J. Anim. Sci.* **2009**, *87*, 136–147. [CrossRef] [PubMed]
86. Theil, P.K.; Lauridsen, C.; Quesnel, H. Neonatal piglet survival: Impact of sow nutrition around parturition on fetal glycogen deposition and production and composition of colostrum and transient milk. *Animal* **2014**, *8*, 1021–1030. [CrossRef] [PubMed]
87. Zhuo, Y.; Feng, B.; Xuan, Y.; Che, L.; Fang, Z.; Lin, Y.; Wu, D. Inclusion of purified dietary fiber during gestation improved the reproductive performance of sows. *J. Anim. Sci. Biotechnol.* **2020**, *11*, 47. [CrossRef]
88. Li, Y.; Zhang, L.; Liu, H.; Yang, Y.; He, J.; Cao, M.; Wu, D. Effects of the ratio of insoluble fiber to soluble fiber in gestation diets on sow performance and offspring intestinal development. *Animals* **2019**, *9*, 422. [CrossRef]
89. Tian, M.; Chen, J.; Liu, J.; Chen, F.; Guan, W.; Zhang, S. Dietary fiber and microbiota interaction regulates sow metabolism and reproductive performance. *Anim. Nutr.* **2020**, *6*, 397–403. [CrossRef]
90. Loisel, F.; Farmer, C.; Ramaekers, P.; Quesnel, H. Effects of high fiber intake during late pregnancy on sow physiology, colostrum production, and piglet performance. *J. Anim. Sci.* **2013**, *91*, 5269–5279. [CrossRef] [PubMed]
91. Lv, D.; Xiong, X.; Yang, H.; Wang, M.; He, Y.; Liu, Y.; Yin, Y. Effect of dietary soy oil, glucose, and glutamine on growth performance, amino acid profile, blood profile, immunity, and antioxidant capacity in weaned piglets. *Sci. China Life Sci.* **2018**, *61*, 1233–1242. [CrossRef] [PubMed]
92. Fernández, L.; Langa, S.; Martín, V.; Maldonado, A.; Jiménez, E.; Martín, R.; Rodríguez, J.M. The human milk microbiota: Origin and potential roles in health and disease. *Pharmacol. Res.* **2013**, *69*, 1–10. [CrossRef] [PubMed]

93. Rodríguez, J.M. The origin of human milk bacteria: Is there a bacterial entero-mammary pathway during late pregnancy and lactation? *Adv. Nutr.* **2014**, *5*, 779–784. [CrossRef] [PubMed]
94. Avershina, E.; Angell, I.L.; Simpson, M.; Storrø, O.; Øien, T.; Johnsen, R.; Rudi, K. Low maternal microbiota sharing across gut, breast milk and vagina, as revealed by 16S rRNA gene and reduced metagenomic sequencing. *Genes* **2018**, *9*, 231. [CrossRef] [PubMed]
95. Chen, W.; Mi, J.; Lv, N.; Gao, J.; Cheng, J.; Wu, R.; Liao, X. Lactation stage-dependency of the sow milk microbiota. *Front. Microbiol.* **2018**, *9*, 945. [CrossRef] [PubMed]
96. Peng, J.; Tang, Y.; Huang, Y. Gut health: The results of microbial and mucosal immune interactions in pigs. *Anim. Nutr.* **2021**, *7*, 282–294. [CrossRef] [PubMed]
97. Sun, C.; Song, R.; Zhou, J.; Jia, Y.; Lu, J. Fermented Bamboo Fiber Improves Productive Performance by Regulating Gut Microbiota and Inhibiting Chronic Inflammation of Sows and Piglets during Late Gestation and Lactation. *Microbiol. Spectr.* **2023**, *11*, e04084-22. [CrossRef]
98. Leal, D.F.; Muro, B.B.; Nichi, M.; Almond, G.W.; Viana, C.H.; Vioti, G.; Garbossa, C.A. Effects of post-insemination energy content of feed on embryonic survival in pigs: A systematic review. *Anim. Reprod. Sci.* **2019**, *205*, 70–77. [CrossRef] [PubMed]
99. Athorn, R.Z.; Stott, P.; Bouwman, E.G.; Chen, T.Y.; Kennaway, D.J.; Langendijk, P. Effect of feeding level on luteal function and progesterone concentration in the vena cava during early pregnancy in gilts. *Reprod. Fertil. Dev.* **2013**, *25*, 531–538. [CrossRef] [PubMed]
100. Hoving, L.L.; Soede, N.M.; Graat, E.A.M.; Feitsma, H.; Kemp, B. Effect of live weight development and reproduction in first parity on reproductive performance of second parity sows. *Anim. Reprod. Sci.* **2010**, *122*, 82–89. [CrossRef]
101. Mallmann, A.L.; Camilotti, E.; Fagundes, D.P.; Vier, C.E.; Mellagi, A.P.G.; Ulguim, R.R.; Bortolozzo, F.P. Impact of feed intake during late gestation on piglet birthweight and reproductive performance: A dose-response study performed in gilts. *J. Anim. Sci.* **2019**, *97*, 1262–1272. [CrossRef]
102. Pedersen, T.F.; Van Vliet, S.; Bruun, T.S.; Theil, P.K. Feeding sows during the transition period—Is a gestation diet, a simple transition diet, or a lactation diet the best choice? Transl. *Anim. Sci.* **2020**, *4*, 34–48. [CrossRef] [PubMed]
103. Feyera, T.; Zhou, P.; Nuntapaitoon, M.; Sørensen, K.U.; Krogh, U.; Bruun, T.S.; Theil, P.K. Mammary metabolism and colostrogenesis in sows during late gestation and the colostral period. *J. Anim. Sci.* **2019**, *97*, 231–245. [CrossRef]
104. Koketsu, Y.; Iida, R. Sow housing associated with reproductive performance in breeding herds. *Mol. Reprod. Dev.* **2017**, *84*, 979–986. [CrossRef]
105. Gianluppi, R.D.F.; Lucca, M.S.; Mellagi, A.P.G.; Bernardi, M.L.; Orlando, U.A.D.; Ulguim, R.R.; Bortolozzo, F.P. Effects of different amounts and type of diet during weaning-to-estrus interval on reproductive performance of primiparous and multiparous sows. *Animal* **2020**, *14*, 1906–1915. [CrossRef]
106. Gaillard, C.; Quiniou, N.; Gauthier, R.; Cloutier, L.; Dourmad, J.Y. Evaluation of a decision support system for precision feeding of gestating sows. *J. Anim. Sci.* **2020**, *98*, skaa255. [CrossRef]
107. Gaillard, C.; Gauthier, R.; Cloutier, L.; Dourmad, J.Y. Exploration of individual variability to better predict the nutrient requirements of gestating sows. *J. Anim. Sci.* **2019**, *97*, 4934–4945. [CrossRef]
108. Gaillard, C.; Durand, M.; Largouët, C.; Dourmad, J.Y.; Tallet, C. Effects of the environment and animal behavior on nutrient requirements for gestating sows: Future improvements in precision feeding. *Anim. Feed Sci. Technol.* **2021**, *279*, 115034. [CrossRef]
109. Gaillard, C.; Dourmad, J.Y. Application of a precision feeding strategy for gestating sows. *Anim. Feed Sci. Technol.* **2022**, *287*, 115280. [CrossRef]
110. Gauthier, R.; Largouët, C.; Rozé, L.; Dourmad, J.Y. Online forecasting of daily feed intake in lactating sows supported by offline time-series clustering, for precision livestock farming. *Comput. Electron. Agric.* **2021**, *188*, 106329. [CrossRef]

animals

MDPI

Review

Nutritional Strategies to Mitigate Post-Weaning Challenges in Pigs: A Focus on Glucans, Vitamin D, and Selenium

John O'Doherty [1,*], Alison Dowley [1], Eadaoin Conway [1] and Torres Sweeney [2]

1 School of Agriculture and Food Science, University College Dublin, Belfield, D04 W6F6 Dublin, Ireland; alison.dowley@ucdconnect.ie (A.D.); eadaoin.conway@ucdconnect.ie (E.C.)

2 School of Veterinary Medicine, University College Dublin, Belfield, D04 W6F6 Dublin, Ireland; torres.sweeney@ucd.ie

* Correspondence: john.vodoherty@ucd.ie

Simple Summary: The pig-farming industry faces significant challenges in ensuring the health and growth of piglets, particularly during the weaning phase. This critical period involves multiple stressors, such as environmental changes, dietary shifts, and social separation, which can adversely affect the piglet's digestive health, immune system, and overall well-being. One of the primary hurdles during weaning is the transition from a milk-based diet to a more complex cereal-based diet. This abrupt dietary change can lead to reduced food intake, digestive issues, gut inflammation, and nutrient absorption difficulties, resulting in diarrhea and poor growth. To tackle these issues, researchers are exploring innovative nutritional strategies. One promising area is the utilization of specific types of fiber, known as glucans, derived from sources like cereals, mushrooms, seaweed, and yeast. Additionally, there is a growing focus on the roles of Vitamin D and selenium, with Vitamin D and selenium-enriched mushrooms serving as natural sources of these vital nutrients. In conclusion, addressing the challenges faced by piglets during weaning necessitates the development of effective nutritional strategies, including the incorporation of glucans, Vitamin D, selenium, and enriched mushrooms. These approaches align with sustainable and responsible pig-farming practices, prioritizing the welfare of the animals and reducing the need for additives and antibiotics.

Abstract: This review examines the challenges faced by the pig industry, with a specific focus on improving the health and growth of weaned pigs. It emphasizes the immediate necessity of investigating alternative approaches to managing pig nutrition and health due to restrictions on the use of antibiotics and the prohibition of zinc oxide in weaned pig diets. The weaning phase is identified as a critical stage in piglet development, characterized by stressors that affect their gastrointestinal health, immune responses, and overall physiology. The primary challenge during weaning arises from transitioning piglets from a digestible milk-based diet to a less digestible cereal-based feed, causing nutritional stress. This manifests as reduced feed intake, leading to gastrointestinal disturbances, intestinal inflammation, and adverse effects on intestinal structure and microbiota. To address these challenges and optimize piglet development, various nutritional strategies have been explored. Notably, glucans, particularly β-glucans from fungi, cereals, algae, and yeast, show promise in alleviating weaning-related issues. Furthermore, it is important to highlight the critical roles played by Vitamin D and selenium in piglet nutrition. These essential nutrients can be sourced naturally from enriched mushrooms that are specifically enriched with Vitamin D and selenium, providing a sustainable dietary option. In conclusion, effective nutritional strategies, including glucans, Vitamin D, selenium, and enriched mushrooms, are beneficial for addressing weaning-related challenges.

Keywords: pig; weaning; β-glucans; mushrooms; Vitamin D; selenium; dam

1. Introduction

Weaning is a critical phase in the development of piglets, with profound effects on their gastrointestinal health, immune responses, and overall physiology [1]. This stage introduces a complex interplay of stressors, including environmental shifts, dietary changes, and social separations, all of which collectively impact production efficiency [2,3]. The nutritional transition during weaning, which involves a shift from a milk-based to a cereal-based diet, poses a significant challenge to the piglet's digestive capacities [4]. Additionally, the move from farrowing rooms to weaner houses exposes them to novel environments and increased pathogen exposure. Social stress is further intensified as piglets are abruptly separated from their mothers and integrated with new piglets. These multifaceted stressors result in decreased feed intake and a range of gastrointestinal issues [5]. In contrast, wild piglets undergo a gradual weaning process over 10–14 weeks, allowing for more robust gastrointestinal development [6]. Commercial weaning, typically occurring between 3 and 4 weeks, coincides with the peak development of the gastrointestinal barrier but leaves the piglet's immune systems underdeveloped. Delaying weaning can enhance disease resistance but comes at the cost of increased production costs.

Following weaning, piglets undergo significant alterations in intestinal morphology, which affect nutrient absorption [7,8]. These changes are linked to an increased incidence of post-weaning diarrhea (PWD) and reduced enzymatic activity, which is essential for nutrient digestion [9]. Consequently, disruptions in the intestinal barrier lead to heightened immune system responses [10]. The weaning process triggers inflammation in piglets due to their immature immune systems and exposure to various antigens [11].

To address these intricate issues, various nutritional strategies have been explored in piglet rearing to enhance health and growth. Glucans, specifically β-glucans derived from sources like seaweed, mushrooms, cereals, and yeast, have gained attention for their potential to alleviate some of the weaning-related challenges. Additionally, the roles of Vitamin D and selenium, two essential nutrients, are emerging as important factors in piglet nutrition. Casein hydrolysates, which are derived from milk protein casein, are also gaining attention as potential alternatives to antibiotic growth promoters in pig nutrition. Understanding the underlying physiological changes during this critical phase is vital for developing effective nutritional strategies. Until recently, the main focus on finding alternatives to in-feed antibiotic growth promoters and zinc oxide has been on dietary manipulations in pigs post-weaning using feed additives in the post-weaning diet. However, other strategies, such as maternal nutrition to improve growth and health in her offspring, also hold promise. This review aims to delve into these strategies, with a focus on glucans, casein hydrolysates, Vitamin D, and selenium, and assess their potential to mitigate the complexities of weaning-related challenges.

2. The Physiological and Immunological Implications of Weaning on Piglets

Weaning is a critical period characterized by considerable stress, influencing the piglet's gut microbiology, immunology, and physiology, which in turn impacts growth, intake, and health [3]. Environmental changes, dietary shifts, and social reorganization contribute to this stress. Unlike the gradual weaning process observed in the wild, commercial weaning procedures are abrupt, potentially increasing health risks due to the piglet's reliance on maternal antibodies [12]. During the suckling period, the piglet's gut microbiota is significantly influenced by the constituents of milk, which promotes the growth of bacterial families such as *Bacteroidaceae*, *Clostridicaceae*, *Lachnospiraceae*, and *Lactobacillicaeae* [13]. However, with the introduction of solid feed and cessation of milk at weaning, substantial changes occur in the intestinal microbiota composition, demonstrated by the emergence of substantial shifts in the population dynamics of gut bacteria [14]. The weaning transition, coupled with a decline in lactose availability, results in higher gastric pH levels, reducing the natural barrier to enteric infections [1,15]. Concomitantly, there is a decrease in beneficial microbes such as *Lactobacillus*, alongside an increase in opportunistic pathogens, including *Clostridium* and *E. coli*, the latter of which plays a pivotal role in the etiology of PWD [16,17]. As the piglet matures, the GIT microbiota diversifies and stabilizes, approximating an adult-like composition by three weeks post-weaning, characterized by a higher abundance of *Prevotellaceae*, *Ruminococcaceae*, *Lactobacillaceae* and *Veillonellaceae*, with a concurrent reduction in *Enterobacteriaceae* and *Bacteroidaceae* [18].

Following weaning, intestinal morphological changes, such as fluctuations in villous height and crypt depth, affect the organ's absorptive function and overall health [19]. Optimal nutrient uptake is compromised by the reduced activity of digestive enzymes such as lactase, sucrase, and maltase, which are critical for carbohydrate digestion [7,20]. These physiological alterations are further aggravated by disruptions to the intestinal barrier, mediated by the inflammatory response, and characterized by weakened tight junctions between epithelial cells [10,21]. Additionally, stress mediators, including cortisol and corticotropin-releasing factors, have been implicated in neuro–immune interactions that adversely affect gastrointestinal functionality [6]. Post-weaning also sees a marked reduction in feed intake, with subsequent negative repercussions for gastrointestinal health and growth [7]. Furthermore, PWD, primarily caused by *E. coli*, poses a significant challenge in pig production. The enterotoxigenic strains of this bacterium bind to the intestinal epithelium, facilitated by fimbriae such as F4 and F18, leading to infection and diarrheal outbreaks [22,23]. Various management strategies that focus on nutrition and the weaning process itself are critical for minimizing these episodes [17]. In summary, weaning represents a developmental stage with profound implications for the gastrointestinal health of piglets. Effective management of this phase, including the understanding of associated stressors and implementation of nutritional interventions, is essential to promote health and growth in piglets, with the added benefit of reducing the need for medicinal supplements such as zinc oxide and in-feed medication in pig production.

3. Exploring Natural Dietary Interventions to Address Dysbiosis in Pig Nutrition

This review delves into the evolving dynamics and challenges within the pig industry, placing a strong emphasis on advancing the health and growth of weaned piglets. A central theme of the review revolves around exploring natural compounds and nutrients as potential dietary supplements for pigs, with a particular focus on the advantages associated with incorporating β-glucans, casein hydrolysates, mushrooms, Vitamin D, and selenium and their synergisms into piglet and sow diets. These natural compounds are acknowledged for their multifaceted properties, encompassing anti-inflammatory, antimicrobial, antioxidant, immunomodulatory, and prebiotic effects. The review emphasizes the major role of maternal nutrition in shaping the health and well-being of piglets, examining how maternal dietary interventions significantly impact piglet development and resilience, both before and after weaning. Another aspect of the review lies in its examination of the unique capabilities of mushrooms, specifically their ability to synthesize Vitamin D and convert inorganic selenium into highly bioavailable organic forms. This positions mushrooms as invaluable additions to pig diets, presenting opportunities to enhance piglet health and performance post-weaning. The review stresses the importance of understanding the underlying mechanisms governing the functional properties of these feed ingredients. Key aspects of gut functionality, including digestive and absorptive capacity, villi architecture, nutrient transporter expression, chemical and physical barriers, microbiota diversity, and immune function, are considered to achieve an optimal response to these dietary interventions [11,24].

4. Structural Specificity and Health Impact of Beta-Glucans in Pig Nutrition

The role of β-glucans in pig nutrition is increasingly recognized for its profound impact on enhancing the immune response and overall health of pigs. These naturally occurring polysaccharides, comprised of glucose molecules linked by β-glycosidic bonds, display a structural diversity that is crucial to their functionality. Depending on their botanical or microbial origin, ranging from cereals like barley and oats to yeasts and mushrooms—the arrangement of these linkages can be primarily 1,3, 1,4, or 1,6, each conferring different physical properties and biological activities. For example, the β-glucans from mushrooms have a 1,3 backbone with 1,6-linked side chains, a structure that has been shown to potentiate immune response through various mechanisms, including the activation of macrophages and other immune cells [25]. Yeast-derived β-glucans, with a higher proportion of 1,3 linkages, have been found to enhance pathogen recognition by the immune system [26]. Conversely, the β-glucans found in cereals, primarily with 1,3 and 1,4 linkages, exhibit a solubility that can influence gut health, which in turn can have a systemic effect on the immune status of pigs [27]. This specificity necessitates a nuanced approach to dietary inclusion, where factors like source, purity, solubility, synergisms, and molecular mass must be carefully considered to ensure the β-glucans incorporated into pig feeds are optimized for the best possible health outcomes [28]. In essence, the structural complexity of β-glucans dictates their potential as a functional feed additive in pig nutrition.

Importantly, all three types of β-glucans have been linked to improved growth performance and feed efficiency in pigs, operating through distinct mechanisms. Yeast and seaweed-derived β-glucans likely enhance growth by boosting health and disease resistance, while cereal-derived β-glucans improve growth through better gut health and nutrient absorption. The selection of β-glucan source should be guided by specific production goals, like enhancing immune function, improving gut health, or optimizing growth performance in pigs.

5. Yeast-Derived β-Glucans: Enhancing Swine Nutrition and Health during Weaning

5.1. Immunomodulatory Potential of Yeast-Derived β-Glucans

β-Glucans, a class of vital polysaccharides, have gained increasing recognition for their health-promoting roles in functional foods and animal nutrition. These molecules are diverse and can be found in various sources, including bacteria, fungi, algae, and cereals, each displaying unique structural attributes. Yeast-derived β-glucans are characterized by having β (1,6)-linked branches on a β (1,3) backbone, in contrast to the primarily linear β (1,4) linkages interspersed with β (1,3) chains typically found in cereal β-glucans [25]. One of the prominent features of yeast-derived β-glucans is their capacity to stimulate the immune system, as they activate a wide range of immune cells, including macrophages, T helper cells, and natural killer cells [29]. This immune activation occurs through their interaction with pattern recognition receptors, which identify these polysaccharides as pathogen-associated molecular patterns, therefore triggering innate immune responses [30,31]. Among these receptors, Dectin-1 plays a crucial role in recognizing β-glucans and initiating immune responses, encompassing processes such as phagocytosis, oxidative burst, and cytokine production [32–34].

The immunomodulatory activity of β-glucans is significantly influenced by their structure and solubility. Insoluble β-glucans activate the dectin-1 pathway, while soluble forms often interact with the complement system, with their action being dependent on specific antibodies [33]. The effectiveness of dectin-1 receptor activation by β-glucans requires a β (1,3) backbone of sufficient length, typically consisting of at least seven glucose units, and often necessitates at least one β (1,6)-side-chain branch [33].

5.2. Enhancing Growth Performance with Yeast-Derived β-Glucans

In the field of pig nutrition, β-glucans, particularly those derived from yeast, have been associated with improved growth performance in weaned pigs. Multiple studies have demonstrated that β-glucans sourced from organisms like *Agrobacterium* spp. and yeast can positively influence the gut microbiota, increasing the abundance of beneficial bacterial taxa such as *Lactobacillus* while reducing harmful species like *E. coli* [28,35–39]. Moreover, yeast β-glucans are known for their anti-inflammatory and antioxidant properties, which may provide added resilience against chronic inflammation and oxidative stress. A recent meta-analysis [40] quantified these benefits, revealing a 7.6% increase in weight gain and a 5.3% increase in feed intake among nursery pigs fed with yeast β-glucans sourced from Saccharomyces cerevisiae. Even more substantial growth improvements were observed with Agrobacterium-derived β-glucans. The meta-analysis recommended an optimal use of 50 mg/kg of Saccharomyces cerevisiae-derived β-glucans in nursery pig diets based on these findings [40].

Kim et al. [39] observed that dietary β-glucans led to a substantial improvement in the daily gain of weaned pigs, with this improvement attributed to enhanced nutrient absorption and improved intestinal health. These findings align with previous research [37], which noted an enhancement in feed conversion ratios with β-glucan supplementation. The pivotal role of β-glucans in promoting gut health has been well emphasized [41]. β-glucans contribute to maintaining the integrity of the gut barrier and support the establishment of a healthy gut microbiome [41], which is critical for preventing post-weaning diarrhea. Additionally, β-glucans also have the potential to alleviate weaning stress in pigs [40]. This study revealed that β-glucans reduced stress-related behaviors and improved feed intake in post-weaned pigs, a factor crucial for ensuring steady growth during the post-weaning period. Research on β-glucans from Agrobacterium sp. ZX09, known for its high purity compared to yeast, has shown positive effects on weaned piglet growth and intestinal health [42]. Both low and high molecular weight β-glucans from *Agrobacterium* sp. ZX09 enhanced piglet growth, emphasizing the importance of source and purity. Low molecular weight β-glucans had strong antioxidant effects, improved mucosal barrier function, and positively affected gut microbiota. High molecular weight β-glucans specifically benefited hindgut bacteria [42].

In summary, current research provides compelling evidence for the significant advantages of incorporating microbial β-glucans, particularly those derived from yeast and bacteria, into the diets of nursery pigs. These β-glucans contribute to enhanced growth, intestinal health, and immune function, ultimately resulting in improved growth performance and overall well-being.

6. The Role of Cereal-Derived β-Glucans in Piglet Diets

6.1. Impact on Gastrointestinal Microbiota

Cereal β-glucans have a significant impact on the gastrointestinal microbiota of pigs, enhancing beneficial bacteria populations such as *Lactobacillus* and *Bifidobacteria*, which are vital to pig health [43]. These polysaccharides also modulate the cycling of nutrients within the pig's system, as indicated by their effect on the excretion patterns of urinary and fecal nitrogen, suggesting improved nutrient utilization [44]. Moreover, β-glucans have been shown to contribute to the reduction of emissions from pig manure, aligning pig-farming practices with environmental sustainability goals [45,46]. However, the inclusion of high levels of intact β-glucans may impede nutrient digestibility, potentially affecting the economic efficiency of pig nutrition [44]. Studies by Metzler and Zebeli [46] have correlated high β-glucan content with a decrease in the apparent ileal digestibility and the total tract digestibility of crude protein and energy, highlighting the necessity for careful dietary inclusion to prevent negative effects on nutrient utilization. At optimal concentrations, cereal β-glucans can increase caecal volatile fatty acids and butyrate levels, conferring a gut health advantage [46]. However, their water-soluble nature can also lead to increased digesta viscosity, which may interfere with feed digestion and nutrient absorption, a concern particularly relevant to the growth and health of nursery pigs [46].

6.2. Structural Configurations and Physiological Impact

The structural configurations of β-glucans, especially the ratios of β (1,3) to β (1,4) linkages, are key determinants in their physicochemical properties, influencing their solubility and the extent of microbial degradation in the pig's gastrointestinal tract [47]. A higher ratio of β (1,3) linkages is known to increase solubility and viscosity, impacting the digestive process. The concentration and solubility of β-glucans differ among cereal grains; for instance, barley has a higher β-glucan content in its endosperm and is characterized by a greater proportion of water-insoluble β-glucans compared to oats [48].

Meta-analytic insights indicate inconsistent growth and feed intake responses to dietary cereal β-glucans [40], suggesting that while certain levels can promote intestinal health, demonstrated by prebiotic effects that bolster beneficial gut microbiota and reduce pathogenic bacteria adhesion, the responses are not uniform. The source of β-glucans plays a significant role; oat-derived β-glucans affect gut microbiota differently from barley-derived β-glucans [48,49]. Therefore, formulating pig diets with β-glucans requires a strategic approach that considers both the type and amount, particularly in cases of soluble β-glucans from barley that could heighten digesta viscosity and associate with digestive issues such as PWD [40]. By contrast, the research of Bach Knudsen and Jørgensen [50] underlines the importance of soluble fiber β-glucans in improving the gut environment of weaned pigs. These fibers increase the viscosity of gut contents, which slows the passage rate and facilitates better nutrient absorption, which is beneficial during the weaning phase when pigs are prone to nutritional upsets and often exhibit inefficient digestion. Therefore, the formulation of pig diets with cereal β-glucans calls for a strategic approach that weighs the type and amount, particularly in the case of soluble β-glucans from barley that could increase digesta viscosity and be associated with digestive issues such as PWD [40].

7. Seaweed: A Sustainable β-Glucan Source for Swine Nutrition

7.1. Immunomodulatory Benefits of Algae β-Glucan

Algae-derived β-glucans are attracting increasing attention as a promising dietary supplement for weaned pigs, offering unique advantages for both growth and health during this crucial developmental stage. One of the primary advantages of incorporating algae β-glucans into the diets of piglets lies in their substantial immunomodulatory capabilities. Algae-derived β-glucans play a vital role in boosting the immune response of weaned pigs [24,51]. This enhancement of the immune system is of paramount importance as it contributes to reducing the occurrence of common post-weaning issues such as diarrhea and respiratory infections. By fortifying the immune defenses, algae β-glucans provide piglets with increased resilience against the microbial threats they encounter in the post-weaning phase [24].

7.2. Laminarin: Antibacterial and Prebiotic Effects

Laminarin, a low molecular weight β-glucan found in various seaweeds, is particularly noteworthy for its antibacterial activity. It is characterized by a linear backbone of (1,3)-β-linked glucopyranose residues with varying β-(1,6)-branching [52,53]. The water solubility of laminarin is influenced by its branching levels, and it accumulates in algal cells during specific seasons to support survival and growth during adverse conditions, such as winter. Laminarin has demonstrated antibacterial properties against a wide range of bacteria in vitro, including pathogenic strains like *E. coli*, *S. Typhimurium*, *L. monocytogenes*, and *St. Aureus* [52,54]. This antibacterial activity extends to purified laminarin extracted from various seaweed species, with a more pronounced effect observed against Gram-negative bacteria [55,56].

When applied in pig nutrition, the inclusion of laminarin-rich extracts in the diet has been associated with reduced populations of *Enterobacteriaceae* and *attaching-effacing Escherichia coli (AEEC)* in the caecum and colon of weaned pigs [57,58]. Moreover, laminarin exhibits prebiotic activity, as demonstrated by an increase in the populations of beneficial *Lactobacillus* species in pig colonic and fecal microbiota following supplementation with both crude and highly purified laminarin-rich extracts [59]. Additionally, investigations in weaned pigs have revealed an increased relative abundance of *Prevotella* spp. following laminarin supplementation, which has been correlated with improved pig performance and maturity [60].

The impact of laminarin supplementation goes beyond microbiota modulation and extends to the production and profiles of short-chain fatty acids (SCFAs) in the gastrointestinal microbiota, particularly affecting butyrate production [35,61]. The SCFAs, including butyrate, are crucial for gut health and have been linked to various physiological benefits. In terms of immunomodulatory activity, dietary supplementation with crude or highly purified laminarin-rich extracts has demonstrated an anti-inflammatory effect on the small intestine and colon of weaned and growing pigs [35]. This anti-inflammatory effect is characterized by the reduced expression of pro-inflammatory cytokine genes, pattern recognition receptors, and the transcription factor NFKB1 [62]. In the colon, laminarin has an immunosuppressive effect, primarily down-regulating genes associated with the Th17 pathway [63]. Furthermore, laminarin-rich extracts have been associated with several performance improvements in weaned pigs, including enhanced final body weight, daily gain, feed intake, and gain-to-feed ratio [63,64]. Additionally, laminarin supplementation has proven effective in reducing diarrhea, particularly in the post-weaning period, as indicated by lower fecal scores in supplemented weaned pigs [57,64]. Importantly, under both hygienic and unsanitary conditions, laminarin-rich extracts have shown promise in reducing the incidence of post-weaning diarrhea and improving daily gains, making them a potential dietary alternative to antibiotic growth promoters and zinc oxide for managing post-weaning diarrhea in pigs [61].

It is essential to acknowledge that the quantitative, structural, and functional variability of laminarin can significantly depend on factors such as extraction methodologies, conditions, and the types of seaweed used [65]. The effectiveness of these β-glucans as bioactive compounds can vary based on parameters such as solubility, molecular weight, and structural characteristics. Although new extraction techniques offer more efficient ways to obtain laminarin from seaweeds, the choice of extraction method and seaweed variety plays a crucial role in determining the quality and properties of the extracted polysaccharides [65]. Understanding these factors is vital for harnessing the full potential of laminarin as a bioactive compound in weaned pig diets.

8. The Role of Vitamin D in Nutrition and Immunity

The pivotal role of Vitamin D in swine health is increasingly evident in the context of modern indoor farming practices, which often limit natural sunlight exposure, essential for the endogenous synthesis of this fat-soluble nutrient. Consequently, dietary Vitamin D supplementation becomes crucial to maintaining pig health, supporting a range of physiological processes from bone development to immune function [66].

Although regulatory guidelines, such as those from the European Food Safety Authority [67], stipulate a maximum dietary Vitamin D content of 50 μg/kg/feed, recent research argues that these standards may not suffice in light of intensified agricultural practices [68,69]. Emerging studies propose that enhanced Vitamin D fortification can contribute to improved immune responses, growth rates, and feed conversion efficiency in pigs, challenging the adequacy of current recommendations by authorities like the National Research Council [70]. Furthermore, the interaction of Vitamin D with the gut microbiota—a relationship that is well-established in human research—shows promising implications in swine, where increased levels of 25(OH)D3, particularly in low-calcium and low-phosphorus diets, have been associated with beneficial shifts in fecal microbial compositions [71].

The classical functions of Vitamin D in mineral regulation and bone health are well-known, yet its immunomodulatory capacity is gaining recognition. Vitamin D mediates immune responses through its active form, calcitriol, acting on the Vitamin D receptor (VDR) present in various immune cells, thus influencing both innate and adaptive immunity [72,73]. Notably, Vitamin D is involved in regulating the expression of antimicrobial peptides and the balance between pro-inflammatory and anti-inflammatory cytokine production. Its deficiency is linked to autoimmune diseases in humans, highlighting its systemic relevance [74]. Although more extensively studied in avian species, where Vitamin D is known to exert antioxidative and immune effects [75–78], the investigation into its effects on pigs suggests similar benefits. For instance, high-dose 25(OH)D3 supplementation has been associated with reduced severity of diarrhea in weaned pigs challenged with pathogens like the porcine epidemic diarrhea virus [79,80]. In light of this evidence, Vitamin D's function in pig health appears to be multifaceted, offering benefits that extend far beyond its traditional roles. This highlights a clear directive for further research to explore the full potential of Vitamin D in enhancing both the health and performance of the post-weaned pig.

9. The Role of Selenium in Nutrition and Immunity

9.1. Selenium's Crucial Role in Pig Health

Selenium is an indispensable trace mineral for swine, playing a crucial role in enhancing immune function, reproductive health, growth, and meat quality. It exercises its biological roles chiefly through selenoproteins, which incorporate the amino acid selenocysteine into their structure at active sites [81]. These selenoproteins, notably glutathione peroxidases (GPXs), thioredoxin reductases (TXNRDs), and iodothyronine deiodinases (DIOs), are essential in modulating the immune system and protecting cells from oxidative harm [82,83]. The dietary form of selenium significantly affects its bioavailability, with organic selenium from sources like enriched yeast showing superior absorption and utilization in the animal's body compared to inorganic forms such as sodium selenite, which are less bioavailable and can be toxic [84]. Studies have shown that organic selenium more effectively improves selenium status in various tissues and animal products, like colostrum and milk, than inorganic sources [85]. The management of selenium intake is critical, balancing a fine line between deficiency, which can subtly impact growth and reproduction, and toxicity, which may result in severe health repercussions [86].

9.2. Selenium's Impact on Gut Microbiota and Immunity

Emerging research has shed light on the impact of selenium on the gut microbiota. Selenium supplementation is linked to a healthier gut flora composition, increasing beneficial bacteria such as *Lactobacillus* while reducing pathogenic species like *E. coli* [87–90]. This supplementation is also associated with reduced inflammatory markers and bolstered immune responses, showcasing selenium's role in modulating inflammation [91,92]. Specifically, in swine, selenium demonstrates the potential to reinforce immunity, particularly under environmental stressors. Providing weaned pigs with selenium-enriched yeast has been shown to enhance growth metrics and immune responses, suggesting a significant role in improving post-weaning resilience [93]. These findings are paralleled in poultry, where selenium has beneficially influenced immune and oxidative stress parameters [94].

At the heart of selenium's role in gut immunity is its integration into selenoproteins with powerful antioxidant capabilities. Selenoproteins like GPXs and TXNRDs are critical in shielding the gut mucosa from oxidative stress caused by free radicals and reactive oxygen species (ROS) during metabolic activities and immune reactions. By neutralizing these reactive molecules, selenium is vital for the maintenance of gut integrity, therefore averting tissue damage and inflammation that can be induced by oxidative stress [95], a factor crucial for ensuring the health of the gastrointestinal system during the post-weaning period. Selenium is also pivotal in augmenting the gut's immune response. It is necessary for the proper activation and functioning of immune cells such as T lymphocytes and natural killer (NK) cells, which are key in fighting off pathogens and preserving immune equilibrium in the gut. Selenium has been implicated in enhancing antibody production, vital to the adaptive immune response, thus facilitating the immune system's capacity to detect and eliminate pathogens, ensuring extensive immune protection in the gastrointestinal tract [96]. Furthermore, selenium's ability to modulate gut inflammation is paramount. Deficiencies in selenium have been connected to a heightened susceptibility to inflammatory conditions like inflammatory bowel disease (IBD). Selenium exerts anti-inflammatory effects by curtailing the production of pro-inflammatory cytokines such as TNF-α and IL-6, helping to manage inflammation and maintain gut health [97], a factor crucial for improved gut health during the post-weaning period.

The intriguing link between selenium and gut microbiota has been uncovered in recent studies [89,90]. Selenium has been shown to affect the composition and diversity of the gut microbial community, a relationship that has significant implications for gut immunity. The balance and diversity of gut bacteria are instrumental in the immune responses within the gut. Thus, selenium's influence on the microbiota composition is a critical factor in promoting a balanced and harmonious gut ecosystem, contributing to a robust and responsive gut immune system in the weaned pig.

10. Mushrooms as a Source of β-Glucan, Vitamin D and Selenium

Mushrooms, especially the globally consumed *Agaricus bisporus* or white button mushroom, present a unique opportunity in animal nutrition, especially for pigs. The agricultural practice of utilizing mushroom by-products stems from a growing awareness of the sustainability issues related to feed production. These by-products, primarily composed of mushrooms unsuitable for the consumer market, embody a resource for nutrient-rich, bioactive compounds that can significantly contribute to pig nutrition and health [98]. *Agaricus bisporus* is endowed with an array of bioactive compounds that offer considerable health benefits. The presence of β-glucan polysaccharides is well-noted for their role in immune modulation. Moreover, an array of polyphenols, amino acids like ergothioneine, and the presence of chitin, terpenoids, Vitamin D2, and ergosterol broaden the potential health benefits. These compounds are collectively known for their anticancer, antioxidant, antiviral, antimicrobial, antibacterial, antifungal, anti-inflammatory, and immunomodulatory effects [98,99].

Ergosterol plays a particularly interesting role in the nutritional value of mushrooms. When exposed to UV light, ergosterol converts to Vitamin D2, a vital nutrient for various physiological functions, including cell growth, neuromuscular and immune function, and inflammation regulation. This is a crucial feature for indoor-reared pigs that lack natural sunlight exposure, as Vitamin D sourced from mushrooms can help in mitigating deficiency [100,101].

Research has indicated that dietary inclusion of *Agaricus bisporus* can influence the pig gut microbiota favorably and exert an anti-inflammatory effect, although no significant changes were noted in piglet performance [102]. This suggests that the mushrooms may contribute to long-term health benefits that are not immediately apparent in growth metrics [102]. Conversely, Duffy et al. [68] demonstrated that finisher pigs fed with Vitamin D2-enriched mushrooms showed improvements in performance, antioxidant status, and pork color stability, indicating the direct benefits of mushroom-derived Vitamin D2 on pig growth and meat quality.

The biofortification of mushrooms with selenium has become an area of great interest due to selenium's critical role in pig health. Organic forms of selenium present in mushrooms, such as selenomethionine, have higher bioavailability compared to inorganic selenium sources, such as selenite or selenate, which are commonly used in pig diets. Selenium is a key component of glutathione peroxidase, an enzyme that protects cells from oxidative damage, and it plays a role in thyroid hormone metabolism and immune response [103]. Selenium-enriched mushrooms have been shown to improve pig health by increasing antioxidant capacity, enhancing immune responses, and improving meat quality, which is indicative of selenium's incorporation into body tissues [104,105]. Furthermore, the supplementation of pig diets with selenium-enriched mushrooms has yielded results such as enhanced gut microbiota, reduced diarrhea scores, and improvements in volatile fatty acids profiles in the caecum [104,105]. These shifts in the gut environment suggest a role for selenium-enriched mushrooms in promoting a gut microbiota that is favorable for pig health, potentially reducing the need for medical interventions and contributing to the resilience of the animals [104,105] Given these multi-dimensional benefits, the integration of mushroom by-products into pig diets extends beyond mere nutrient supplementation.

11. Synergistic Effects of β-Glucan with Casein Hydrolysates

Casein hydrolysates, which are derived from milk protein casein, are gaining attention as potential alternatives to antibiotic growth promoters in pig nutrition [106]. They are valued for their high nutritive content and the presence of bioactive peptides [107,108]. These bioactive peptides are naturally embedded within the structure of casein and can be released through enzymatic hydrolysis during digestion or food processing [107,108]. Extensive research has demonstrated the significant bioactivity of casein hydrolysates in various experimental settings, including in vitro, ex vivo, and in vivo studies [109,110].

However, a common challenge with bioactive compounds is their often-limited bioavailability in vivo, which can restrict their effectiveness as health-promoting agents [111].

One potential issue is that the bioactivity of casein hydrolysates may be compromised in the stomach during digestion. To address this concern, microencapsulation techniques can be employed to protect these bioactive compounds and enhance their bioavailability [106,112]. For example, studies have shown that microencapsulation using substances like yeast β-glucan can preserve the bioactivity of casein hydrolysates in vivo [106,112]. It is worth noting that β-glucans themselves have been extensively studied for their antioxidant, immunological, and anti-inflammatory effects [113–115], but research on the use of casein hydrolysates in pig diets is relatively limited. Some studies have explored their potential benefits. For instance, piglets supplemented with yeast β-glucan and casein hydrolysate exhibited reduced inflammation, characterized by the upregulation of tight junction protein CLDN3 and the downregulation of pro-inflammatory cytokine genes [116], factors crucial for improved gut health during the post-weaning period. Furthermore, in post-weaning pigs, the combination of casein hydrolysate and yeast β-glucan was found to improve gastrointestinal function and health [106]. These findings suggest that the use of casein hydrolysates in combination with microencapsulation techniques, such as those involving β-glucans, holds promise for enhancing the bioavailability and effectiveness of these bioactive compounds in improving pig health and performance.

12. The Benefits of Incorporating β-Glucans into Sow Diets

The development of the GIT and immune system in piglets is profoundly influenced by maternal factors during gestation and lactation. Supplementing the diets of gestating and lactating sows has the potential to positively impact the health of piglets, particularly during the pre-weaning phase, which may lead to improved post-weaning performance [24]. The GIT colonization in piglets starts immediately post-birth, with bacteria from the sow and the environment crucial in this process [117]. The microbiota composition of piglets can be influenced by altering the sow's diet to encourage beneficial bacteria and decrease pathogenic species, potentially reducing piglet susceptibility to PWD [118,119]. The sow's vaginal and fecal microbiota are significant contributors to the piglet's intestinal microbiota. Crespo-Piazuelo et al. [120] showed that sows supplemented with a probiotic containing *Bacillus altitudinis* resulted in piglets displaying fecal shedding of the probiotic strain, suggesting a vertical transmission of beneficial bacteria. Additionally, laminarin and fucoidan from seaweed extracts given to gestating sows reduced Enterobacteriaceae levels in sow feces and decreased colonic Escherichia coli in piglets at weaning [119,121]. Furthermore, yeast β-glucan combined with casein hydrolysate in sow diets during late pregnancy has been linked to a more favorable fecal microbiota composition, with increases in beneficial bacteria like Lactobacillus [116]. Such piglets weaned from these supplemented sows had increased villus height in the duodenum and increased villus height to crypt depth ratio in the jejunum, as well as a decreased expression of the pro-inflammatory cytokine genes, the tight junction gene *CLDN3* and the mucin gene *MUC2* in the duodenum and jejunum.

The quality of colostrum and milk is vital for delivering antimicrobial and immune-enhancing properties to piglets [118]. Colostrum intake is essential for stimulating intestinal growth and function, facilitating the absorption of immunoglobulin G (IgG) for systemic immunity, and providing energy for thermoregulation [122–124]. Colostrum and milk contain immunoglobulins and other antimicrobial compounds that support the establishment of a beneficial commensal microbiota [118]. The IgG, abundant in colostrum, decreases after birth, but its early presence is critical for protection against infections during weaning [125,126]. Similarly, IgA provides a crucial defense against GIT pathogens and plays a role in preventing PWD [127]. Dietary supplementation of gestating and lactating sows with β-glucans and other bioactive compounds can enhance immunoglobulin concentrations in colostrum and milk, potentially improving piglet health outcomes [128–130]. Piglets from sows fed with laminarin and fucoidan-rich diets showed improved immune function, such as enhanced leukocyte phagocytosis capacity (Leonard et al., 2010) and increased resistance to ETEC infections [119].

In summary, the strategic dietary supplementation of sows with β-glucans and other bioactive compounds can significantly influence piglet GIT colonization, immunity, and resilience against post-weaning gastrointestinal challenges. Through these nutritional interventions, it is possible to reduce the presence of pathogenic bacteria, enhance immune function, and improve piglet growth and health outcomes.

13. Conclusions

The multifaceted roles of β-glucans, derived from diverse sources such as yeast, mushrooms, cereals, and seaweeds, are becoming increasingly recognized for their potential in swine nutrition. These polysaccharides, through their immunomodulatory, antimicrobial, and gut health-promoting activities, offer a natural alternative to traditional feed additives. Yeast-derived β-glucans, with their robust impact on immune cell activation, have shown promising effects on the growth performance and gut microbiota composition of weaned pigs, enhancing the presence of beneficial bacteria while suppressing pathogenic strains. Similarly, mushroom-derived β-glucans, along with an array of other bioactive compounds, contribute to antioxidative capacity and overall animal well-being, with added benefits from Vitamin D and selenium fortification. Laminarin from seaweed adds to the complexity of β-glucans in swine diets by providing distinctive antibacterial and prebiotic effects that could support intestinal health and improve post-weaning growth metrics. Moreover, the incorporation of β-glucans into sow diets may impart long-term health benefits to piglets, emphasizing the importance of maternal nutrition on offspring development.

Although the benefits of β-glucans are evident, the picture is nuanced. It is essential to select appropriate sources and doses of β-glucans and to consider synergy with other compounds, such as casein hydrolysates, for maximum efficacy. Furthermore, the influence of Vitamin D and selenium, particularly from mushroom sources, extends the potential health benefits for post-weaned pigs, reinforcing the need for a strategic, well-balanced approach to dietary supplementation. By enhancing immunity, promoting healthy gut microbiota, and improving growth and resilience, β-glucans stand as a significant contributor to the advancement of sustainable and productive pig nutrition practices post-weaning.

Author Contributions: J.O., A.D., E.C. and T.S. designed the review; J.O., A.D., E.C. and T.S. wrote, reviewed, and edited the manuscript. All authors have read and agreed to the published version of the manuscript.

Funding: A.D. and E.C. were funded by the Science Foundation Ireland (SFI) [Grant number: 16/RC/3889].

Data Availability Statement: Not applicable.

Conflicts of Interest: None of the authors had a financial or personal conflict of interest in relation to the present review.

References

1. Heo, J.M.; Opapeju, F.O.; Pluske, J.R.; Kim, J.C.; Hampson, D.J.; Nyachoti, C.M. Gastrointestinal health and function in weaned pigs: A review of feeding strategies to control post-weaning diarrhoea without using in-feed antimicrobial compounds. *J. Anim. Physiol. Anim. Nutr.* **2013**, *97*, 207–237. [CrossRef] [PubMed]
2. Lallès, J.P.; Boudry, G.; Favier, C.; Le Floc'h, N.; Luron, I.L.; Montagne, L.; Sève, B. Gut Function and Dysfunction in Young Pigs: Physiology. *Anim. Res.* **2004**, *53*, 301–316. [CrossRef]
3. Campbell, J.M.; Crenshaw, J.D.; Polo, J. The biological stress of early weaned piglets. *J. Anim. Sci. Biotechnol.* **2013**, *4*, 19. [CrossRef] [PubMed]
4. Lallès, J.-P.; Bosi, P.; Smidt, H.; Stokes, C.R. Nutritional management of gut health in pigs around weaning. *Proc. Nutr. Soc.* **2007**, *66*, 260–268. [CrossRef] [PubMed]
5. McCracken, B.A.; Spurlock, M.E.; Roos, M.A.; Zuckermann, F.A.; Gaskins, H.R. Weaning Anorexia May Contribute to Local Inflammation in the Piglet Small Intestine. *J. Nutr.* **1999**, *129*, 613–619. [CrossRef] [PubMed]
6. Moeser, A.J.; Pohl, C.S.; Rajput, M. Weaning Stress and Gastrointestinal Barrier Development: Implications for Lifelong Gut Health in Pigs. *Anim. Nutr.* **2017**, *3*, 313–321. [CrossRef]

7. Pluske, J.R.; Hampson, D.J.; Williams, I.H. Factors influencing the structure and function of the small intestine in the weaned pig: A review. *Livest. Prod. Sci.* **1997**, *51*, 215–236. [CrossRef]
8. Dong, G.Z.; Pluske, J.R. The Low Feed Intake in Newly-weaned Pigs: Problems and Possible Solutions. *Asian-Australas. J. Anim. Sci.* **2007**, *20*, 440–452. [CrossRef]
9. Boudry, G.; Péron, V.; Le Huërou-Luron, I.; Lallès, J.-P.; Sève, B. Weaning Induces Both Transient and Long-Lasting Modifications of Absorptive, Secretory, and Barrier Properties of Piglet Intestine. *J. Nutr.* **2004**, *134*, 2256–2262. [CrossRef]
10. Hu, C.H.; Xiao, K.; Luan, Z.S.; Song, J. Early weaning increases intestinal permeability, alters expression of cytokine and tight junction proteins, and activates mitogen-activated protein kinases in pigs. *J. Anim. Sci.* **2013**, *91*, 1094–1101. [CrossRef]
11. Pluske, J.R. Feed and feed additives-related aspects of gut health and development in weanling pigs. *J. Anim. Sci. Biotechnol.* **2013**, *4*, 1. [CrossRef] [PubMed]
12. Castillo, M.; Martín-Orúe, S.M.; Nofrarías, M.; Manzanilla, E.G.; Gasa, J. Changes in Caecal Microbiota and Mucosal Morphology of Weaned Pigs. *Vet. Microbiol.* **2007**, *124*, 239–247. [CrossRef] [PubMed]
13. Guevarra, R.B.; Lee, J.H.; Lee, S.H.; Seok, M.-J.; Kim, D.W.; Na Kang, B.; Johnson, T.J.; Isaacson, R.E.; Kim, H.B. Piglet gut microbial shifts early in life: Causes and effects. *J. Anim. Sci. Biotechnol.* **2019**, *10*, 1. [CrossRef] [PubMed]
14. Bian, G.; Ma, S.; Zhu, Z.; Su, Y.; Zoetendal, E.G.; Mackie, R.; Zhu, W. Age, Introduction of Solid Feed and Weaning are More Important Determinants of Gut Bacterial Succession in Piglets Than Breed and Nursing Mother as Revealed by a Reciprocal Cross-Fostering Model. *Environ. Microbiol.* **2016**, *18*, 1566–1577. [CrossRef]
15. Canibe, N.; Højberg, O.; Kongsted, H.; Vodolazska, D.; Lauridsen, C.; Nielsen, T.S.; Schönherz, A.A. Review on Preventive Measures to Reduce Post-Weaning Diarrhoea in Piglets. *Animals* **2022**, *12*, 2585. [CrossRef]
16. Gresse, R.; Chaucheyras-Durand, F.; Fleury, M.A.; Van de Wiele, T.; Forano, E.; Blanquet-Diot, S. Gut Microbiota Dysbiosis in Postweaning Piglets: Understanding the Keys to Health. *Trends Microbiol.* **2017**, *25*, 851–873. [CrossRef]
17. Fairbrother, J.M.; Nadeau, É.; Gyles, C.L. *Escherichia coli* in postweaning diarrhea in pigs: An update on bacterial types, pathogenesis, and prevention strategies. *Anim. Heal. Res. Rev.* **2005**, *6*, 17–39. [CrossRef]
18. Frese, S.A.; Parker, K.; Calvert, C.C.; Mills, D.A. Diet shapes the gut microbiome of pigs during nursing and weaning. *Microbiome* **2015**, *3*, 28. [CrossRef]
19. Pluske, J.; Williams, I.; Aherne, F. Maintenance of Villous Height and Crypt Depth in Piglets by Providing Continuous Nu-trition After Weaning. *Anim. Sci.* **1996**, *62*, 131–144. [CrossRef]
20. Xiong, X.; Tan, B.; Song, M.; Ji, P.; Kim, K.; Yin, Y.; Liu, Y. Nutritional Intervention for the Intestinal Development and Health of Weaned Pigs. *Front. Veter Sci.* **2019**, *6*, 46. [CrossRef]
21. Spreeuwenberg, M.A.M.; Verdonk, J.M.A.J.; Gaskins, H.R.; Verstegen, M.W.A. Small Intestine Epithelial Barrier Function Is Compromised in Pigs with Low Feed Intake at Weaning. *J. Nutr.* **2001**, *131*, 1520–1527. [CrossRef] [PubMed]
22. Sun, Y.; Kim, S.W. Intestinal Challenge with Enterotoxigenic Escherichia coli in Pigs, and Nutritional Intervention to Prevent Postweaning Diarrhea. *Anim. Nutr.* **2017**, *3*, 322–330. [CrossRef] [PubMed]
23. Sterndale, S.O.; Evans, D.J.; Mansfield, J.P.; Clarke, J.; Sahibzada, S.; Abraham, S.; Pluske, J.R. Effect of Mucin 4 Allele on Sus-ceptibility to Experimental Infection with Enterotoxigenic F4 *Escherichia coli* in Pigs Fed Experimental Diets. *J. Anim. Sci. Biotechnol.* **2019**, *10*, 56.
24. O'Doherty, J.V.; Venardou, B.; Rattigan, R.; Sweeney, T. Feeding Marine Polysaccharides to Alleviate the Negative Effects Associated with Weaning in Pigs. *Animals* **2021**, *11*, 2644. [CrossRef] [PubMed]
25. Jayachandran, M.; Chen, J.; Chung, S.S.M.; Xu, B. A Critical Review on the Impacts of Beta-Glucans on Gut Microbiota and Human Health. *J. Nutr. Biochem.* **2018**, *61*, 101–110. [CrossRef] [PubMed]
26. Cerletti, C.; Esposito, S.; Iacoviello, L. Edible Mushrooms and Beta-Glucans: Impact on Human Health. *Nutrients* **2021**, *13*, 2195. [CrossRef] [PubMed]
27. Bashir, K.M.I.; Choi, J.-S. Clinical and Physiological Perspectives of β-Glucans: The Past, Present, and Future. *Int. J. Mol. Sci.* **2017**, *18*, 1906. [CrossRef] [PubMed]
28. Kim, K.; Ehrlich, A.; Perng, V.; Chase, J.A.; Raybould, H.; Li, X.; Liu, Y. Algae-Derived β-Glucan Enhanced Gut Health and Immune Responses of Weaned Pigs Experimentally Infected with a Pathogenic *E. coli*. *Anim. Feed Sci. Technol.* **2019**, *248*, 114–125. [CrossRef]
29. Choi, E.Y.; Lee, S.S.; Hyeon, J.Y.; Choe, S.H.; Keum, B.R.; Lim, J.M.; Cho, K.K. Effects of β-Glucan on the Release of Nitric Oxide by Macrophages Stimulated with Lipopolysaccharide. *Asian-Australas. J. Anim. Sci.* **2016**, *29*, 1664–1674. [CrossRef]
30. Kogan, G.; Kocher, A. Role of yeast cell wall polysaccharides in pig nutrition and health protection. *Livest. Sci.* **2007**, *109*, 161–165. [CrossRef]
31. Samuelsen, A.B.C.; Schrezenmeir, J.; Knutsen, S.H. Effects of orally administered yeast-derived beta-glucans: A review. *Mol. Nutr. Food Res.* **2013**, *58*, 183–193. [CrossRef] [PubMed]
32. Brown, J.; O'Callaghan, C.A.; Marshall, A.S.J.; Gilbert, R.J.C.; Siebold, C.; Gordon, S.; Jones, E.Y. Structure of the Fungal Be-ta-Glucan-Binding Immune Receptor Dectin-1: Implications for Function. *Protein Sci.* **2007**, *16*, 1042–1052. [CrossRef] [PubMed]
33. Stier, H.; Ebbeskotte, V.; Gruenwald, J. Immune-modulatory effects of dietary Yeast Beta-1,3/1,6-D-glucan. *Nutr. J.* **2014**, *13*, 38. [CrossRef] [PubMed]
34. Tsoni, S.V.; Brown, G.D. β-Glucans and Dectin-1. *Ann. N. Y. Acad. Sci.* **2008**, *1143*, 45–60. [CrossRef] [PubMed]

35. Sweeney, T.; Collins, C.; Reilly, P.; Pierce, K.; Ryan, M.; O'Doherty, J. Effect of Purified β-Glucans Derived from Laminaria Digitata, Laminaria Hyperborea, and Saccharomyces Cerevisiae on Piglet Performance, Selected Bacterial Populations, Volatile Fatty Acids and Pro-Inflammatory Cytokines in the Gastrointestinal Tract of Pigs. *Br. J. Nutr.* **2012**, *108*, 1226–1234. [CrossRef] [PubMed]
36. Luo, J.; Zeng, D.; Cheng, L.; Mao, X.; Yu, J.; Yu, B.; Chen, D. Dietary β-Glucan Supplementation Improves Growth Performance, Carcass Traits and Meat Quality of Finishing Pigs. *Anim. Nutr.* **2019**, *5*, 380–385. [CrossRef] [PubMed]
37. Dritz, S.S.; Shi, J.; Kielian, T.L.; Goodband, R.D.; Nelssen, J.L.; Tokach, M.D.; Chengappa, M.M.; Smith, J.E.; Blecha, F. Influence of dietary β-glucan on growth performance, nonspecific immunity, and resistance to *Streptococcus* suis infection in weanling pigs. *J. Anim. Sci.* **1995**, *73*, 3341–3350. [CrossRef]
38. Li, J.; Li, D.F.; Xing, J.J.; Cheng, Z.B.; Lai, C.H. Effects of β-Glucan Extracted from Saccharomyces Cerevisiae on Growth Performance, and Immunological and Somatotropic Responses of Pigs Challenged with *Escherichia coli* Lipopolysaccharide. *J. Anim. Sci.* **2006**, *84*, 2374–2381. [CrossRef]
39. Singh, R.P.; Bhardwaj, A. β-glucans: A potential source for maintaining gut microbiota and the immune system. *Front Nutr.* **2023**, *10*, 1143682. [CrossRef]
40. Choi, H.; Kim, S.W. Characterization of β-Glucans from Cereal and Microbial Sources and Their Roles in Feeds for Intestinal Health and Growth of Nursery Pigs. *Animals* **2023**, *13*, 2236. [CrossRef]
41. Kim, K.; Ehrlich, A.; Perng, V.; Chase, J.; Raybould, H.; Li, X.; Atwill, E.R.; Whelan, R.; Sokale, A.; Liu, Y. 069 Effects of dietary β-glucan on growth performance, diarrhea, and gut permeability of weanling pigs experimentally infected with a pathogenic *Escherichia coli*. *J. Anim. Sci.* **2017**, *95*, 34–35. [CrossRef]
42. Luo, J.; Chen, D.; Mao, X.; He, J.; Yu, B.; Cheng, L.; Zeng, D. Purified β-glucans of Different Molecular Weights Enhance Growth Performance of LPS-challenged Piglets via Improved Gut Barrier Function and Microbiota. *Animals* **2019**, *9*, 602. [CrossRef] [PubMed]
43. O'Shea, C.J.; Sweeney, T.; Lynch, M.B.; Gahan, D.A.; Callan, J.J.; O'Doherty, J.V. Effect of -glucans contained in barley- and oat-based diets and exogenous enzyme supplementation on gastrointestinal fermentation of finisher pigs and subsequent manure odor and ammonia emissions. *J. Anim. Sci.* **2009**, *88*, 1411–1420. [CrossRef] [PubMed]
44. O'Connell, J.; Sweeney, T.; Callan, J.J.; O'Doherty, J.V. The effect of cereal type and exogenous enzyme supplementation in pig diets on nutrient digestibility, intestinal microflora, volatile fatty acid concentration and manure ammonia emissions from finisher pigs. *Anim. Sci.* **2005**, *81*, 357–364. [CrossRef]
45. Garry, B.; Fogarty, M.; Curran, T.; O'Connell, M.; O'Doherty, J. The effect of cereal type and enzyme addition on pig performance, intestinal microflora, and ammonia and odour emissions. *Animal* **2007**, *1*, 751–757. [CrossRef] [PubMed]
46. Metzler-Zebeli, B.U.; Zebeli, Q. Cereal β-glucan alters nutrient digestibility and microbial activity in the intestinal tract of pigs, and lower manure ammonia emission: A meta-analysis. *J. Anim. Sci.* **2013**, *91*, 3188–3199. [CrossRef] [PubMed]
47. Johansen, H.N.; Knudsen, K.E.B.; Wood, P.J.; Fulcher, R.G. Physico-Chemical Properties and the Degradation of Oat Bran Polysaccharides in the Gut of Pigs. *J. Sci. Food Agric.* **1997**, *73*, 81–92. [CrossRef]
48. Virkki, L.; Johansson, L.; Ylinen, M.; Maunu, S.; Ekholm, P. Structural Characterization of Water-Insoluble Non-Starchy Polysaccharides of Oats and Barley. *Carbohydr. Polym.* **2005**, *59*, 357–366. [CrossRef]
49. Reilly, P.; Sweeney, T.; O'Shea, C.; Pierce, K.M.; Figat, S.; Smith, A.G.; Gahan, D.A.; O'Doherty, J.V. The Effect of Cereal-Derived Beta-Glucans and Exogenous Enzyme Supplementation on Intestinal Microflora, Nutrient Digestibility, Mineral Metabolism and Volatile Fatty Acid Concentrations in Finisher Pigs. *Anim. Feed Sci. Technol.* **2010**, *158*, 165–176. [CrossRef]
50. Bach Knudsen, K.E.; Jørgensen, H. Intestinal Degradation of Dietary Carbohydrates—From Birth to Maturity. In *Digestive Physiology in Pigs*; Lindberg, J.E., Ogle, B., Eds.; CABI Publishing: Wallingford, UK, 2001; pp. 109–120.
51. Corino, C.; Modina, S.C.; Di Giancamillo, A.; Chiapparini, S.; Rossi, R. Seaweeds in Pig Nutrition. *Animals* **2019**, *9*, 1126. [CrossRef]
52. Adams, J.; Ross, A.; Anastasakis, K.; Hodgson, E.; Gallagher, J.; Jones, J.; Donnison, I. Seasonal variation in the chemical composition of the bioenergy feedstock *Laminaria digitata* for thermochemical conversion. *Bioresour. Technol.* **2011**, *102*, 226–234. [CrossRef]
53. Kadam, S.U.; O'Donnell, C.P.; Rai, D.K.; Hossain, M.B.; Burgess, C.M.; Walsh, D.; Tiwari, B.K. Laminarin from Irish Brown Seaweeds *Ascophyllum nodosum* and *Laminaria hyperborea*: Ultrasound Assisted Extraction, Characterization and Bioactivity. *Mar. Drugs* **2015**, *13*, 4270–4280. [CrossRef] [PubMed]
54. Venardou, B.; O'doherty, J.V.; Garcia-Vaquero, M.; Kiely, C.; Rajauria, G.; McDonnell, M.J.; Ryan, M.T.; Sweeney, T. Evaluation of the Antibacterial and Prebiotic Potential of *Ascophyllum nodosum* and Its Extracts Using Selected Bacterial Members of the Pig Gastrointestinal Microbiota. *Mar. Drugs* **2021**, *20*, 41. [CrossRef] [PubMed]
55. Liu, Z.; Xiong, Y.; Yi, L.; Dai, R.; Wang, Y.; Sun, M.; Shao, X.; Zhang, Z.; Yuan, S. Endo-β-1,3-glucanase digestion combined with the HPAEC-PAD-MS/MS analysis reveals the structural differences between two laminarins with different bioactivities. *Carbohydr. Polym.* **2018**, *194*, 339–349. [CrossRef] [PubMed]
56. Sellimi, S.; Maalej, H.; Rekik, D.M.; Benslima, A.; Ksouda, G.; Hamdi, M.; Sahnoun, Z.; Li, S.; Nasri, M.; Hajji, M. Antioxidant, antibacterial and in vivo wound healing properties of laminaran purified from *Cystoseira barbata* seaweed. *Int. J. Biol. Macromol.* **2018**, *119*, 633–644. [CrossRef] [PubMed]

57. Walsh, A.; Sweeney, T.; O'Shea, C.; Doyle, D.; O'Doherty, J. Effect of supplementing varying inclusion levels of laminarin and fucoidan on growth performance, digestibility of diet components, selected faecal microbial populations and volatile fatty acid concentrations in weaned pigs. *Anim. Feed. Sci. Technol.* **2013**, *183*, 151–159. [CrossRef]
58. Bouwhuis, M.A.; Sweeney, T.; Mukhopadhya, A.; Thornton, K.; McAlpine, P.O.; O'Doherty, J.V. Zinc methionine and laminarin have growth-enhancing properties in newly weaned pigs influencing both intestinal health and diarrhoea occurrence. *J. Anim. Physiol. Anim. Nutr.* **2016**, *101*, 1273–1285. [CrossRef] [PubMed]
59. Murphy, P.; Bello, F.D.; O'Doherty, J.; Arendt, E.K.; Sweeney, T.; Coffey, A. Analysis of bacterial community shifts in the gastrointestinal tract of pigs fed diets supplemented with β-glucan from *Laminaria digitata*, *Laminaria hyperborea* and *Saccharomyces cerevisiae*. *Animal* **2013**, *7*, 1079–1087. [CrossRef]
60. Vigors, S.; O'doherty, J.V.; Rattigan, R.; McDonnell, M.J.; Rajauria, G.; Sweeney, T. Effect of a Laminarin Rich Macroalgal Extract on the Caecal and Colonic Microbiota in the Post-Weaned Pig. *Mar. Drugs* **2020**, *18*, 157. [CrossRef]
61. Rattigan, R.; Sweeney, T.; Maher, S.; Ryan, M.T.; Thornton, K.; O'doherty, J.V. Effects of reducing dietary crude protein concentration and supplementation with either laminarin or zinc oxide on the growth performance and intestinal health of newly weaned pigs. *Anim. Feed. Sci. Technol.* **2020**, *270*, 114693. [CrossRef]
62. Ryan, M.T.; O'Shea, C.J.; Collins, C.B.; O'Doherty, J.V.; Sweeney, T. Effects of dietary supplementation with *Laminaria hyperborea*, *Laminaria digitata*, and *Saccharomyces cerevisiae* on the IL-17 pathway in the porcine colon. *J. Anim. Sci.* **2012**, *90*, 263–265. [CrossRef] [PubMed]
63. Walsh, A.M.; Sweeney, T.; O'Shea, C.J.; Doyle, D.N.; O'Doherty, J.V. Effect of dietary laminarin and fucoidan on selected microbiota, intestinal morphology and immune status of the newly weaned pig. *Br. J. Nutr.* **2013**, *110*, 1630–1638. [CrossRef] [PubMed]
64. McDonnell, P.; Figat, S.; O'Doherty, J.V. The effect of dietary laminarin and fucoidan in the diet of the weanling piglet on performance, selected faecal microbial populations and volatile fatty acid concentrations. *Animal* **2010**, *4*, 579–585. [CrossRef] [PubMed]
65. Venardou, B.; O'doherty, J.V.; Garcia-Vaquero, M.; Kiely, C.; Rajauria, G.; McDonnell, M.J.; Ryan, M.T.; Sweeney, T. In Vitro Evaluation of Brown Seaweed *Laminaria* spp. as a Source of Antibacterial and Prebiotic Extracts That Could Modulate the Gastrointestinal Microbiota of Weaned Pigs. *Animals* **2023**, *13*, 823. [CrossRef] [PubMed]
66. Yang, P.; Ma, Y. Recent advances of vitamin D in immune, reproduction, performance for pig: A review. *Anim. Health Res. Rev.* **2021**, *22*, 85–95. [CrossRef] [PubMed]
67. European Food Safety Authority (EFSA). Safety and Efficacy of 25-Hydroxycholecalciferol as a Feed Additive for Poultry and Pigs. *EFSA J.* **2009**, *7*, 969. [CrossRef]
68. Duffy, S.K.; Kelly, A.K.; Rajauria, G.; Jakobsen, J.; Clarke, L.C.; Monahan, F.J.; O'Doherty, J.V. The Use of Synthetic and Natural Vitamin D Sources in Pig Diets to Improve Meat Quality and Vitamin D Content. *Meat Sci.* **2018**, *143*, 60–68. [CrossRef]
69. Yang, J.; Tian, G.; Chen, D.; Zheng, P.; Yu, J.; Mao, X.; Yu, B. Effects of Dietary 25-Hydroxyvitamin D3 Supplementation on Growth Performance, Immune Function and Antioxidative Capacity in Weaned Piglets. *Arch. Anim. Nutr.* **2019**, *73*, 44–51. [CrossRef]
70. NRC. *Nutrient Requirements of Swine*, 11th Revised ed.; The National Academies Press: Washington, DC, USA, 2012.
71. Zhang, L.; Yang, M.; Piao, X. Effects of 25-Hydroxyvitamin D3 on Growth Performance, Serum Parameters, Fecal Microbiota, and Metabolites in Weaned Piglets Fed Diets with Low Calcium and Phosphorus. *J. Sci. Food Agric.* **2022**, *102*, 597–606. [CrossRef]
72. Adams, J.S.; Liu, P.T.; Chun, R.; Modlin, R.L.; Hewison, M. Vitamin D in Defense of the Human Immune Response. *Ann. N. Y. Acad. Sci.* **2007**, *1117*, 94–105. [CrossRef]
73. Baeke, F.; Takiishi, T.; Korf, H.; Gysemans, C.; Mathieu, C. Vitamin D: Modulator of the immune system. *Curr. Opin. Pharmacol.* **2010**, *10*, 482–496. [CrossRef] [PubMed]
74. Colotta, F.; Jansson, B.; Bonelli, F. Modulation of inflammatory and immune responses by vitamin D. *J. Autoimmun.* **2017**, *85*, 78–97. [CrossRef] [PubMed]
75. Lu, L.; Li, S.; Zhang, L.; Liu, X.; Li, D.; Zhao, X.; Liu, Y. Expression of β-Defensins in Intestines of Chickens Injected with Vitamin D. *Genet. Mol. Res.* **2015**, *14*, 3330–3337. [CrossRef] [PubMed]
76. Morris, A.; Shanmugasundaram, R.; Lilburn, M.S.; Selvaraj, R.K. 25-Hydroxycholecalciferol Supplementation Improves Growth Performance and Decreases Inflammation during an Experimental Lipopolysaccharide Injection. *Poult. Sci.* **2014**, *93*, 1951–1956. [CrossRef] [PubMed]
77. Morris, A.; Shanmugasundaram, R.; McDonald, J.; Selvaraj, R. Effect of In Vitro and In Vivo 25-Hydroxyvitamin D Treatment on Macrophages, T Cells, and Layer Chickens during a Coccidia Challenge. *J. Anim. Sci.* **2015**, *93*, 2894–2903. [CrossRef] [PubMed]
78. Shojadoost, B.; Behboudi, S.; Villanueva, A.; Brisbin, J.; Ashkar, A.; Sharif, S. Vitamin D3 modulates the function of chicken macrophages. *Res. Veter- Sci.* **2015**, *100*, 45–51. [CrossRef]
79. Yang, J.; Tian, G.; Chen, D.; Zheng, P.; Yu, J.; Mao, X.; He, J.; Luo, Y.; Luo, J.; Huang, Z.; et al. Dietary 25-Hydroxyvitamin D3 Supplementation Alleviates Porcine Epidemic Diarrhea Virus Infection by Improving Intestinal Structure and Immune Response in Weaned Pigs. *Animals* **2019**, *9*, 627. [CrossRef]
80. Zhang, L.; Liu, S.; Piao, X. Dietary 25-Hydroxycholecalciferol Supplementation Improves Performance, Immunity, Antioxi-dant Status, Intestinal Morphology, and Bone Quality in Weaned Piglets. *J. Sci. Food Agric.* **2021**, *101*, 2592–2600. [CrossRef]
81. Thomson, C.D. Selenium | Physiology. In *Encyclopedia of Food Sciences and Nutrition*, 2nd ed.; Caballero, B., Ed.; Academic Press: Oxford, UK, 2003; pp. 5117–5124.

82. Avery, J.C.; Hoffmann, P.R. Selenium, Selenoproteins, and Immunity. *Nutrients* **2018**, *10*, 1203. [CrossRef]
83. Lubos, E.; Loscalzo, J.; Handy, D.E.; Mailloux, R.J.; Treberg, J.R.; Spiers, J.G.; Chen, H.-J.C.; Cuffe, J.S.; Sernia, C.; Lavidis, N.A.; et al. Glutathione Peroxidase-1 in Health and Disease: From Molecular Mechanisms to Therapeutic Opportunities. *Antioxid. Redox Signal.* **2011**, *15*, 1957–1997. [CrossRef]
84. Verma, A.K.; Kumar, A.; Rahal, A.; Kumar, V.; Roy, D. Inorganic Versus Organic Selenium Supplementation: A Review. *Pak. J. Biol. Sci.* **2012**, *15*, 418–425. [CrossRef]
85. Mahan, D.C.; Peters, J.C. Long-term effects of dietary organic and inorganic selenium sources and levels on reproducing sows and their progeny. *J. Anim. Sci.* **2004**, *82*, 1343–1358. [CrossRef] [PubMed]
86. Nathues, H.; Boehne, I.; Grosse Beilage, T.; Gerhauser, I.; Hewicker-Trautwein, M.; Wolf, P.; Grosse Beilage, E. Peracute Se-lenium Toxicosis Followed by Sudden Death in Growing and Finishing Pigs. *Can. Vet. J.* **2010**, *51*, 515–518. [PubMed]
87. Victor, H.; Zhao, B.; Mu, Y.; Dai, X.; Wen, Z.; Gao, Y.; Chu, Z. Effects of Se-chitosan on the Growth Performance and Intestinal Health of the Loach *Paramisgurnus dabryanus* (Sauvage). *Aquaculture* **2019**, *498*, 263–270. [CrossRef]
88. Ren, Z.; Zhao, Z.; Wang, Y.; Huang, K. Preparation of Selenium/Zinc-Enriched Probiotics and Their Effect on Blood Selenium and Zinc Concentrations, Antioxidant Capacities, and Intestinal Microflora in Canine. *Biol. Trace Elem. Res.* **2010**, *141*, 170–183. [CrossRef]
89. Molan, A.L.; Liu, Z.; Tiwari, R. The Ability of Green Tea to Positively Modulate Key Markers of Gastrointestinal Function in Rats. *Phytother. Res.* **2010**, *24*, 1614–1619. [CrossRef]
90. Dalia, A.M.; Loh, T.C.; Sazili, A.Q.; Jahromi, M.F.; Samsudin, A.A. Effects of vitamin E, inorganic selenium, bacterial organic selenium, and their combinations on immunity response in broiler chickens. *BMC Veter. Res.* **2018**, *14*, 249. [CrossRef]
91. Kaushal, N.; Kudva, A.K.; Patterson, A.D.; Chiaro, C.; Kennett, M.J.; Desai, D.; Prabhu, K.S. Crucial Role of Macrophage Se-lenoproteins in Experimental Colitis. *J. Immunol.* **2014**, *193*, 3683–3692. [CrossRef]
92. Gîlcă-Blanariu, G.-E.; Diaconescu, S.; Ciocoiu, M.; Ştefănescu, G. New Insights into the Role of Trace Elements in IBD. *BioMed Res. Int.* **2018**, *2018*, 1813047. [CrossRef]
93. Liu, L.; Chen, D.; Yu, B.; Luo, Y.; Huang, Z.; Zheng, P.; Mao, X.; Yu, J.; Luo, J.; Yan, H.; et al. Influences of Selenium-Enriched Yeast on Growth Performance, Immune Function, and Antioxidant Capacity in Weaned Pigs Exposure to Oxidative Stress. *BioMed Res. Int.* **2021**, *2021*, 5533210. [CrossRef]
94. Khan, A.Z.; Khan, I.U.; Khan, S.; Afzal, S.; Hamid, M.; Tariq, M.; Liu, R. Selenium-Enriched Probiotics Improve Hepatic Protection by Regulating Pro-Inflammatory Cytokines and Antioxidant Capacity in Broilers Under Heat Stress Conditions. *J. Adv. Vet. Anim. Res.* **2019**, *6*, 355–361. [CrossRef] [PubMed]
95. Rayman, M.P. The importance of selenium to human health. *Lancet* **2000**, *356*, 233–241. [CrossRef] [PubMed]
96. Hoffmann, P.R.; Berry, M.J. The influence of selenium on immune responses. *Mol. Nutr. Food Res.* **2008**, *52*, 1273–1280. [CrossRef] [PubMed]
97. Ala, M.; Kheyri, Z. The rationale for selenium supplementation in inflammatory bowel disease: A mechanism-based point of view. *Nutrition* **2021**, *85*, 111153. [CrossRef] [PubMed]
98. Papoutsis, K.; Grasso, S.; Menon, A.; Brunton, N.P.; Lyng, J.G.; Jacquier, J.C.; Bhuyan, D.J. Recovery of Ergosterol and Vitamin D2 from Mushroom Waste—Potential Valorization by Food and Pharmaceutical Industries. *Trends Food Sci. Technol.* **2020**, *99*, 351–366. [CrossRef]
99. Öztürk, M.; Tel-Çayan, G.; Muhammad, A.; Terzioğlu, P.; Duru, M.E. Mushrooms: A Source of Exciting Bioactive Compounds. In *Studies in Natural Products Chemistry*; Atta-ur-Rahman, Ed.; Elsevier: New York, NY, USA, 2015; Volume 45, pp. 363–456.
100. Cardwell, G.; Bornman, J.F.; James, A.P.; Black, L.J. A Review of Mushrooms as a Potential Source of Dietary Vitamin D. *Nutrients* **2018**, *10*, 1498. [CrossRef]
101. Hewison, M. Vitamin D and immune function: An overview. *Proc. Nutr. Soc.* **2012**, *71*, 50–61. [CrossRef]
102. Solano-Aguilar, G.I.; Jang, S.; Lakshman, S.; Gupta, R.; Beshah, E.; Sikaroodi, M.; Vinyard, B.; Molokin, A.; Gillevet, P.M.; Urban, J.F. The Effect of Dietary Mushroom *Agaricus bisporus* on Intestinal Microbiota Composition and Host Immunological Function. *Nutrients* **2018**, *10*, 1721. [CrossRef]
103. Surai, P.F. *Selenium in Nutrition and Health*; Nottingham University Press: Nottingham, UK, 2006.
104. Conway, E.; Sweeney, T.; Dowley, A.; Vigors, S.; Ryan, M.; Yadav, S.; Wilson, J.; O'doherty, J.V. Selenium-Enriched Mushroom Powder Enhances Intestinal Health and Growth Performance in the Absence of Zinc Oxide in Post-Weaned Pig Diets. *Animals* **2022**, *12*, 1503. [CrossRef]
105. Dowley, A.; Sweeney, T.; Conway, E.; Vigors, S.; Ryan, M.T.; Yadav, S.; Wilson, J.; O'doherty, J.V. The effects of dietary supplementation with mushroom or selenium enriched mushroom powders on the growth performance and intestinal health of post-weaned pigs. *J. Anim. Sci. Biotechnol.* **2023**, *14*, 12. [CrossRef]
106. Mukhopadhya, A.; O'Doherty, J.V.; Sweeney, T. A Combination of Yeast Beta-Glucan and Milk Hydrolysate is a Suitable Alternative to Zinc Oxide in the Race to Alleviate Post-Weaning Diarrhea in Piglets. *Sci. Rep.* **2019**, *9*, 616. [CrossRef] [PubMed]
107. Hou, Y.; Wu, Z.; Dai, Z.; Wang, G.; Wu, G. Protein hydrolysates in animal nutrition: Industrial production, bioactive peptides, and functional significance. *J. Anim. Sci. Biotechnol.* **2017**, *8*, 24. [CrossRef] [PubMed]
108. Mohanty, D.; Mohapatra, S.; Misra, S.; Sahu, P. Milk derived bioactive peptides and their impact on human health—A review. *Saudi J. Biol. Sci.* **2015**, *23*, 577–583. [CrossRef] [PubMed]

109. Mukhopadhya, A.; Sweeney, T. Milk Proteins: Processing of Bioactive Fractions and Effects on Gut Health. In *Milk Proteins*; IntechOpen: London, UK, 2016; p. 83.
110. Nagpal, R.; Behare, P.; Rana, R.; Kumar, A.; Kumar, M.; Arora, S.; Yadav, H. Bioactive Peptides Derived from Milk Proteins and Their Health Beneficial Potentials: An Update. *Food Funct.* **2011**, *2*, 18–27. [CrossRef]
111. Toopcham, T.; Mes, J.J.; Wichers, H.J.; Yongsawatdigul, J. Immunomodulatory Activity of Protein Hydrolysates Derived from Virgibacillus Halodenitrificans SK1-3-7 Proteinase. *Food Chem.* **2017**, *224*, 320–328. [CrossRef] [PubMed]
112. Yao, M.; McClements, D.J.; Xiao, H. Improving Oral Bioavailability of Nutraceuticals by Engineered Nanoparticle-Based Delivery Systems. *Curr. Opin. Food Sci.* **2015**, *2*, 14–19. [CrossRef]
113. Cao, H.; Yu, R.; Zhang, Y.; Hu, B.; Jian, S.; Wen, C.; Yang, G. Effects of Dietary Supplementation with β-Glucan and Bacillus Subtilis on Growth, Fillet Quality, Immune Capacity, and Antioxidant Status of Pengze Crucian Carp (Carassius Auratus Var. Pengze). *Aquaculture* **2019**, *508*, 106–112. [CrossRef]
114. Du, B.; Lin, C.; Bian, Z.; Xu, B. An Insight into Anti-Inflammatory Effects of Fungal Beta-Glucans. *Trends Food Sci. Technol.* **2015**, *41*, 49–59. [CrossRef]
115. Suchecka, D.; Harasym, J.; Wilczak, J.; Gajewska, M.; Oczkowski, M.; Gudej, S.; Gromadzka-Ostrowska, J. Antioxidative and Anti-Inflammatory Effects of High Beta-Glucan Concentration Purified Aqueous Extract from Oat in Experimental Model of LPS-Induced Chronic Enteritis. *J. Funct. Foods* **2015**, *14*, 244–254. [CrossRef]
116. Dowley, A.; O'doherty, J.V.; Mukhopadhya, A.; Conway, E.; Vigors, S.; Maher, S.; Ryan, M.T.; Sweeney, T. Maternal supplementation with a casein hydrolysate and yeast beta-glucan from late gestation through lactation improves gastrointestinal health of piglets at weaning. *Sci. Rep.* **2022**, *12*, 17407. [CrossRef]
117. Gomez de Agüero, M.; Ganal-Vonarburg, S.C.; Fuhrer, T.; Rupp, S.; Uchimura, Y.; Li, H.; Steinert, A.; Heikenwalder, M.; Hapfelmeier, S.; Sauer, U.; et al. The maternal microbiota drives early postnatal innate immune development. *Science* **2016**, *351*, 1296–1302. [CrossRef] [PubMed]
118. Bauer, E.; Williams, B.A.; Smidt, H.; Verstegen, M.W.; Mosenthin, R. Influence of the Gastrointestinal Microbiota on Devel-opment of the Immune System in Young Animals. *Curr. Issues Intest. Microbiol.* **2006**, *7*, 35–52. [PubMed]
119. Heim, G.; Sweeney, T.; O'Shea, C.J.; Doyle, D.N.; O'Doherty, J.V. Effect of maternal supplementation with seaweed extracts on growth performance and aspects of gastrointestinal health of newly weaned piglets after challenge with enterotoxigenic *Escherichia coli* K88. *Br. J. Nutr.* **2014**, *112*, 1955–1965. [CrossRef] [PubMed]
120. Crespo-Piazuelo, D.; Gardiner, G.E.; Ranjitkar, S.; Bouwhuis, M.A.; Ham, R.; Phelan, J.P.; Lawlor, P.G. Maternal Supplemen-tation with *Bacillus altitudinis* Spores Improves Porcine Offspring Growth Performance and Carcass Weight. *Br. J. Nutr.* **2022**, *127*, 403–420. [CrossRef] [PubMed]
121. Leonard, S.G.; Sweeney, T.; Bahar, B.; O'Doherty, J.V. Effect of maternal seaweed extract supplementation on suckling piglet growth, humoral immunity, selected microflora, and immune response after an ex vivo lipopolysaccharide challenge1. *J. Anim. Sci.* **2012**, *90*, 505–514. [CrossRef] [PubMed]
122. Xu, R.; Wang, F.; Zhang, S. Postnatal adaptation of the gastrointestinal tract in neonatal pigs: A possible role of milk-borne growth factors. *Livest. Prod. Sci.* **2000**, *66*, 95–107. [CrossRef]
123. Rooke, J.A.; Bland, I.M. The acquisition of passive immunity in the new-born piglet. *Livest. Prod. Sci.* **2002**, *78*, 13–23. [CrossRef]
124. Le Dividich, J.; Rooke, J.A.; Herpin, P. Nutritional and immunological importance of colostrum for the new-born pig. *J. Agric. Sci.* **2005**, *143*, 469–485. [CrossRef]
125. Klobasa, F.; Werhahn, E.; Butler, J.E. Composition of Sow Milk During Lactation. *J. Anim. Sci.* **1987**, *64*, 1458–1466. [CrossRef]
126. Quesnel, H. Colostrum production by sows: Variability of colostrum yield and immunoglobulin G concentrations. *Animal* **2011**, *5*, 1546–1553. [CrossRef]
127. Theil, P.K.; Lauridsen, C.; Quesnel, H. Neonatal Piglet Survival: Impact of Sow Nutrition Around Parturition on Fetal Gly-cogen Deposition and Production and Composition of Colostrum and Transient Milk. *Animal* **2014**, *8*, 1021–1030. [CrossRef] [PubMed]
128. Jang, Y.; Kang, K.; Piao, L.; Jeong, T.; Auclair, E.; Jonvel, S.; Kim, Y. Effects of Live Yeast Supplementation to Gestation and Lactation Diets on Reproductive Performance, Immunological Parameters and Milk Composition in Sows. *Livest. Sci.* **2013**, *152*, 167–173. [CrossRef]
129. Zanello, G.; Meurens, F.; Serreau, D.; Chevaleyre, C.; Melo, S.; Berri, M.; Salmon, H. Effects of Dietary Yeast Strains on Immunoglobulin in Colostrum and Milk of Sows. *Vet. Immunol. Immunopathol.* **2013**, *152*, 20–27. [CrossRef] [PubMed]
130. Leonard, S.; Sweeney, T.; Bahar, B.; Lynch, B.; O'doherty, J. Effect of Maternal Fish Oil and Seaweed Extract Supplementation on Colostrum and Milk Composition, Humoral Immune Response, and Performance of Suckled Piglets. *J. Anim. Sci.* **2010**, *88*, 2988–2997. [CrossRef]

animals

MDPI

Review

Characterization of β-Glucans from Cereal and Microbial Sources and Their Roles in Feeds for Intestinal Health and Growth of Nursery Pigs

Hyunjun Choi and Sung Woo Kim *

Department of Animal Science, North Carolina State University, Raleigh, NC 27695, USA
* Correspondence: sungwoo_kim@ncsu.edu

Simple Summary: The use of antibiotics in animal feeds has been phased out due to concerns surrounding microbial resistance to antibiotics. β-glucans have been shown to improve the intestinal health and growth performance of nursery pigs. β-glucans are non-starch polysaccharides originating from the cell walls of various sources including yeast, bacteria, fungi, and cereal grains. Depending on the sources and dose levels of β-glucans, however, their impacts on intestinal health and growth were not consistent due to the quantitative, compositional, and structural differences of β-glucans. Cereal grains-based diets provide high amounts of soluble fractions of β-glucans, causing digesta viscosity in the GIT of pigs and interfering with the nutrient digestion and intestinal health of pigs. Microbial β-glucans, however, showed positive effects on the intestinal health and growth of nursery pigs. Microbial β-glucans affect the intestinal immune system through activating dectin-1 and toll-like receptors related to the intestinal health of nursery pigs. Therefore, this review investigated the quantitative, compositional, and structural differences of β-glucans and the functional roles of β-glucans in the intestinal health and growth efficiency of nursery pigs.

Abstract: The objectives of this review are to investigate the quantitative, compositional, and structural differences of β-glucans and the functional effects of β-glucans on the intestinal health and growth of nursery pigs. Banning antibiotic feed supplementation increased the research demand for antibiotic alternatives to maintain the intestinal health and growth of nursery pigs. It has been proposed that β-glucans improve the growth efficiency of nursery pigs through positive impacts on their intestinal health. However, based on their structure and source, their impacts can be extensively different. β-glucans are non-starch polysaccharides found in the cell walls of yeast (*Saccharomyces cerevisiae*), bacteria, fungi (*Basidiomycota*), and cereal grains (mainly barley and oats). The total β-glucan content from cereal grains is much greater than that of microbial β-glucans. Cereal β-glucans may interfere with the positive effects of microbial β-glucans on the intestinal health of nursery pigs. Due to their structural differences, cereal β-glucans also cause digesta viscosity, decreasing feed digestion, and decreasing nutrient absorption in the GIT of nursery pigs. Specifically, cereal β-glucans are based on linear glucose molecules linked by β-(1,3)- and β-(1,4)-glycosidic bonds with relatively high water-soluble properties, whereas microbial β-glucans are largely linked with β-(1,3)- and β-(1,6)-glycosidic bonds possessing insoluble properties. From the meta-analysis, the weight gain and feed intake of nursery pigs increased by 7.6% and 5.3%, respectively, through the use of yeast β-glucans (from *Saccharomyces cerevisiae*), and increased by 11.6% and 6.9%, respectively, through the use of bacterial β-glucans (from *Agrobacterium* sp.), whereas the use of cereal β-glucans did not show consistent responses. The optimal use of yeast β-glucans (*Saccharomyces cerevisiae*) was 50 mg/kg in nursery pig diets based on a meta-analysis. Collectively, use of microbial β-glucans can improve the intestinal health of nursery pigs, enhancing immune conditions, whereas the benefits of cereal β-glucans on intestinal health were not consistent.

Keywords: β-glucan; growth performance; intestinal health; prebiotics; swine

1. Introduction

Weaning is considered the most critical period for nursery pigs, as piglets are exposed to a new environment, are separated from their dam, struggle with new pen mates, and transition from milk to solid feeds, which all negatively affect their overall health, intestinal immune status, and growth performance [1,2]. Antibiotics have been used in nursery feeds to mitigate the negative effects of weaning stress and to improve the intestinal health and growth of nursery pigs. Due to concerns about antibiotic-resistant bacteria, however, the use of antibiotics in feeds has been phased out in many countries [3]. Thus, there is a demand for the investigation of feed additives to reduce the usage of antibiotics and to improve the growth rate of pigs [4]. β-glucans, non-starch polysaccharides (NSP) in cereal grains and microorganisms, have been proposed as a potential means of improving the intestinal health and growth of nursery pigs [5,6]. However, cereal and microbial β-glucans (yeast, bacteria, and other origins) have compositional and structural differences [7].

Cereal β-glucans are based on linear glucose molecules linked by β-(1,3)- and β-(1,4)-glycosidic bonds with relatively high water-soluble properties, whereas yeast β-glucans (from *Saccharomyces cerevisiae*) are largely linked with β-(1,3)- and β-(1,6)-glycosidic bonds possessing insoluble properties [8,9]. Moreover, the total β-glucan content from microbial β-glucans (yeast, bacteria, and algae) is lower compared with the levels found in cereal grain-based diets. Due to these differences, cereal β-glucans can cause increased viscosity of digesta and negatively impact feed digestion in nursery pigs [10], whereas microbial β-glucans may not have those effects. Therefore, the objectives of this review are to investigate the compositional and structural differences between cereal and microbial β-glucans, to provide an overview of the functional effects of microbial β-glucans on intestinal health and growth of nursery pigs, and to investigate the potential application of microbial β-glucans as a feed additive for growth of nursery pigs.

2. Difference of Composition and Structure of β-Glucans Influence Viscosity of Digesta in GIT of Nursery Pigs

β-glucans are NSP that make up a component of cell walls. β-glucans are derived from yeast (*Saccharomyces cerevisiae*), bacteria, fungi, and cereal grains (mainly from barley and oats) [7]. Those β-glucan sources can cause increased viscosity of digesta in the gastrointestinal tract (GIT) of pigs. Viscosity of digesta in the GIT of nursery pigs, however, can be influenced by the structure, amounts, purity, and molecular weight of β-glucans [9]. Therefore, understanding the compositional and structural differences in β-glucan sources is critical to investigating their effects on the intestinal health and growth of nursery pigs.

Structural and Compositional Difference of β-Glucans

Barley and oats contain generally higher content of β-glucans than other cereal feedstuffs [11]. The β-glucan content from barley was 5 to 11%, and 3 to 7% from oats [12]. In cereal grains, β-glucans are present in endosperm and sub-aleurone cell walls [7], which require breakdown during the digestion process in pigs.

Cereal β-glucans are based on linear glucose molecules linked with β-(1,3)- and β-(1,4)-glycosidic bonds with relatively high water-soluble properties in the digesta of animals [7] (Figure 1). However, the β-(1,3)- to β-(1,4)-glycosidic bonds ratio of barley is greater than that of oats. In β-glucans, the β-(1,3)-glycosidic bonds are relatively more fermentable than β-(1,4)-glycosidic bonds in the digesta, and the lower molecular weight of β-glucans also increases the fermentation in the digesta of pigs [13]. Barley had a greater proportion of β-(1,3)-glycosidic bonds and a lower molecular weight than oats [14], which may result in higher water-soluble digesta in pigs fed barley-based diets than that in pigs fed oat-based diets [15]. A previous study showed that the β-glucans of barley are already 80% depolymerized in the small intestine of pigs [13]. Moreover, the ileal digestibility of barley β-glucans ranged from 63 to 72%, and the total tract digestibility ranged from 89 to 93%, indicating that most of the β-glucans in barley are digested in the small intestine of pigs [16]. Thus, β-glucans in barley may have greater water solubility in the GIT of pigs compared with that in oats.

Unlike cereal β-glucans, yeast β-glucans (from *Saccharomyces cerevisiae*) are largely linked with β-(1,3)- and β-(1,6)-glycosidic bonds, which contain 53 to 83% of the insoluble fraction [7]. However, structural differences also exist within the microbial β-glucans, which can affect the viscosity in the digesta of nursery pigs. The β-(1,3)-glycosidic bonds are relatively soluble, whereas β-(1,6)-glycosidic bonds are less soluble in the digesta of pigs [17]. Laminarin, a β-glucan derived from algae, is extensively linked with β-(1,3)-glycosidic bonds randomly attached to β-(1,6)-glycosidic bonds, making it relatively soluble and thus causing viscosity in the digesta of pigs. However, laminarin from *Laminaria hyperborean* is interestingly less fermentable due to fewer β-(1,3)-bonds not causing viscous digesta in pigs [18]. The β-glucans from yeast (*Saccharomyces cerevisiae*) mainly consist of branched β-(1,3)-linkage bonds and generally have greater molecular weight compared with Laminarin [19]. The structure of the bacterial β-glucan (from *Agrobacterium* sp.) mainly consists of linear β-(1,3)-glycosidic bonds. Therefore, considering the structural difference among the microbial β-glucans, yeast β-glucans (from *Saccharomyces cerevisiae*) generally have less soluble properties than algal and bacterial β-glucans in the GIT of nursery pigs.

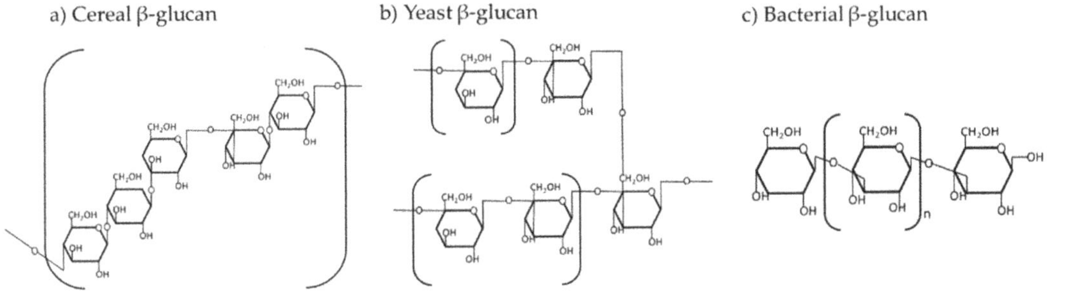

Figure 1. Structural and branching degree of β-glucans from different sources: (**a**) cereal β-glucans (linked with β-(1,3)- and β-(1,4)-glycosidic bonds); (**b**) yeast β-glucans (linked with β-(1,3)- and β-(1,6)-glycosidic bonds); and (**c**) bacterial β-glucans (linked with β-(1,3)-glycosidic bonds). The concept used in this figure was adapted from Du et al. [20].

Quantitative contributions of β-glucans in typical feed fed to pigs are mainly from cereal grains (~30 g/kg feed) [21–23] rather than microbial feed additives (~1 g/kg feed) (Tables 1–3). Considering the property of β-glucans from microorganisms, the use of microbial feed additives would not cause viscosity issues in the GIT of pigs. Viscosity refers to the ability of mixed fluids (digesta) and soluble polysaccharides such as gums, pectin, and β-glucans to thicken or form gels in the GIT of pigs [24]. In pigs fed diets with highly soluble NSP, the viscosity of digesta was increased [25,26]. Specifically, in pigs fed barley-based diets, the viscosity of digesta in the stomach and ileum was greater when compared with corn-based diets [27]. This is likely due to the high content of soluble NSP in barley [28]. Moreover, barley-based diets also increased the viscosity of digesta in the small intestine of nursery pigs, which can result in a higher incidence of enterotoxigenic *Escherichia coli* (ETEC) infections [29] and reduced feed digestion [30]. Exogenous enzymes can degrade the NSP fractions to reduce viscosity of digesta of nursery pigs [30,31]. However, the viscosity of digesta was not decreased by enzyme supplementation of nursery pigs fed diets containing 50% barley [32]. Additionally, 10% oat-derived β-glucans did not affect the viscosity of digesta, except in the stomach [33]. The possibility of diverse outcomes is likely due to the high depolymerization of the β-glucans from various sources in the GIT of pigs [34]. The depolymerization of cereal β-glucans occurs in the stomach [35], and a high proportion of β-glucans are hydrolyzed in the small intestine of pigs [16,36]. Therefore, some β-glucans in cereal grain-based diet such as processed barley could decrease the intestinal health of nursery pigs by increasing viscosity, whereas microbial β-glucans may not cause increased viscosity and decreased feed digestion.

3. Effects of Dietary β-Glucans on Intestinal Microbiota and Intestinal Health of Nursery Pigs

The intestinal tract is where feed digestion and absorption occur. Intestinal health is inclusive of seven major criteria: (1) mucosal and luminal microbiota; (2) mucosal inflammation; (3) mucosal oxidative stress; (4) morphological damages and mucosal integrity; (5) crypt stem cell proliferation and tissue repair; (6) effective digestion and absorption of nutrients; and (7) overall well-being and growth efficiency [37]. Among the factors that influence the intestinal health of nursery pigs, feed is highly influential on the intestinal microbiota, intestinal immune responses, and digestion and absorption of nutrients [34,38]. Both cereal and microbial β-glucans (from yeast and bacteria) have been shown to improve the intestinal health of nursery pigs [6,39]. However, some previous studies have not detected the positive effects of dietary β-glucans on intestinal health of nursery pigs, raising questions about the efficacy of β-glucans on the improving intestinal health of pigs. Therefore, this section is focused on the potential of β-glucans to improve the intestinal health of nursery pigs.

3.1. *Effects of Cereal β-Glucans on Intestinal Health of Nursery Pigs*

Supplementation of exogenous β-glucans extracted from cereal grains at 3.5% increased the beneficial microbiota in the ileum, cecum, and colon of pigs [40]. Additionally, barley-derived β-glucans decreased K88-ETEC adhesion to the enterocytes of nursery pigs [41], reducing pathogenic infection in the small intestine. In pigs fed exogenous oat β-glucans, the abundance of *Lactobacillus* spp. and *Bifidobacteria* spp. was greater than in pigs fed exogenous barley β-glucans [40]. Oat β-glucans also increased populations of *Bifidobacteria* spp. and *Lactobacillus* spp. in the stomach and colon of nursery pigs [42]. The reason for different results from β-glucans from cereal grains may be due to the higher insoluble fractions of oats than barley [13]. These studies indicate that cereal β-glucans possess prebiotic effects, modulating the intestinal microbiota and mitigating the negative effects of pathogenic bacterial infection in the GIT of pigs, but the effects of cereal β-glucans could vary. Prebiotics are non-digestible soluble NSP and are fermented by gut microbiota, which potentially enhance the beneficial microbiota in the GIT of pigs [4,43]. However, high levels of soluble β-glucans from barley, especially in processed barley, can cause increased viscosity of digesta and negatively affect microbiota in the GIT of pigs [44]. Moreover, increased viscosity could result in the increased fermentation of pathogenic bacteria related to the post-weaning diarrhea (PWD) of nursery pigs [29,45]. Both barley and oat β-glucan extracts may have beneficial effects on the intestinal microbiota of nursery pigs, but high inclusion rates of high-β-glucan barley in feeds, especially in processed barley, should be used with caution on account of increased digesta viscosity.

3.2. *Effects of Microbial β-Glucans on Intestinal Health and Growth Performance of Nursery Pigs*

Microbial (yeast and bacteria) and algal β-glucans decreased the population of pathogenic bacteria (*Enterobacteria*) in the ileum and colon of pigs [39], indicating the potential role of microbial β-glucans in improving the intestinal health of nursery pigs.

Biological indicators used to determine the inflammation status of the intestine of nursery pigs include decreased levels of pro-inflammatory cytokines (TNF-α, IL-8, IL-6, IL-1β, and IFN-γ) and increased levels of anti-inflammatory cytokines (IL-4, IL-10, and IL-13) [4]. After weaning, the mRNA expression of pro-inflammatory cytokines was increased [46], indicating that weaning stress affects cytokine signaling modulation in the small intestine of nursery pigs [47]. Supplementation of yeast β-glucans reduced pro-inflammatory cytokines and increased anti-inflammatory cytokines in the jejunum of nursery pigs [48]. The potential of microbial β-glucans to improve the immune response may be attributed to the activation of dectin-1 receptor in the intestine through β-(1,3)-

glycosidic bonds present in β-glucans [6,49]. The increased dectin-1 receptor stimulation by microbial β-glucans (yeast and bacteria) increased phagocytosis in the immune cells and increased cytokines, modulating the immune response through humoral immunity in pigs [50]. As a result, microbial β-glucans reduce the energy cost of the immune response through the activation of dectin-1 receptor, decrease inflammation in the GIT, and improve the growth rate of nursery pigs [51,52]. Therefore, supplementation of microbial β-glucans could reduce enteric inflammation in the GIT and improve the growth rate of nursery pigs [34].

The effects of microbial β-glucans include (1) reduced pathogenic microbiota in the GIT; (2) increased immune responses (pro-inflammatory cytokines); (3) increased mucosa protein and tight junction protein of enterocytes; and (4) improved morphology of nursery pigs. The possibility for these effects is mainly due to the prevention of enterocyte inflammation in nursery pigs, which increases growth performance [6,10,53]. However, the optimal use of β-glucans may be variable depending on β-glucan sources due to differences in the purity, molecular weight, conformation, chemical structure, and solubility of β-glucans in nursery diets [6]. Therefore, this section investigates the effects of microbial β-glucans on the intestinal health and growth of nursery pigs.

3.2.1. Yeast (*Saccharomyces cerevisiae*)

The use of yeast β-glucans (from *Saccharomyces cerevisiae*) has positive effects on the intestinal health and growth performance of nursery pigs, with an increase in weight gain of 7.6% and an increase in feed intake of 5.3% (Table 1). Yeast β-glucans decreased Enterobacteria spp. in the ileum and proximal colon [39]. Additionally, yeast β-glucans improved the morphology parameters of nursery pigs such as VH:CD and jejunum goblet cells [54] and increased the digestibility of nutrients for nursery pigs [55]. The reason for the improvement in the intestinal health of nursery pigs is likely due to the activation of the dectin-1 receptor in the small intestine. However, yeast β-glucans (from *Saccharomyces cerevisiae*) did not linearly improve the growth performance of nursery pigs with increasing β-glucan levels [10,53]. The reason for the growth of pigs showing quadratic changes through yeast β-glucans (from *Saccharomyces cerevisiae*) could be due to high immune stimulation increasing energy use for body maintenance [52,56,57]. During the period of high immune stimulation, proinflammatory cytokines such as TNF-α, IL-6, and IL-1 are released to activate macrophages for defense against infection in pigs [10,58,59]. The supplementation of yeast β-glucans (from *Saccharomyces cerevisiae*) showed quadratic responses in the growth performance, IL-1, and TNF-α in broiler chickens [57]. In the case of an in vitro study using macrophages from mice, zymosan (a form of yeast β-glucan) increased TNF-α secretion [60]. The optimal use of yeast β-glucans (from *Saccharomyces cerevisiae*) could be considered to improve the immune responses of nursery pigs related to growth performance. In this review, the optimal use of yeast β-glucans (from *Saccharomyces cerevisiae*) was determined as 50 mg/kg of nursery diets (Figure 2). In summary, yeast β-glucans (from *Saccharomyces cerevisiae*) have the potential to increase the intestinal health and growth performance of nursery pigs, showing decreased pathogenic bacteria in the GIT, improved morphology parameters, and increased nutrient digestibility.

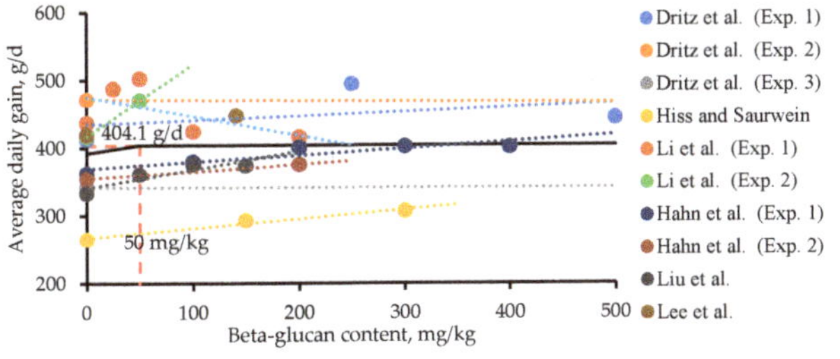

Figure 2. Improvement in body weight gain of nursery pigs fed diets with increasing quantities of

yeast (*Saccharomyces cerevisiae*) β-glucans (0 to 1000 mg/kg) using a linear broken line analysis. The meta-analysis is conducted by Proc NLMIXED to determine the breakpoint on the regression of body-weight gain in nursery pigs based on the data from six published studies (ten experiments with a non-challenged period). The breakpoint (a one-slope broken line analysis) was 50 mg/kg (standard error = 0.561; $p < 0.05$) of β-glucan content in nursery pig diets. The equation for body weight gain in nursery pigs was ADG, g/d = 404.1 − 0.235 × z1 (β-glucan content, mg/kg), $R^2 = 0.87$ if β-glucan content is ≥breakpoint, then z1 = 0. Due to the lack of published data for other microbial β-glucans, a meta-analysis was not conducted [10,55,58,61,62].

A meta-analysis was conducted to determine the optimal use of yeast β-glucans (from *Saccharomyces cerevisiae*) in feeds based on the growth performance data of nursery pigs. A total of 29 datasets with body weight (BW), average daily gain (ADG), average daily feed intake (ADFI), and gain to feed ratio (G:F) from six published research papers with ten experiments were used. For the literature search in PubMed, Web of Science, and Google Scholar, the used keywords were β-glucans, growth performance, intestinal health, and nursery pigs. The found papers were manually screened based on the title and experimental procedures. During this screening process, data from growing pigs or sows were excluded. Additionally, papers which did not contain information about specific levels of β-glucans in the test product were not included in the meta-analysis. For the meta-analysis, the inclusion rate of yeast β-glucans (from *Saccharomyces cerevisiae*) with respect to the growth response was evaluated with a one-slope broken line analysis using the Proc NLMIXED procedure in SAS (SAS Inst. Inc., Cary, NC, USA) [63]. Using a one-slope broken line analysis, the optimal use of yeast β-glucans in feeds for the ADG of nursery pigs was obtained. Statistical significance and tendency were declared at $p < 0.05$ and $0.05 \leq p < 0.10$, respectively. The optimal use of yeast (*Saccharomyces cerevisiae*) β-glucans in nursery pig diets was 50 mg/kg (Figure 2). For other microbial β-glucans, a meta-analysis of their optimal use was not conducted due to the limited amount of data.

Table 1. Effects of the use of yeast β-glucans (from *Saccharomyces cerevisiae*) on the intestinal health and growth performance of nursery pigs [1,2].

Item	Initial BW (kg) or Age (d)	Experimental Period (d)	β-Glucan Compound (mg/kg)	β-Glucan (mg/kg)	Results	Reference
Intestinal health	8.0 kg	28	500	141	Increased jejunal goblet cells, tended to decrease diarrhea during d 0 to 14, tended to increase VH:CD, and tended to increase apparent ileal and total tract digestibility of energy	[54]
	6.4 kg	35	-	100, 200, 300, and 400	Linearly increased apparent total tract digestibility of nutrients	[55]
	5.8 kg	21	-	50, 100, and 150	Increased villus height and VH:CD on the jejunum	[61]
	15.3 kg	28	-	250	Decreased *Enterobacteria* spp. In ileum and proximal colon	[39]

Item	Initial BW (kg) or age (d)	Experimental period (d)	β-glucan compound (mg/kg)	β-glucan (mg/kg)	ADG (% change)	ADFI (% change)	G:F (% change)	Reference
Growth performance	4.9 kg	28	-	250	19.9 **	23.2 **	−1.1	[58]
				500	7.7	11.6	−1.1	
	5.0 kg	28	-	1000	−1.9	−7.1 **	2.6	
				1000	0	−1.5 **	0	
	28 d	28	-	150	10.6	7.4	0	[62]
				300	15.8	15.4 *	0	
	8.7 [3] kg	28	-	25	11.4	7.5	3.1	[10]
				50	14.8	11.6	2.7	
				100	−3	−4.6	0.6	
				200	−4.8	−3.7	−1.3	
	8.2 kg	28	-	50	12.7 **	11.5 **	1.3	

Table 1. *Cont.*

Item	Initial BW (kg) or age (d)	Experimental period (d)	β-glucan compound (mg/kg)	β-glucan (mg/kg)	ADG (% change)	ADFI (% change)	G:F (% change)	Reference
Growth performance	6.4 [4] kg	35	-	100	4.7	3.2	1.6	[55]
				200	10.5	9	0	
				300	11	6.8	3.2	
				400	10.7	5.4	4.8	
	6.2 kg	35	-	200	5.9	1.8	4.2	
	5.8 kg	21	-	50	8.4 **	2.4	5.9	[61]
				100	12.9 **	6.0 **	6.6	
				150	12.3 **	8.8	3.2	
	8.0 kg	28	500	141	6.7 *	2.8	3.8	[54]
	6.0 kg	35	2000	NA	7.4 **	6.5 **	0.9	[64]
	6.0 kg	48	2000	NA	−5.5	−6.9	1.6	[65]
	Average % change:				7.6	5.3	1.9	

BW, body weight; NA, not available; VH:CD, villus height to crypt depth ratio. [1] Asterisk marks (*, **) represent statistical tendency ($p < 0.10$) and significant difference ($p < 0.05$), respectively. [2] The percentage increase or decrease in the average daily gain (ADG), average daily feed intake (ADFI), and gain-to-feed ratio (G:F) was determined in beta-glucan supplementation groups relative to the control group. [3] β-glucan supplementation contents quadratically increased ($p < 0.05$) the ADG of nursery pigs. [4] β-glucan supplementation contents linearly tended to increase ($p < 0.10$) the ADG of nursery pigs.

3.2.2. Bacteria (*Agrobacterium* sp.)

The supplementation of bacterial β-glucans (from *Agrobacterium* sp.) showed positive effects on the intestinal health and growth performance of nursery pigs, resulting in an 11.6% increase in weight gain and a 6.9% increase in the feed intake of nursery pigs (Table 2). Specifically, the supplementation of 50 mg/kg bacterial β-glucans (from *Agrobacterium* sp. *ZX09*) in feeds increased villus height, decreased crypt depth, and increased VH:CD after lipopolysaccharide (LPS) challenge [6]. Moreover, the 50 mg/kg of bacterial β-glucans (from *Agrobacterium* sp. *ZX09*) decreased the intestinal permeability of the small intestine of nursery pigs [61]. The intestinal permeability function can be determined by tight junction proteins such as occludin, claudin, and MUC1 and 2. High tight junction protein complexes between intestinal cells inhibit the paracellular flow, thus enhancing pathogen prevention [4]. Additionally, the highly viscous mucus in the intestine, consisting of cross-linked mucins, antimicrobial factors, and trefoil peptides, acts as an additional physical and chemical intestinal barrier and prevents microorganisms from making contact with the intestinal epithelium [48]. The reason for the decrease in intestinal permeability is likely the activation of dectin-1 receptor. The increase in dectin-1 receptor in the intestine can increase phagocytosis in immune cells and cytokine production, which can improve the intestinal health of nursery pigs. Lastly, bacterial β-glucans (from *Agrobacterium* sp.) linearly increased IL-10 and linearly decreased TNF-α in the jejunum mucosa of nursery pigs. As prebiotics effects of β-glucans in the intestinal microbiota of pigs, supplementation with 200 mg/kg of bacterial β-glucans (from *Agrobacterium* sp.) increased the relative abundance of *Fournierella*, *Parabacteria*, and *Alistipes* in the ileum, providing growth substrates with alpha-glucosidase activity, and increased *Oscillospira*, a butyrate-producing bacteria [66]. Additionally, supplementation with 300 mg/kg of bacterial β-glucans (from *Agrobacterium* sp.) showed interaction with morphological parameters (villus height), the expression genes related to intestinal integrity (Z0-1, Occludin-1, and MUC2), and the growth performance of nursery pigs challenged with ETEC [67], indicating that bacterial β-glucans can be highly effective under challenged conditions in mitigating pathogenic bacteria infection [48]. In terms of the growth of nursery pigs, bacterial β-glucans (from *Agrobacterium* sp.) also showed a quadratic response (as was shown in the yeast β-glucans (from *Saccharomyces cerevisiae*)) [6], which indicates that bacterial β-glucans also require optimal usage in order to improve intestinal health and growth. However, due to the lack of published data, the optimal use of bacterial β-glucans cannot be determined. In summary, bacterial β-glucans can decrease intestinal permeability, which can prevent pathogenic bacteria infections and improve the intestinal health and growth performance of nursery pigs.

Table 2. Effects of the use of bacterial β-glucans (from *Agrobacterium* sp. and *Paenibacillus polymyxa*) on the intestinal health and growth performance of nursery pigs [1,2].

Item	Initial BW (kg) or Age (d)	Experimental Period (d)	β-Glucan Compound (mg/kg)	β-Glucan (mg/kg)	Results	Reference
Intestinal health Bacteria (*Agrobacterium* sp.)	21 d	28	-	50	Increased villus height, decreased crypt depth, and increased VH:CD after LPS challenge; increased mRNA abundance representing intestinal permeability (Z0-1, occludin, claudin, and MUC1 and 2), and decreased malondialdehyde in the jejunal mucosa after LPS challenge	[48]
	7.0 kg	28	-	50, 100, and 200	Linearly increased IL-10 and linearly decreased TNF-α level of jejunal mucosa	[6]
				100	Increased MUC1 and 2 to β-actin mRNA ratio	[6]
	6.1 kg	21	-	200	Increased VH:CD in jejunum and increased mRNA abundance of an intestinal permeability parameter (occludin)	[66]
	6.1 kg	21	500	300	Increased VH:CD in jejunum, increased mRNA abundance of intestinal permeability parameter in jejunum (Z0-1, claudin-1, and MUC2), and increased *Lactobacillus* spp. and propionic acid in cecum digesta after ETEC challenge	[67]
					Decreased malondialdehyde, TNF-α, and IL-6 in jejunum after ETEC challenge	[68]

Item	Initial BW (kg) or age (d)	Experimental period (d)	β-glucan compound (mg/kg)	β-glucan (mg/kg)	ADG (% change)	ADFI (% change)	G:F (% change)	Reference
Growth performance	21 d	21	-	50	21.6 **	11.0 **	9.2	[48]
Bacteria (*Agrobacterium* sp.)				50	14.1	8.2	6.6	
	7.0 [3] kg	28	-	25	2.5	2.8	−0.6	[6]
				50	10.4	8.0	2.4	[6]
				100	15.7	10.2	4.9	
				200	−0.9	1.0	−2.3	
	6.1 kg	21	-	200	17.6	6.9	4.3	[66]
		Average % change			11.6	6.9	3.5	
Bacteria (*Paenibacillus polymyxa*)	5.6 kg	28		400	5.8 *	−0.8	6.6	[69]

BW, body weight; NA, not available; MUC, mucin; LPS, lipopolysaccharide; VH:CD, villus height to crypt depth ratio; IL-10, interleukin-10; TNF-α, tumor necrosis factor-α; Z0-1, zonula occludens-1. [1] Asterisk marks (*, **) represent statistical tendency ($p < 0.10$) and significant difference ($p < 0.05$), respectively. [2] The percentage increase or decrease in the average daily gain (ADG), average daily feed intake (ADFI), and gain to feed ratio (G:F) is determined in beta-glucan supplementation groups relative to the control group. [3] β-glucan supplementation contents linearly ($p < 0.05$) and quadratically ($p < 0.05$) increased the ADG of nursery pigs.

3.2.3. Algae (*Euglena gracilis*, *Laminaria digitata*, and *Laminaria hyperborea*)

The use of algal β-glucans has been shown to improve the intestinal health of nursery pigs by decreasing intestinal permeability in jejunal mucosa and decreasing pathogenic bacteria such as *Enterobacteria* spp. (Table 3), but it did not improve growth performance [54]. Specifically, 54 mg/kg of algal β-glucans increased mRNA abundance, representing a decrease in intestinal permeability (claudin, occludin, and MUC2) in the jejunal mucosa of nursery pigs [49]. Additionally, 108 mg/kg of algal β-glucans increased the mRNA abundance of dectin-1 receptors in the jejunal mucosa, and 141 mg/kg of β-glucans also increased the relative gene expression of tight junction proteins (claudin, occludin, and MUC1) in the jejunum of nursery pigs. Lastly, microbiota data showed that 250 mg/kg of 2 algal β-glucans (from *Laminaria digitata* and *Laminaria hyperborea*) decreased *Enterobacteria* spp. In the ileum and proximal colon of nursery pigs. Several studies showed improvements in the growth performance and intestinal health in pigs fed seaweed extract-supplemented diets (from *Laminaria* spp.) [70–72]. However, information on the algal β-glucans content in the seaweed extract was not available. Further research is needed to investigate the effects of algal β-glucans on the growth performance of nursery pigs.

Table 3. Effects of the use of algal β-glucans (from *Euglena gracilis*, *Laminaria digitata*, and *Laminaria hyperborea*) on the intestinal health of nursery pigs.

Item	Initial BW (kg) or Age (d)	Experimental Period (d)	β-Glucan (mg/kg)	Results	Reference
Algae (*Euglena gracilis*)	7.7 kg	17	54	Increased mRNA abundance representing intestinal permeability (claudin, occludin, and MUC2) in jejunal mucosa on d 12	[65]
			108	Increased mRNA abundance representing intestinal permeability (dectin) in jejunal mucosa on d 5 and 12. Decreased transcellular permeability.	
Algae (*Laminaria digitata*)	15.3 kg	28	250	Decreased *Enterobacteria* spp. in ileum and proximal colon; increased acetic acid and decreased propionic acid in ileum	[39]
Algae (*Laminaria hyperborea*)	15.3 kg	28	250	Decreased *Enterobacteria* spp. in ileum and proximal colon; decreased total volatile fatty acid in the ileum	

MUC, mucin; mRNA, messenger ribonucleic acid.

4. Conclusions

Due to their quantitative, compositional, and structural differences, cereal β-glucans have relatively high water-soluble properties, whereas microbial β-glucans (yeast and bacteria) have water-insoluble properties in the digesta of nursery pigs. The high water-soluble properties of cereal β-glucans, if fed in ample amounts, are shown to cause digesta viscosity, negatively affecting the intestinal health and nutrient utilization in nursery pigs. In contrast, the use of microbial β-glucans showed positive effects on the intestinal health of nursery pigs at an optimal level through mainly activating the dectin-1 receptor and prebiotic effects without causing digesta viscosity. From this review, it is evident that the use of microbial β-glucans can improve intestinal health and nutrient utilization, which, in turn, can improve the growth efficiency of nursery pigs.

Author Contributions: Conceptualization, H.C. and S.W.K.; methodology, S.W.K.; formal analysis, H.C. and S.W.K.; investigation, H.C.; data curation, H.C. and S.W.K.; writing—original draft preparation, H.C.; writing—review and editing, H.C. and S.W.K.; supervision, S.W.K. All authors have read and agreed to the published version of the manuscript.

Funding: North Carolina Agricultural Foundation (#660101, Raleigh, NC, USA) and USDA-NIFA Hatch (#02893, Washington DC, USA).

Institutional Review Board Statement: Not applicable.

Informed Consent Statement: Not applicable.

Data Availability Statement: Not applicable.

Acknowledgments: All the members of Kim Lab at North Carolina State University.

Conflicts of Interest: The authors declare no conflict of interest.

References

1. Flis, M.; Sobotka, W.; Antoszkiewicz, Z. Fiber substrates in the nutrition of weaned piglets—A review. *Ann. Anim. Sci.* **2017**, *17*, 627–644. [CrossRef]
2. Zheng, L.; Duarte, M.E.; Sevarolli Loftus, A.; Kim, S.W. Intestinal health of pigs upon weaning: Challenges and nutritional intervention. *Front. Vet. Sci.* **2021**, *8*, 628258. [CrossRef] [PubMed]
3. Casewell, M.; Friis, C.; Marco, E.; McMullin, P.; Phillips, I. The European ban on growth-promoting antibiotics and emerging consequences for human and animal health. *J. Antimicrob. Chemother.* **2003**, *52*, 159–161. [CrossRef] [PubMed]
4. Kim, S.W.; Duarte, M.E. Understanding intestinal health in nursery pigs and the relevant nutritional strategies. *Anim. Biosci.* **2021**, *34*, 338. [CrossRef] [PubMed]
5. Li, J.; Xing, J.; Li, D.; Wang, X.; Zhao, L.; Lv, S.; Huang, D. Effects of β-glucan extracted from *Saccharomyces cerevisiae* on humoral and cellular immunity in weaned piglets. *Arch. Anim. Nutr.* **2005**, *59*, 303–312. [CrossRef]
6. Luo, J.; Liu, S.; Yu, B.; He, J.; Mao, X.; Cheng, L.; Chen, D. Beta-glucan from *Agrobacterium sp. ZX09* improves growth performance and intestinal function in weaned piglets. *J. Anim. Physiol. Anim. Nutr.* **2019**, *103*, 1818–1827. [CrossRef]
7. Suchecka, D.; Gromadzka-Ostrowska, J.; Żyła, E.; Harasym, J.; Oczkowski, M. Selected physiological activities and health promoting properties of cereal beta-glucans. A review. *J. Anim. Feed Sci.* **2017**, *26*, 183–191. [CrossRef]
8. Volman, J.J.; Ramakers, J.D.; Plat, J. Dietary modulation of immune function by β-glucans. *Physiol. Behav.* **2008**, *94*, 276–284. [CrossRef] [PubMed]
9. Kaur, R.; Sharma, M.; Ji, D.; Xu, M.; Agyei, D. Structural features, modification, and functionalities of beta-glucan. *Fibers* **2019**, *8*, 1. [CrossRef]
10. Li, J.; Li, D.F.; Xing, J.J.; Cheng, Z.B.; Lai, C.H. Effects of β-glucan extracted from *Saccharomyces cerevisiae* on growth performance, and immunological and somatotropic responses of pigs challenged with *Escherichia coli* lipopolysaccharide. *J. Anim. Sci.* **2006**, *84*, 2374–2381. [CrossRef]
11. Jha, R.; Rossnagel, B.; Pieper, R.; Van Kessel, A.; Leterme, P. Barley and oat cultivars with diverse carbohydrate composition alter ileal and total tract nutrient digestibility and fermentation metabolites in weaned piglets. *Animal* **2010**, *4*, 724–731. [CrossRef] [PubMed]
12. Skendi, A.; Biliaderis, C.G.; Lazaridou, A.; Izydorczyk, M.S. Structure and rheological properties of water soluble β-glucans from oat cultivars of Avena sativa and Avena bysantina. *J. Cereal Sci.* **2003**, *38*, 15–31. [CrossRef]
13. Holtekjølen, A.K.; Vhile, S.G.; Sahlstrøm, S.; Knutsen, S.H.; Uhlen, A.K.; Åssveen, M.; Kjos, N.P. Changes in relative molecular weight distribution of soluble barley beta-glucan during passage through the small intestine of pigs. *Livest. Sci.* **2014**, *168*, 102–108. [CrossRef]
14. Lambo, A.M.; Öste, R.; Nyman, M.E.L. Dietary fibre in fermented oat and barley β-glucan rich concentrates. *Food Chem.* **2005**, *89*, 283–293. [CrossRef]
15. Wood, P.J. Oat and rye β-glucan: Properties and function. *Cereal Chem.* **2010**, *87*, 315–330. [CrossRef]
16. Högberg, A.; Lindberg, J.E. Influence of cereal non-starch polysaccharides and enzyme supplementation on digestion site and gut environment in weaned piglets. *Anim. Feed Sci. Technol.* **2004**, *116*, 113–128. [CrossRef]
17. MacArtain, P.; Gill, C.I.; Brooks, M.; Campbell, R.; Rowland, I.R. Nutritional value of edible seaweeds. *Nutr. Rev.* **2007**, *65*, 535–543. [CrossRef]
18. Read, S.M.; Currie, G.; Bacic, A. Analysis of the structural heterogeneity of laminarin by electrospray-ionisation-mass spectrometry. *Carbohydr. Res.* **1996**, *281*, 187–201. [CrossRef]
19. Manners, D.J.; Masson, A.J.; Patterson, J.C. The structure of a β-(1→3)-D-glucan from yeast cell walls. *Biochem. J.* **1973**, *135*, 19–30. [CrossRef]
20. Du, B.; Meenu, M.; Liu, H.; Xu, B. A concise review on the molecular structure and function relationship of β-glucan. *Int. J. Mol. Sci.* **2019**, *20*, 4032. [CrossRef]
21. Zhou, X.; Beltranena, E.; Zijlstra, R.T. Effect of feeding wheat- or barley-based diets with low or and high nutrient density on nutrient digestibility and growth performance in weaned pigs. *Anim. Feed Sci. Technol.* **2016**, *218*, 93–99. [CrossRef]
22. Nasir, Z.; Wang, L.F.; Young, M.G.; Swift, M.L.; Beltranena, E.; Zijlstra, R.T. The effect of feeding barley on diet nutrient digestibility and growth performance of starter pigs. *Anim. Feed Sci. Technol.* **2015**, *210*, 287–294. [CrossRef]
23. Che, T.M.; Perez, V.G.; Song, M.; Pettigrew, J.E. Effect of rice and other cereal grains on growth performance, pig removal, and antibiotic treatment of weaned pigs under commercial conditions1. *J. Anim. Sci.* **2012**, *90*, 4916–4924. [CrossRef]
24. Dikeman, C.L.; Fahey, G.C., Jr. Viscosity as related to dietary fiber: A review. *Crit. Rev. Food Sci. Nutr.* **2006**, *46*, 649–663. [CrossRef] [PubMed]
25. Owusu-Asiedu, A.; Patience, J.F.; Laarveld, B.; Van Kessel, A.G.; Simmins, P.H.; Zijlstra, R.T. Effects of guar gum and cellulose on digesta passage rate, ileal microbial populations, energy and protein digestibility, and performance of grower pigs. *J. Anim. Sci.* **2006**, *84*, 843–852. [CrossRef] [PubMed]
26. Hooda, S.; Matte, J.J.; Vasanthan, T.; Zijlstra, R.T. Dietary oat β-glucan reduces peak net glucose flux and insulin production and modulates plasma incretin in portal-vein catheterized grower pigs. *J. Nutr.* **2010**, *140*, 1564–1569. [CrossRef] [PubMed]
27. Willamil, J.; Badiola, I.; Devillard, E.; Geraert, P.A.; Torrallardona, D. Wheat-barley-rye-or corn-fed growing pigs respond differently to dietary supplementation with a carbohydrase complex. *J. Anim. Sci.* **2012**, *90*, 824–832. [CrossRef]

28. Knudsen, K.E.B. Carbohydrate and lignin contents of plant materials used in animal feeding. *Anim. Feed Sci. Technol.* **1997**, *67*, 319–338. [CrossRef]
29. Hopwood, D.E.; Pethick, D.W.; Pluske, J.R.; Hampson, D.J. Addition of pearl barley to a rice-based diet for newly weaned piglets increases the viscosity of the intestinal contents, reduces starch digestibility and exacerbates post-weaning colibacillosis. *Br. J. Nutr.* **2004**, *92*, 419–427. [CrossRef]
30. Passos, A.A.; Park, I.; Ferket, P.; Von Heimendahl, E.; Kim, S.W. Effect of dietary supplementation of xylanase on apparent ileal digestibility of nutrients, viscosity of digesta, and intestinal morphology of growing pigs fed corn and soybean meal based diet. *Anim. Nutr.* **2015**, *1*, 19–23. [CrossRef]
31. Chen, H.; Zhang, S.; Kim, S.W. Effects of supplemental xylanase on health of the small intestine in nursery pigs fed diets with corn distillers' dried grains with solubles. *J. Anim. Sci.* **2020**, *98*, skaa185. [CrossRef]
32. Medel, P.; Baucells, F.; Gracia, M.I.; De Blas, C.; Mateos, G.G. Processing of barley and enzyme supplementation in diets for young pigs. *Anim. Feed Sci. Technol.* **2002**, *95*, 113–122. [CrossRef]
33. Schop, M.; Jansman, A.; de Vries, S.; Gerrits, W. Increased diet viscosity by oat β-glucans decreases the passage rate of liquids in the stomach and affects digesta physicochemical properties in growing pigs. *Animal* **2020**, *14*, 269–276. [CrossRef] [PubMed]
34. Baker, J.T.; Duarte, M.E.; Holanda, D.M.; Kim, S.W. Friend or foe? Impacts of dietary xylans, xylooligosaccharides, and xylanases on intestinal health and growth performance of monogastric animals. *Animals* **2021**, *11*, 609. [CrossRef] [PubMed]
35. Johansen, H.N.; Bach Knudsen, K.E.; Wood, P.J.; Fulcher, R.G. Physico-chemical properties and the degradation of oat bran polysaccharides in the gut of pigs. *J. Sci. Food Agric.* **1997**, *73*, 81–92. [CrossRef]
36. Thacker, P.A.; Campbell, G.L.; Grootwassink, J. The effect of organic acids and enzyme supplementation on the performance of pigs fed barley-based diets. *Can. J. Anim. Sci.* **1992**, *72*, 395–402. [CrossRef]
37. Bischoff, S.C. 'Gut health': A new objective in medicine? *BMC Med.* **2011**, *9*, 24. [CrossRef]
38. Bach Knudsen, K.E.; Hedemann, M.S.; Lærke, H.N. The role of carbohydrates in intestinal health of pigs. *Anim. Feed Sci. Technol.* **2012**, *173*, 41–53. [CrossRef]
39. Sweeney, T.; Collins, C.B.; Reilly, P.; Pierce, K.M.; Ryan, M.; O'doherty, J.V. Effect of purified β-glucans derived from Laminaria digitata, Laminaria hyperborea and Saccharomyces cerevisiae on piglet performance, selected bacterial populations, volatile fatty acids and pro-inflammatory cytokines in the gastrointestinal tract of pigs. *Br. J. Nutr.* **2012**, *108*, 1226–1234. [CrossRef]
40. Reilly, P.; Sweeney, T.; O'Shea, C.; Pierce, K.M.; Figat, S.; Smith, A.G.; Gahan, D.A.; O'Doherty, J.V. The effect of cereal-derived beta-glucans and exogenous enzyme supplementation on intestinal microflora, nutrient digestibility, mineral metabolism and volatile fatty acid concentrations in finisher pigs. *Anim. Feed Sci. Technol.* **2010**, *158*, 165–176. [CrossRef]
41. Ewaschuk, J.B.; Johnson, I.R.; Madsen, K.L.; Vasanthan, T.; Ball, R.; Field, C.J. Barley-derived β-glucans increases gut permeability, ex vivo epithelial cell binding to E. coli, and naïve T-cell proportions in weanling pigs1,2. *J. Anim. Sci.* **2012**, *90*, 2652–2662. [CrossRef]
42. Metzler-Zebeli, B.U.; Zijlstra, R.T.; Mosenthin, R.; Gänzle, M.G. Dietary calcium phosphate content and oat β-glucan influence gastrointestinal microbiota, butyrate-producing bacteria and butyrate fermentation in weaned pigs. *FEMS Microbiol. Ecol.* **2011**, *75*, 402–413. [CrossRef]
43. Liu, Y.; Espinosa, C.D.; Abelilla, J.J.; Casas, G.A.; Lagos, L.V.; Lee, S.A.; Kwon, W.B.; Mathai, J.K.; Navarro, D.M.; Jaworski, N.W. Non-antibiotic feed additives in diets for pigs: A review. *Anim. Nutr.* **2018**, *4*, 113–125. [CrossRef] [PubMed]
44. Metzler-Zebeli, B.U.; Zebeli, Q. Cereal β-glucan alters nutrient digestibility and microbial activity in the intestinal tract of pigs, and lower manure ammonia emission: A meta-analysis. *J. Anim. Sci.* **2013**, *91*, 3188–3199. [CrossRef]
45. McDonald, D.E.; Pethick, D.W.; Mullan, B.P.; Hampson, D.J. Increasing viscosity of the intestinal contents alters small intestinal structure and intestinal growth, and stimulates proliferation of enterotoxigenic *Escherichia coli* in newly-weaned pigs. *Br. J. Nutr.* **2001**, *86*, 487–498. [CrossRef]
46. Hu, C.H.; Xiao, K.; Luan, Z.S.; Song, J. Early weaning increases intestinal permeability, alters expression of cytokine and tight junction proteins, and activates mitogen-activated protein kinases in pigs1. *J. Anim. Sci.* **2013**, *91*, 1094–1101. [CrossRef]
47. Moeser, A.J.; Pohl, C.S.; Rajput, M. Weaning stress and gastrointestinal barrier development: Implications for lifelong gut health in pigs. *Anim. Nutr.* **2017**, *3*, 313–321. [CrossRef]
48. Luo, J.; Chen, D.; Mao, X.; He, J.; Yu, B.; Cheng, L.; Zeng, D. Purified β-glucans of different molecular weights enhance growth performance of LPS-challenged piglets via improved gut barrier function and microbiota. *Animals* **2019**, *9*, 602. [CrossRef] [PubMed]
49. Kim, K.; Ehrlich, A.; Perng, V.; Chase, J.A.; Raybould, H.; Li, X.; Atwill, E.R.; Whelan, R.; Sokale, A.; Liu, Y. Algae-derived β-glucan enhanced gut health and immune responses of weaned pigs experimentally infected with a pathogenic *E. coli*. *Anim. Feed Sci. Technol.* **2019**, *248*, 114–125. [CrossRef]
50. Drummond, R.A.; Brown, G.D. The role of Dectin-1 in the host defence against fungal infections. *Curr. Opin. Microbiol.* **2011**, *14*, 392–399. [CrossRef] [PubMed]
51. Huntley, N.F.; Nyachoti, C.M.; Patience, J.F. Lipopolysaccharide immune stimulation but not β-mannanase supplementation affects maintenance energy requirements in young weaned pigs. *J. Anim. Sci. Biotechnol.* **2018**, *9*, 47. [CrossRef]
52. Van der Meer, Y.; Jansman, A.J.M.; Gerrits, W.J.J. Low sanitary conditions increase energy expenditure for maintenance and decrease incremental protein efficiency in growing pigs. *Animal* **2020**, *14*, 1811–1820. [CrossRef] [PubMed]

53. Shen, Y.B.; Piao, X.S.; Kim, S.W.; Wang, L.; Liu, P.; Yoon, I.; Zhen, Y.G. Effects of yeast culture supplementation on growth performance, intestinal health, and immune response of nursery pigs. *J. Anim. Sci.* **2009**, *87*, 2614–2624. [CrossRef] [PubMed]
54. Lee, J.J.; Kyoung, H.; Cho, J.H.; Choe, J.; Kim, Y.; Liu, Y.; Kang, J.; Lee, H.; Kim, H.B.; Song, M. Dietary yeast cell wall improves growth performance and prevents of diarrhea of weaned pigs by enhancing gut health and anti-Inflammatory immune responses. *Animals* **2021**, *11*, 2269. [CrossRef] [PubMed]
55. Hahn, T.-W.; Lohakare, J.D.; Lee, S.L.; Moon, W.K.; Chae, B.J. Effects of supplementation of β-glucans on growth performance, nutrient digestibility, and immunity in weanling pigs. *J. Anim. Sci.* **2006**, *84*, 1422–1428. [CrossRef] [PubMed]
56. Blecha, F.; Reddy, D.N.; Chitko-McKown, C.G.; McVey, D.S.; Chengappa, M.M.; Goodband, R.D.; Nelssen, J.L. Influence of recombinant bovine interleukin-1β and interleukin-2 in pigs vaccinated and challenged with *Streptococcus suis*. *Vet. Immunol. Immunopathol.* **1995**, *44*, 329–346. [CrossRef]
57. Zhang, B.; Guo, Y.; Wang, Z. The modulating effect of β-1, 3/1, 6-glucan supplementation in the diet on performance and immunological responses of broiler chickens. *Asian Australas. J. Anim. Sci.* **2008**, *21*, 237–244. [CrossRef]
58. Dritz, S.S.; Shi, J.; Kielian, T.L.; Goodband, R.D.; Nelssen, J.L.; Tokach, M.D.; Chengappa, M.M.; Smith, J.E.; Blecha, F. Influence of dietary β-glucan on growth performance, nonspecific immunity, and resistance to *Streptococcus suis* infection in weanling pigs. *J. Anim. Sci.* **1995**, *73*, 3341–3350. [CrossRef]
59. Duarte, M.E.; Stahl, C.H.; Kim, S.W. Intestinal damages by F18+ *Escherichia coli* and its amelioration with an antibacterial bacitracin fed to nursery pigs. *Antioxidants* **2023**, *12*, 1040. [CrossRef]
60. Young, S.-H.; Ye, J.; Frazer, D.G.; Shi, X.; Castranova, V. Molecular mechanism of tumor necrosis factor-α production in 1→3-β-glucan (zymosan)-activated macrophages. *J. Biol. Chem.* **2001**, *276*, 20781–20787. [CrossRef]
61. Liu, G.; Yu, L.; Martínez, Y.; Ren, W.; Ni, H.; Abdullah Al-Dhabi, N.; Duraipandiyan, V.; Yin, Y. Dietary *Saccharomyces cerevisiae* cell wall extract supplementation alleviates oxidative stress and modulates serum amino acids profiles in weaned piglets. *Oxid. Med. Cell. Longev.* **2017**, *2017*, 3967439. [CrossRef]
62. Hiss, S.; Sauerwein, H. Influence of dietary β-glucan on growth performance, lymphocyte proliferation, specific immune response and haptoglobin plasma concentrations in pigs. *J. Anim. Physiol. Anim. Nutr.* **2003**, *87*, 2–11. [CrossRef] [PubMed]
63. Choi, H.; Kim, B.G. A low-fiber diet requires a longer adaptation period before collecting feces of pigs compared with a high-fiber diet in digestibility experiments using the inert marker method. *Anim. Feed Sci. Technol.* **2019**, *256*, 114254. [CrossRef]
64. Sun, Y.; Park, I.; Guo, J.; Weaver, A.C.; Kim, S.W. Impacts of low level aflatoxin in feed and the use of modified yeast cell wall extract on growth and health of nursery pigs. *Anim. Nutr.* **2015**, *1*, 177–183. [CrossRef]
65. Kim, S.W.; Holanda, D.M.; Gao, X.; Park, I.; Yiannikouris, A. Efficacy of a yeast cell wall extract to mitigate the effect of naturally co-occurring mycotoxins contaminating feed ingredients fed to young pigs: Impact on gut health, microbiome, and growth. *Toxins* **2019**, *11*, 633. [CrossRef]
66. Wu, Y.; Li, X.; Liu, H.; Du, Y.; Zhou, J.; Zou, L.; Xiong, X.; Huang, H.; Tan, Z.; Yin, Y. A water-soluble β-glucan improves growth performance by altering gut microbiome and health in weaned pigs. *Anim. Nutr.* **2021**, *7*, 1345–1351. [CrossRef] [PubMed]
67. Zhou, Y.; Luo, Y.; Yu, B.; Zheng, P.; Yu, J.; Huang, Z.; Mao, X.; Luo, J.; Yan, H.; He, J. Effect of β-glucan supplementation on growth performance and intestinal epithelium functions in weaned pigs challenged by enterotoxigenic *Escherichia coli*. *Antibiotics* **2022**, *11*, 519. [CrossRef]
68. Zhou, Y.; Luo, Y.; Yu, B.; Zheng, P.; Yu, J.; Huang, Z.; Mao, X.; Luo, J.; Yan, H.; He, J. *Agrobacterium sp. ZX09* β-glucan attenuates Enterotoxigenic *Escherichia coli*-induced disruption of intestinal epithelium in weaned pigs. *Int. J. Mol. Sci.* **2022**, *23*, 10290. [CrossRef]
69. Hwang, Y.H.; Park, B.K.; Lim, J.H.; Kim, M.S.; Song, I.B.; Park, S.C.; Jung, H.K.; Hong, J.H.; Yun, H.I. Effects of β-glucan from *Paenibacillus polymyxa* and L-theanine on growth performance and immunomodulation in weanling piglets. *Asian Australas. J. Anim. Sci.* **2008**, *21*, 1753–1759. [CrossRef]
70. Gahan, D.A.; Lynch, M.B.; Callan, J.J.; O'sullivan, J.T.; O'Doherty, J.V. Performance of weanling piglets offered low-, medium-or high-lactose diets supplemented with a seaweed extract from *Laminaria* spp. *Animal* **2009**, *3*, 24–31. [CrossRef]
71. Heim, G.; Sweeney, T.; O'Shea, C.J.; Doyle, D.N.; O'Doherty, J.V. Effect of maternal supplementation with seaweed extracts on growth performance and aspects of gastrointestinal health of newly weaned piglets after challenge with enterotoxigenic *Escherichia coli* K88. *Br. J. Nutr.* **2014**, *112*, 1955–1965. [CrossRef] [PubMed]
72. Zou, T.; Yang, J.; Guo, X.; He, Q.; Wang, Z.; You, J. Dietary seaweed-derived polysaccharides improve growth performance of weaned pigs through maintaining intestinal barrier function and modulating gut microbial populations. *J. Anim. Sci. Biotechnol.* **2021**, *12*, 28. [CrossRef] [PubMed]

animals

MDPI

Systematic Review

Evaluating the Effects of Non-Nutritive Sweeteners on Pigs: A Systematic Review

Mariah R. Jansen and Kwangwook Kim *

Department of Animal Science, Michigan State University, East Lansing, MI 48824, USA; jansenm2@msu.edu
* Correspondence: kkim@msu.edu

Simple Summary: This systematic review examines the effects of non-nutritive sweeteners (NNS) on pigs, focusing on growth performance, feed preference, gut health, and other clinical indicators. Sweeteners such as stevia, sucralose, and neotame have been tested in various studies to evaluate their influence on swine production. Results show that NNS supplementation generally improves growth performance and feed intake in pigs, with some studies reporting reduced diarrhea rates and improved gut health. However, the effects of NNS on gut microbiota are inconsistent, with some sweeteners promoting beneficial bacteria growth while others show minimal changes in microbial diversity. Despite these outcomes, research on the long-term effects of NNS on gut health and the immune system remains limited. This review highlights the need for further studies to explore the mechanisms behind NNS effects, especially in diverse dietary and environmental conditions. Identifying optimal types and dosages of NNS, along with understanding their interactions with the gut microbiome, will be crucial in determining their role as a dietary supplement in swine production.

Abstract: Non-nutritive sweeteners (NNS) have been investigated for their potential to improve feed palatability and growth performance in pigs, although their use in swine production remains limited. This systematic review evaluates the effects of NNS on pigs, drawing from 18 studies published between 1990 and 2024. Following the PRISMA guidelines and using the PICOS framework, a total of 448 papers were initially identified, of which 18 met the inclusion criteria for review. The results are mixed: some studies suggest that NNS like stevioside, sucralose, and neotame may improve performance and reduce diarrhea, while others show limited or no effects. The impact of NNS on gut microbiota is similarly inconsistent, with some sweeteners promoting beneficial bacterial growth, while others show minimal changes in microbial diversity. This review emphasizes the need for more research to clarify the effects of NNS in pigs, particularly the mechanisms behind their influence on growth and gut health. Additionally, further studies are needed to determine optimal dosages and assess the long-term impacts of NNS on pig immune function and overall health. The findings highlight the current gaps in knowledge and suggest that more evidence is needed to understand the role of NNS in swine nutrition.

Keywords: growth performance; gut health; non-nutritive sweetener; nutrition; pig; sugar substitutes; swine nutrition

1. Introduction

Non-nutritive sweeteners (NNS), also known as high-intensity sweeteners or sugar substitutes, have minimal caloric value and are 10 to 1000 times sweeter than sucrose [1–3], which is the sweetener commonly found in foods like candy and soda [4]. The discovery of NNS began with saccharin in 1879, followed by significant advancements from the 1960s to the 1980s with the development of acesulfame-K, aspartame, sucralose, neotame, advantame, and steviol glycosides [5,6]. Their use surged in the 2000s, particularly in low-calorie foods and medications, with several sweeteners gaining approval from the FDA and the EU [7,8].

NNS have become increasingly popular in both human nutrition and livestock feeds. As their use expanded in livestock, it became essential to determine the preference and palatability of these sweeteners among different species. Various preference tests conducted on livestock such as cattle, sheep, and goats showed positive results [9–11]. In ruminants, particularly dairy cattle, NNS have been used to activate sweet taste receptors in the small intestine, potentially increasing glucose uptake and influencing rumen microbiota [12–14]. Studies on calves have also assessed the impact of these sweeteners on growth performance and feed preference during stress and production [15,16]. Similarly, in poultry, NNS have been tested for their effects on growth performance and feed preference, with research investigating their impact on intestinal morphology and immune responses to understand the mechanisms behind these improvements [17,18]. The intense sweetness of these sweeteners offered benefits such as increased feed palatability and intake, reduced inflammation, and positive changes to the gut microbiota, which positively impacted livestock health at a lower cost than sucrose [19–22]. Consequently, NNS became valuable feed additives to reduce calorie depletion and promote health during critical periods such as weaning [2].

Despite these positive results, the mechanisms underlying the benefits of NNS are not yet thoroughly understood. One potential mechanism is the activation of specific taste receptors by NNS, leading to the secretion of beneficial hormones that regulate appetite, glucose metabolism, and digestive processes [12,23,24]. Another mechanism involves the modification of gut microbiota, which can improve overall gut health and function [13,14,25]. Alterations in the gut microbiome can enhance nutrient absorption, boost immune responses, and reduce inflammation by promoting the growth of beneficial bacteria and inhibiting harmful pathogens. Additionally, NNS may influence the expression of genes related to metabolic pathways and immune function, further contributing to their positive effects on health [18].

However, minimal research has been conducted on the supplementation of NNS, particularly in pigs. While there is growing evidence of the benefits of NNS in other animals, specific studies on pigs are limited, leaving a gap in our understanding of how these sweeteners affect pig physiology, growth, and health. Investigating these aspects in pigs is essential to optimize their use and harness their full potential in improving livestock production.

Therefore, this systematic review aims to provide a comprehensive overview of data concerning feed preference, growth performance, health promotion, and the gut microbiome in pigs. It also seeks to understand the optimal use and underlying mechanisms of NNS supplementation, highlighting areas that require further investigation. By synthesizing existing research, this review aims to identify gaps in the current knowledge and propose directions for future studies to optimize the use of NNS in pig nutrition.

2. Materials and Methods

2.1. Study Protocol

The protocol of this systematic review was registered at the International Prospective Register of Systematic Reviews (PROSPERO) (ID: CRD42024518080 [26]). The systematic review was conducted in accordance with the Preferred Reporting Items for Systematic Reviews and Meta-analysis (PRISMA) guidelines.

2.2. Eligibility Criteria

The summary of the inclusion and exclusion criteria of the study characteristics, based on the PICOS framework (i.e., populations, interventions, comparators, outcomes of interest, and study designs) [27], is presented in Table 1. Peer-reviewed papers that were available in full-text and written in English were included. Review papers, abstracts, protocols, editorials, opinion pieces, and dissertations were excluded.

Table 1. Summary of inclusion and exclusion criteria.

	Inclusion Criteria	Exclusion Criteria
Population (P)	Pigs (*Sus scrofa* and/or *sus domesticus*) Breeds: Commonly industrial used	Non-pig species (e.g., guinea pig)
Intervention (I)	Dietary supplementation with various non-nutritive sweeteners	Non-nutritive sweeteners are combined with the treatment
Comparators (C)	Control groups fed a basal or commercial diet without non-nutritive sweeteners Additional comparators: Different concentrations/types of non-nutritive sweeteners and combinations in both feed and water	Studies without a control group
Outcomes (O)	Growth performance Incidence of diarrhea and overall health status Feed palatability and preference Gut health and microbiota composition Intestinal development Blood biochemical parameters	Studies only measured in vitro or ex vivo
Study designs (S)	Controlled experimental trials with random allocation to treatment and control groups	Observational studies, including cross-sectional, cohort, and case-control studies Studies without a clear intervention, detailed methodology, or outcomes measurement procedures

2.3. Information Sources and Search Strategy

The published research studies included in this review were found through searches in scientific databases, performed using the PubMed Advanced Search Builder, Scopus, Web of Science Core Collection, and AGRICOLA-USDA. A series of developed keywords utilizing Boolean search terms were used. A combination of keyword searches was employed to identify studies on non-nutritive sweeteners in pigs, and comprehensive search strategies for each database are provided in Table 2. The searches were completed in March 2024, with date limits applied to studies published between 1990 and 2024.

Table 2. Search strategies for PubMed, Scopus, Web of Science, and Agricola.

	PubMed
Query	((“non-nutritive sweeteners”[All Fields] OR “artificial sweeteners”[All Fields] OR “sugar substitutes”[All Fields] OR “sweetener”[All Fields] OR “sweeteners”[All Fields] OR “Sweetening Agents”[MeSH Terms]) AND (“Piglet”[All Fields] OR “Piglets”[All Fields] OR “Pig”[All Fields] OR “Pigs”[All Fields] OR “Swine”[All Fields] OR “Porcine”[All Fields] OR “Sus scrofa”[All Fields])) AND (1990/1/1:2024/12/12[pdat])
Language	Limited by English
Range	Year 1990–2024
	Scopus
Query	(“non-nutritive sweeteners” OR “artificial sweeteners” OR “sugar substitutes” OR “sweetener” OR “sweeteners” OR “Sweetening Agents”) AND (“Piglet” OR “Piglets” OR “Pig” OR “Pigs” OR “Swine” OR “Porcine” OR “Sus scrofa”) AND PUBYEAR > 1989 AND PUBYEAR < 2026 AND (LIMIT-TO (EXACTKEYWORD, “Swine”) OR LIMIT-TO (EXACTKEYWORD, “Pig”)) AND (LIMIT-TO (LANGUAGE, “English”))
Language	Limited by English
Range	Year 1990–2024

Table 2. *Cont.*

	Web of Science
Query	TS = ("non-nutritive sweeteners" OR "artificial sweeteners" OR "sugar substitutes" sweetener OR "Sweetening Agents" OR sweeteners) AND TS = (pig OR pigs OR piglet OR piglets OR swine OR porcine or "Sus scrofa") ǀ Timespan: 1 January 1990 to 31 December 2024 (Publication Date)
Language	Limited by English
Range	Year 1990–2024
	Agricola
Query	"non-nutritive sweeteners" OR "non-nutritive sweetener" OR "artificial sweeteners" OR "artificial sweetener" OR "sugar substitutes" OR "sugar substitute "OR "sweetener" OR "sweeteners" OR "Sweetening Agents" AND "Piglet" OR "Piglets" OR "Pig" OR "Pigs" OR "Swine" OR "Porcine" OR "Sus scrofa"
Language	Limited by English
Range	Year 1990–2024

2.4. Study Selection

All retrieved references were imported into Zotero reference management software (version 6.0.37), and duplicates were initially removed. The remaining references were then imported into Covidence systematic review software [28], where additional duplicates were removed. Prior to article screening, two researchers (M.R.J. and K.K.) developed a procedure for title and abstract screening using 20 randomly selected papers. In the first phase of screening, M.R.J. and K.K. independently assessed all study titles and abstracts against the eligibility criteria in Table 1. The agreement in the abstract and title screening between the two reviewers was 84.8% (Cohen's Kappa = 0.615). Then, a full-text review was performed. The agreement for full-text screening was 80.0% (Cohen's Kappa = 0.636). Discrepancies at each stage were resolved through discussion with two reviewers.

2.5. Data Collection Process and Data Items

Based on the selected study criteria "Effects of Non-Nutritive Sweeteners in Pigs", data extraction was performed. The data extraction forms were initially drafted by M.R.J. and discussed with K.K. Data extracted from each study included the following items: the names of the authors, the year of publication, and the country where the study was conducted; the number and species of animals used in the study; the mean age of the animals at the start of the experiment; the type and dosage of sweetener administered, along with the number of animals per treatment group; the total duration of the experiment; specific outcomes and parameters measured during the study, such as growth performance, feed intake, feed preference, gut health, and biochemical markers; and key findings from the study, including the effects of different sweeteners on the evaluated characteristics. These data items were systematically extracted to ensure consistency and comprehensiveness in capturing the relevant details of each study, facilitating a thorough comparison and synthesis of the results.

2.6. Quality Assessment and Risk of Bias

Two reviewers (M.R.J. and K.K.) independently evaluated the risk of bias of included studies using the SYRCLE's risk of bias tool for animal experimental studies. The checklist comprises ten domains, categorized into six types of biases: sequence generation, baseline characteristics, and allocation concealment (selection bias); random housing and blinding of caregivers and/or investigators (performance bias); random outcome assessment and blinding of outcome assessors (detection bias); incomplete outcome data (attrition bias); selective outcome reporting (reporting bias); and other sources of bias (other). Each item in the tool was assessed as "low risk of bias", "high risk of bias", or "unclear risk of bias". Disagreements between the reviewers were resolved through discussion.

2.7. Synthesis of Results

Given the heterogeneity in study outcomes, outcome measures, and trial designs, a qualitative evaluation and synthesis of the study results were performed. Consequently, a meta-analysis was not conducted, and publication bias was not assessed.

3. Results

3.1. Study Selection Process

Figure 1 shows the study selection process. The search yielded 448 references in total, from which 124 duplicates were removed. A total of 324 abstracts were then screened, among which 249 were judged ineligible, leaving 74 papers to be read in full text. In total, 18 papers met the eligibility criteria and were included.

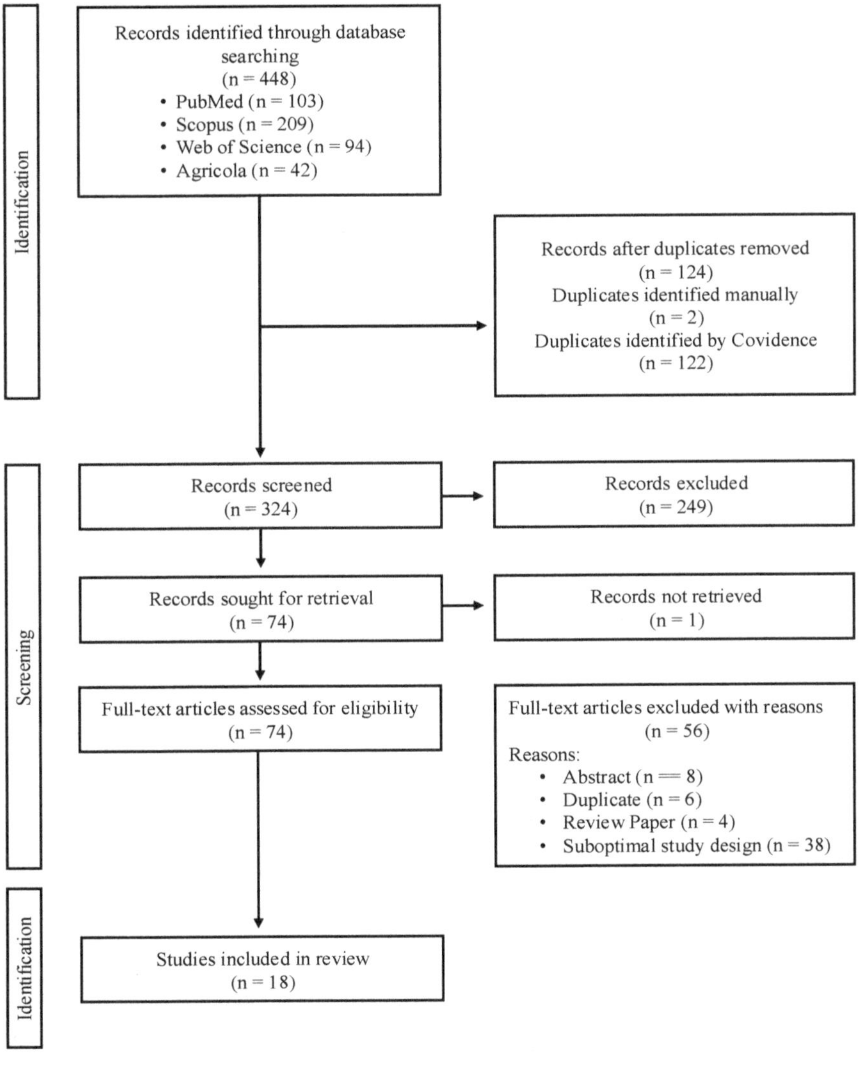

Figure 1. PRISMA flow diagram for the Covidence literature screening on non-nutritive sweetener supplementation in pigs.

3.2. Summary of Study Designs and Sample Characteristics

Table 3 provides details of the characteristics of each study. Four papers were published between 2020 and 2024, nine papers were published between 2010 and 2019, four papers were published between 2000 and 2009, and one paper was published before 2000. The total sample size ranged from 12 pigs [29] to 216 pigs [30,31].

Most of the studies used crossbred pigs as the sample population. These crossbreds included Large White/Landrace × Pietrain pigs [32], Great Yorkshire × Dutch Landrace × D-line [33], Landrace × Large White [34–37], Duroc × Landrace × Large White [30,31], Duroc × Landrace × Yorkshire [38–40], and Large White/Landrace × Large White [41]. Two studies experimented with purebred pigs, specifically Yorkshire [42] and Gloucestershire Old Spot pigs [22]. Four studies did not report the breed of pigs used [29,43–45].

Most studies used young pigs ranging from 21 to 28 days of age or weighing between 7.01 ± 0.3 to 9.05 ± 0.04 kg [22,30–33,35–39,42,43,45]. Other studies used pigs of different weights and ages, including those weighing approximately 23 kg [29], 68.08 ± 0.74 kg [40], and 34.1 ± 2.5 kg [41], as well as pigs aged between 2 to 4 months [44].

Table 3. The summary of study and main results of non-nutritive sweeteners in pigs [1].

Authors, Year, Country	Number of Animals/Species	Age	Type of Sweetener/Number Per Group	Experiment Duration	Characteristics Evaluated	Main Results
Kelly et al. (2017), UK [22]	16, Gloucestershire Old Spot pigs	28 days	0.015% NHDC + Saccharin	2 weeks	• Mucosa-associated microbiota	• Relative abundances of 76 small intestinal and 81 large intestinal mucosa associated OTU's were different among the NHDC + Saccharin treatment, respectively • Relative abundance levels of total *Campylobacteraceae* were lower throughout the small intestine, and large intestine, with NHDC + Saccharin treatment • *Helicobacteraceae* abundance increased within the small intestine of the NHDC + Saccharin-treated group • A *Lactobacillus gasseri* related OTU abundance was increased in the large intestine of NHDC + Saccharin-treated group
Geuns et al. (2003), Belgium [29]	12 female pigs	Not specified; initial body weight: approximately 23 kg	1.67 g/kg stevioside (n = 6)	14 day	• Metabolism of stevioside to steviol in the feces • Absorption of stevioside and steviol in the blood	• Stevioside was completely converted into steviol in feces • Steviol was noted to have concentrations around 853 ± 48 µg/g dry weight in the feces • No stevioside or steviol was detected in the blood

Table 3. *Cont.*

Authors, Year, Country	Number of Animals/Species	Age	Type of Sweetener/Number Per Group	Experiment Duration	Characteristics Evaluated	Main Results
Wang et al. (2014), China [30]	*Experiment 1 and 2:* 216. Duroc × Landrace × Large White pigs, respectively	28 ± 1 days, respectively	*Experiment 1:* • 100, 150, 200, 250, or 300 mg/kg stevioside (n = 36, respectively) *Experiment 2:* • 100, 150, 200, 250, or 300 mg/kg rebaudioside A (n = 36, respectively)	42 day, respectively	• Growth performance • Diarrhea incidence rate	*Experiment 1:* • Stevioside increased ADG linearly throughout the experiment • ADFI increased linearly with higher doses of stevioside • 251 mg/kg stevioside showed lowest incidence of diarrhea based on the broken-line regression model • Optimal dosage for dietary stevioside supplementation: 200 to 250 mg/kg *Experiment 2:* • Rebaudioside A increased ADG linearly throughout the experiment • Feed:gain decreased linearly as the mg/kg of rebaudioside A treatment increased • 213 mg/kg rebaudioside A showed greatest ADFI according to the broken-1 line regression model • The lowest diarrhea incidence was observed at a dose of 191 mg/kg based on the broken-line regression analysis

Table 3. *Cont.*

Authors, Year, Country	Number of Animals/Species	Age	Type of Sweetener/Number Per Group	Experiment Duration	Characteristics Evaluated	Main Results
Zhu et al. (2016), China [31]	*Experiment 1:* 48, Duroc × Landrace × Large White pigs *Experiment 2:* 216, Duroc × Landrace × Large White pigs *Experiment 3:* 108, Duroc × Landrace × Large White pigs	*Experiment 1:* Not specified; initial body weight: 9.05 ± 0.04 kg *Experiment 2:* Not specified; initial body weight: 7.35 ± 0.06 kg *Experiment 3:* Not specified; initial body weight: 7.34 ± 0.08 kg	*Experiment 1:* • 30 mg/kg neotame *Experiment 2:* • 10, 20, 30, 40, or 50 mg/kg neotame *Experiment 3:* • 50 or 500 mg/kg neotame	*Experiment 1:* 15 days *Experiment 2:* 35 days *Experiment 3:* 35 days	*Experiment 1:* • Feed intake • Diet preference percentage *Experiment 2:* • Growth performance *Experiment 3:* • Hematological parameters • Serum biochemical parameters • Histopathological parameters	*Experiment 1:* • 30 mg/kg neotame significantly increased feed intake during day 7 and 10, and 1 to 10 • 30 mg/kg neotame increased diet preference percentage during day 3, 6, 7, 10, and 1 to 10 *Experiment 2:* • A linear increase in ADFI was observed when increasing neotame levels during day 1 to 22 and 1 to 35 • ADG and ADFI increased quadratically during day 1 to 22, 23 to 35, and 1 to 35 when neotame levels increased • Optimal concentrations for maximum ADG and ADFI during the entire experimental period were 21.7 mg/kg and 20.7 mg/kg, respectively *Experiment 3:* • No adverse effects on hematological or serum biochemical parameters were observed • Normal histological structures were present in the liver and kidney at up to 500 mg/kg neotame

Table 3. *Cont.*

Authors, Year, Country	Number of Animals/Species	Age	Type of Sweetener/Number Per Group	Experiment Duration	Characteristics Evaluated	Main Results
Clouard and Val-Laillet (2014), France [32]	32 female, Large White/Landrace × Pietrain pigs	26.38 ± 0.09 days	• 10 to 20% stevia rebaudiana, minimum of 90% steviol glycosides, 5 to 10% high-saponin plants extract, and 70 to 85% excipients (n = 8, respectively)	28 days	• Growth performance • Feed preference	• No significant effects of feed additives on growth performance • Feed consumption of 10 to 20% stevia rebaudiana was increased on day 23 (1-h preference test)
Sterk et al. (2008), The Netherlands [33]	198, Great Yorkshire × Dutch Landrace × D-line pigs	26 days	• 150 mg/kg of NHDC + Saccharin C-150 (n = 66) • 150 mg/kg of NHDC + Saccharin 3D (n = 66)	19 days	• Growth performance • Feed intake per visit (g) • Latency time, total visits per day • Number of visits with food consumed • Fecal consistency	• No significant differences between day 0 to 11 were observed between feed intake characteristics • 3D treatment had an increased percentage of visits with feed intake between day 12 to 19 and increased feed intake on day 8 and 10 • Sweetener diets decreased the percentage of soft feces
Moran et al. (2010), UK [34]	56, Landrace × Large White pigs	28 days	• 150 mg/kg body weight NHDC + Saccharin in feed (n = 12) • 150 mg/kg body weight NHDC + Saccharin in water (n = 8) • 34.8 mg/d/animal saccharin in water (n = 8) • 8.7 mg/d/animal NHDC in water (n = 8) • NHDC + Saccharin + inhibitor of human sweet taste receptor in water (n = 4)	3 days	• Expression of Na +/glucose co-transporter (SGLT1) • Detection of gut hormones and sweet taste receptors	• Expression of SGLT1 mRNA and relative protein abundance was increased by sweetener treatments • D-Glucose uptake was increased by sweetener treatments • Sweet taste receptor T1R2/T1R3 and gustducin were co-expressed in the intestine when supplemented with sweeteners

Table 3. *Cont.*

Authors, Year, Country	Number of Animals/Species	Age	Type of Sweetener/Number Per Group	Experiment Duration	Characteristics Evaluated	Main Results
Daly et al. (2014), UK [35]	24, Landrace × Large White pigs	28 days	• 5% lactose • 0.015% NHDC + Saccharin	2 weeks	• Gut microbiota • Cecal lactic acid concentration	• NHDC + Saccharin increased populations of *Lactobacillus* in the cecal microbiota • NHDC + Saccharin increased cecal lactic acid concentration
Daly et al. (2016), UK [36]	16, Landrace × Large White pigs	28 days	• 0.015% NHDC + Saccharin	2 weeks	• Gut microbiota • Cecal short chain fatty acid concentration analysis	• NHDC + Saccharin increased *Lactobacillaceae* and reduced *Veillonellaceae* and *Ruminococcaceae* abundance in the cecum • NHDC + Saccharin increased cecal lactic acid concentration • NHDC + Saccharin enhanced the expression of *Lactobacillus* 4228 (sugar transporters) in vitro
Daly et al. (2021), UK [37]	24, Landrace × Large White pigs	28 days	• 2 mM sucralose • 0.25 mM saccharin • 10 mM acesulfame-K • 1 mM aspartame • 10 mM cyclamate (n = 4, respectively)	3 days	• Activation of pig sweet taste receptor T1R2-T1R3 Intestinal capacity of glucose uptake	• Sucralose, saccharin, and acesulfame K increased T1R2-T1R3 receptor activation • Sucralose, saccharin, and acesulfame K resulted in increased expression and activity of intestinal SGLT1, enhancing glucose absorption.

Table 3. *Cont.*

Authors, Year, Country	Number of Animals/Species	Age	Type of Sweetener/Number Per Group	Experiment Duration	Characteristics Evaluated	Main Results
Lee et al. (2019), South Korea [38]	*Experiment 1:* 30, Landrace × Yorkshire × Duroc pigs *Experiment 2:* 52 pigs.	*Experiment 1:* Not specified; initial body weight: 7.01 ± 0.3 kg *Experiment 2:* Not specified; initial body weight: 18.67 ± 0.52 kg	*Experiment 1:* • 0.05% saccharin (50% saccharin-natrium) • 0.03% saccharin-neotame (50% saccharin-natrium + 2% neotame) • 0.02% neotame (10% neotame) • 0.02% saccharin-neotame mix (10% saccharin-natrium + 10% neotame) *Experiment 2:* • 0.05% saccharin (50% saccharin-natrium) • 0.03% saccharin-neotame mix (50% saccharin-natrium + 2% neotame) • 0.02% neotame (10% neotame) • 0.02% saccharin-neotame mix (10% saccharin-natrium + 10% neotame)	*Experiment 1:* 21 day *Experiment 2:* 14 day	*Experiment 1:* • Growth performance • Feed preference *Experiment 2:* • Growth performance • Nutrient digestibility • Blood biochemical analysis • Fecal bacterial counts	*Experiment 1:* • 0.02% Saccharin-neotame mix treatment had a high tendency to increase gain:feed throughout day 0 to 21 • 0.02% Neotame treatment increased feed preference during day 0 to 21 *Experiment 2:* • ADG was highest in 0.02% saccharin-neotame mix treatment during the first week • ADFI was highest in 0.02% saccharin-neotame mix treatment throughout the experiment • Fecal *Lactobacillus* counts were higher in the 0.02% saccharin-neotame mix reduced blood cholesterol, while 0.02% neotame groups
Liu et al. (2022), China [39]	108, Duroc × Landrace × Yorkshire pigs	21 days	• 100, 200, or 400 mg/kg stevia residue extract (n = 36, respectively)	42 days	• Growth performance • Diarrhea rate • Antioxidant capacity • Intestinal health • Gut microbiota	• 100 and 200 mg/kg stevia residue extract decreased the rate of diarrhea during day 29 to 42 • 200 mg/kg stevia residue extract reduced the rate of diarrhea throughout the experiment • Stevia residue extract reduced malondialdehyde content and increased total superoxide dismutase in serum and liver • 200 and 400 mg/kg stevia residue extract increased total antioxidant capacity • Relative abundance of *Prevotellaceae* (Family) and *Roseburia* and *Prevotella* (Genus) increased with 400 mg/kg stevia residue extract

Table 3. *Cont.*

Authors, Year, Country	Number of Animals/Species	Age	Type of Sweetener/Number Per Group	Experiment Duration	Characteristics Evaluated	Main Results
Xiong et al. (2022), China [40]	48, Duroc × Landrace × Yorkshire pigs	Not specified; initial body weight: 68.08 ± 0.74 kg	• 100, 200, 400, 600, or 800 mg/kg stevia residue extract (n = 8, respectively)	75 days	• Growth performance • Carcass traits • Meat quality • Antioxidant capacity • Gut microbiota	• Stevia residue extract supplementation linearly increased body weight on day 35 • Stevia residue extract supplementation increased ADG and tended to increase ADFI from day 1 to 75, with the highest gain observed at 100 mg/kg • Hot carcass weight, and gastric index increased, and tended to increase circumference with the supplementation of 100 mg/kg stevia residue extract. • Triglyceride content within serum was increased in the 800 mg/kg stevia residue extract treatment • Malondialdehyde content was lower in 100 mg/kg Stevia residue extract compared to other treatments • 600 and 800 mg/kg stevia residue extract reduced malondialdehyde content within the *longissimus thoracis* compared to the control • Stevia residues extract increased serum total superoxide dismutase activity • Complexity of species diversity of 400, 600, and 800 mg/kg was decreased within Chao1 and observed indexes and tended to decrease the Shannon index
Clouard et al. (2012), France [41]	27, Large White/Landrace × Large White pigs	Not specified; initial body weight: 34.1 ± 2.5 kg	• 2.25% maltodextrin (n = 3/experiment) • 0.37% saccharin (n = 3/experiment)	1 week habituation + 2 weeks conditioning sessions	• Feed and water consumption • Two-choice drinking and feeding test	• Maltodextrin supplementation increased water consumption during the conditioning sessions

Table 3. *Cont.*

Authors, Year, Country	Number of Animals/Species	Age	Type of Sweetener/Number Per Group	Experiment Duration	Characteristics Evaluated	Main Results
Munro et al. (2000), Canada [42]	209, Purebred Yorkshire pigs	24 ± 4 days	• 5% sucrose • 83.3 mg/kg stevia • 167 mg/kg stevia • 334 mg/kg stevia	3 weeks	• Growth performance	• 167 mg/kg stevia reduced body weight during week 1 post-weaning • 5% sucrose increased feed intake and feed:gain throughout the trial
Maenz et al. (1993), Canada [43]	12 pigs/trial with 4 replicate trials.	28 days	• Palasweet (saccharin-based sweetener)	10 days	• Growth performance • Water intake • Diarrhea score	• No significant effects of sweetener were observed
Glaser et al. (2000), Switzerland [44]	75 pigs	2 to 4 months	• 0.01, 0.05, or 0.35 g/L acesulfame-K (n = 4, respectively) • 0.15 or 0.3 g/L alitame (n = 4, respectively) • 1.5, 3, or 5 g/L aspartame (n = 4, respectively) • 5, 10, or 20 g/L cyclamate (n = 4, respectively) • 0.6, 1.2, 2.4 g/L dulcin (n = 4, respectively) • 0.2 g/L monellin (n = 4) • 0.6 g/L NHDC (n = 4) • 0.05 g/L 5-Nitro-2-propoxy aniline (P-4000) (n = 4) • 2.5 g/L perillartine (n = 4, respectively) • 0.2, 0.4, or 0.8 g/L saccharin (n = 4, respectively) • 0.062 or 0.125 g/L sucralose (n = 5 to 6, respectively) • 0.20 g/L Thaumatin (n = 4)		• Gustatory responses • Preference test • Richter-type drinking test	• Monellin, thaumatin, NHDC, P-4000, perillartine, aspartame, and cyclamate did not elicit preference in pigs • Alitame, sucralose, saccharin, acesulfame-K, and dulcin elicit appeal in pigs

Table 3. *Cont.*

Authors, Year, Country	Number of Animals/Species	Age	Type of Sweetener/Number Per Group	Experiment Duration	Characteristics Evaluated	Main Results
Zhang et al. (2020), China [45]	*Experiment 1:* 48 pigs *Experiment 2:* 180 pigs *Experiment 3:* 108 pigs	*Experiment 1:* Not specified; initial body weight: 8.90 ± 0.09 kg *Experiment 2:* Not specified; initial body weight: 7.95 ± 0.17 kg *Experiment 3:* Not specified; initial body weight: 7.97 ± 0.18 kg	*Experiment 1:* • 150 mg/kg sucralose (n = 48) *Experiment 2:* • 75, 150, 225, or 300 mg/kg sucralose (n = 36, respectively) *Experiment 3:* • 150 or 1500 mg/kg sucralose (n = 36, respectively)	*Experiment 1:* 15 days *Experiment 2:* 28 days *Experiment 3:* 28 days	*Experiment 1:* • Feed intake • Diet preference percentage *Experiment 2:* • Growth performance *Experiment 3:* • Growth performance • Hematological parameters • Serum Parameters • Organ index • Histopathological analysis	*Experiment 1:* • Higher diet preference percentage for sucralose treatment during the entire experimental period *Experiment 2:* • 150 mg/kg sucralose increased ADG throughout day 15 to 28, and the entire experimental period (day 0 to 28) • 150 mg/kg sucralose increased ADFI during day 0 to 14, 15 to 28, and 0 to 28 • 225 and 300 mg/kg sucralose decreased ADFI throughout the experiment • Maximum level of inclusion to maximize ADG was 146.7 mg/kg, 150.1 mg/kg, and 149.6 mg/kg of sucralose for day 0 to 14, 15 to 28, and 0 to 28, respectively • Maximum level of inclusion to maximize ADFI was 137.8 mg/kg, 145.8 mg/kg, and 141.8 mg/kg of sucralose for day 0 to 14, 15 to 28, and 0 to 28, respectively *Experiment 3:* • 150 mg/kg sucralose had higher ADG and ADFI than control and 1500 mg/kg group throughout the experiment • No significant differences of hematological, serum parameters, histopathological analysis, and organ index were noted among treatments

[1] ADG = Average daily gain; ADFI = Average daily feed intake; NHDC = Neohesperidin dihydrochalcone; SGLT 1 = Sodium glucose cotransporter 1; NHDC + Saccharin = A combination of NHDC and saccharin.

3.3. Summary of Non-Nutritive Sweetener Intervention and Evaluated Characteristics

Table 3 provides details of the non-nutritive sweetener interventions of each study. The studies used a variety of non-nutritive sweeteners across different experimental setups, including the variety of parameters to measure in each intervention. Aspartame, acesulfame-K, and cyclamate were added to the drinking water in two studies by Daly [36] and Glaser [43]. These studies measured the activation of the pig sweet taste receptor and conducted preference tests using the Richter-type drinking test. Alitame, Dulcin, Monellin, 5-Nitro-2-propoxy aniline (P-4000), Perillartine, and Thaumatin were added to the drinking water, and gustatory responses, preference, and Richter-type drinking tests were conducted [44]. Maltodextrin was added to both feed and water to measure feed and water consumption and was tested for two-choice drinking and feeding [41]. Neotame was included in the diet in different concentrations to assess the effects on feed intake, diet preference, growth performance, hematological and serum biochemical parameters, and histopathological parameters [31], as well as growth performance, nutrient digestibility, blood biochemical analysis, and fecal bacterial counts [38]. Various studies have tested saccharin in either feed or drinking water to evaluate different parameters: activation of the pig sweet taste receptor [37]; feed and water consumption, and two-choice drinking and feeding [41]; growth performance, feed preferences, nutrient digestibility, blood biochemicals, and fecal bacterial counts [38]; gustatory responses, preference test, Richter-type drinking test [44]; and expression of Na^+/glucose co-transporter (SGLT1), and detection of gut hormones and sweet taste receptors [34]. A combination of saccharin and neotame in different doses was also added to the diet to measure growth performance, feed preferences, nutrient digestibility, blood biochemicals, and fecal bacterial counts [38]. Stevia or stevia residue was added to the diet in various concentrations to measure growth performance [42]; growth performance and feed preferences [32]; growth performance, carcass traits, meat quality, antioxidant capacity, and gut microbiota [40]; and growth performance, diarrhea rate, antioxidant capacity, intestinal health, and gut microbiota [39]. Stevioside was supplemented in the diet in different doses to assess growth performance and diarrhea incidence rate [30], as well as metabolism and absorption mechanisms in feces and blood [29]. Sucralose in various doses was added to the diet and drinking water to test the activation of the pig sweet taste receptor [37]; growth performance, feed intake, diet preference, hematological and serum parameters, organ index, and histopathological analysis [45]; and gustatory responses, preference, and Richter-type drinking test [44]. NHDC + saccharin (a combination of neohesperidin dihydrochalcone (NHDC) and saccharin) was added in various concentrations to the feed or drinking water to measure growth performance, feed intake per visit, latency time and total visits per day, number of visits with feed consumed, and fecal consistency [33]; mucosa-associated microbiota [22]; gut microbiota and cecal lactic acid concentration [35]; gut microbiota and cecal short-chain fatty acid concentration analysis [36]; and the expression of SGLT1, detection of gut hormones, and sweet taste receptors [34]. Sucrose or Rebaudioside A was added to the diet to measure growth performance [42], and growth performance and diarrhea incidence rate [30], respectively.

3.4. Summary of Quality Assessment and Risk of Bias

The results of risk of bias assessments of the 18 studies are reported in Figure 2. For sequence generation, most studies had a low risk of bias, ensuring proper randomization [30–34,36,38–45]. Regarding baseline characteristics, the majority of studies reported a low risk of bias, indicating comparable baseline characteristics among groups [29,31–34,36–43,45]. However, allocation concealment had several studies with unclear or high risk, indicating potential biases in the allocation process [22,29,30,32–34,36,37,41,42,44,45]. In terms of random housing, there were instances of both low and high risks, reflecting variability in whether housing was randomized. The blinding of caregivers and/or investigators was often marked by an unclear risk due to insufficient reporting on blinding procedures. For random outcome assessment, this domain was often marked by an unclear risk, with few studies explicitly mentioning random outcome assessment [30–34,39–45]. Similarly, the blinding of outcome assessors had most studies with an unclear risk. Most studies showed a low risk of bias for incomplete outcome data, indicating proper handling of data acquisition and processing [22,29–31,34,36,37,39–45]. Selective outcome reporting was generally low across studies, suggesting comprehensive reporting of outcomes, except for Wang (2014), which was unclear. All studies were rated as low risk for other sources of bias, indicating minimal other sources of bias.

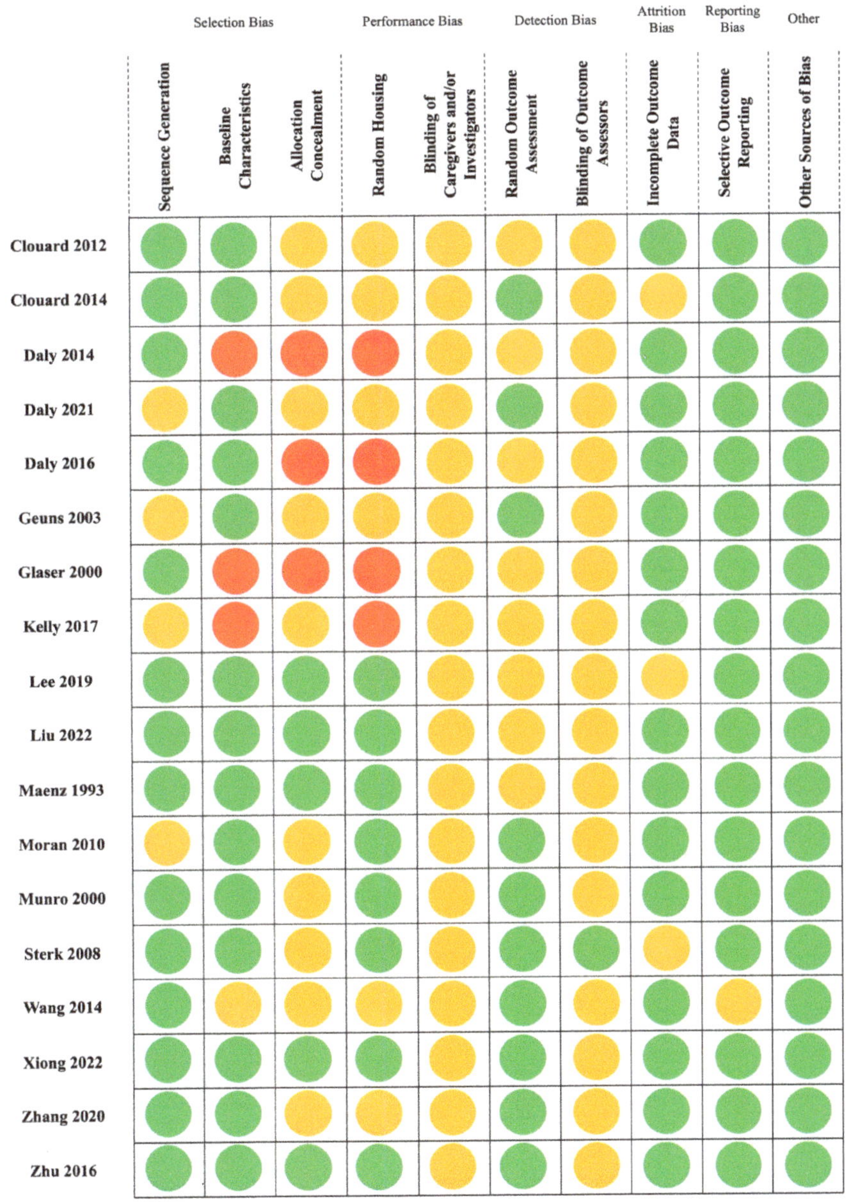

Figure 2. Traffic light plot of risk of bias assessments for included studies. Green = Low risk of bias; Yellow = Unclear; Red = High risk of bias.

3.5. Growth Performance and Feed Preference

3.5.1. Neotame and Neotame + Saccharin

Lee et al. [38] reported that the supplementation of 0.02% neotame significantly increased the average daily gain (ADG) of pigs during the first 7 days of a 14-day feeding period ($p = 0.049$), although this effect was not observed over a 21-day feeding period. Additionally, pigs supplemented with 0.02% neotame had significantly higher average daily feed intake (ADFI) ($p = 0.047$) compared to those supplemented with a 0.02% neotame + saccharin blend during a 14-day period. However, no significant effect on ADFI was seen during week 1 of the 14-day period or throughout a 21-day period. Feed efficiency was not significantly affected by supplementation of 0.02% neotame during the 14- and 21-day feeding periods. Pigs supplemented with a 0.02% neotame + saccharin blend showed a tendency for increased feed efficiency ($p = 0.055$) over the 21-day feeding period, but neither 0.02% nor 0.03% neotame + saccharin blends significantly affected ADG, ADFI, or feed efficiency during the 14- and 21-day periods [38].

Zhu et al. [31] observed that ADG increased quadratically ($p < 0.05$) with increasing dietary neotame levels (10, 20, 30, 40, and 50 mg/kg) during days 1 to 22, 23 to 35, and 1 to 35. ADFI also increased linearly ($p < 0.05$) as neotame levels rose during the same periods and increased quadratically ($p < 0.05$) with increasing neotame levels over the 35-day period. Pigs consuming 30 mg/kg neotame showed significantly higher consumption ($p < 0.05$) compared to a control diet on days 7, 10, and throughout the 15-day feeding period, with a significant increase in feed preference percentage ($p < 0.05$) compared to the control on days 3, 6, 7, 10, and throughout the same period. Zhu et al. [30] also predicted optimal neotame dosages for ADG based on a quadratic plot model: 20.4 and 18.0 mg/kg for days 1 to 22, 22.9 and 22.0 mg/kg for days 23 to 35, and 21.7 and 20.7 mg/kg for the entire 35-day feeding period [31].

3.5.2. Rebaudioside A, Maltodextrin, and Saccharin

Wang et al. [30] demonstrated that rebaudioside A supplementation in 28-day-old weanling pigs led to a linear increase in ADG ($p \leq 0.05$) and a linear decrease in the feed-to-gain ratio ($p < 0.05$) as the dosage increased from 0 to 300 mg/kg over a 42-day feeding period. The highest ADG was achieved with 300 mg/kg rebaudioside A, making it the most effective dosage among those tested (0, 100, 150, 200, and 250 mg/kg). Additionally, a broken-line regression analysis revealed that the optimal dosage of rebaudioside A to maximize average daily feed intake (ADFI) was 213 mg/kg [30].

Clouard et al. [41] observed that a 2.25% maltodextrin inclusion in water tended to be consumed more ($p = 0.09$) during 12 one-tank training sessions from days 14 to 28 of a 28-day trial period compared to the control diet. Furthermore, the maltodextrin inclusion was consumed significantly more ($p = 0.008$) during these sessions than a 0.37% saccharin inclusion treatment. In contrast, the 0.37% saccharin inclusion in water was consumed significantly less ($p < 0.05$) than both the control diet and the maltodextrin inclusion over the same period [41]. Saccharin supplementation did not significantly affect ADG, feed consumption, or average daily water intake in 28-day-old weanling pigs during a 10-day feeding period [43]. Preference was elicited in pigs when saccharin was supplemented in water at concentrations of 0.2, 0.4, and 0.8 g/L [44].

3.5.3. Stevia

Supplementation of stevioside at levels of 100, 150, 200, 250, and 300 mg/kg linearly increased ADG and ADFI ($p \leq 0.05$) in 28-day-old pigs during a 42-day feeding period, while also linearly decreased ($p < 0.05$) the feed-to-gain ratio over the same period [30]. Supplementation of 100 to 800 mg/kg had no significant effect on ADG, ADFI, or feed/gain ratio during days 0 to 35 but did linearly increase body weight on day 35 ($p < 0.05$) [40]. The greatest ADG was observed with 300 mg/kg stevioside supplementation during the 42-day feeding period [30] and with 100 mg/kg stevia supplementation during a 75-day feeding period [40]. Supplementation of 83.3, 167, 334 mg/kg [42], 100, 200, 400 mg/kg [39], and 10 to 20% stevia [32] had no significant effect on ADG, ADFI, or feed-to-gain ratio in 24-, 21-, and 26-day-old pigs during 21-, 42-, and 28-day feeding periods, respectively. Pigs supplemented with 167 mg/kg stevia showed the lowest growth during days 0 to 7, and the highest growth during days 8 to 14 of a 21-day feeding period, compared to diets supplemented with 83.3 or 334 mg/kg stevia, and 5% sucrose [42]. Supplementation of 10 to 20% stevia did not significantly affect the initial and final body weight of the pigs but significantly increased feed consumption on day 28 ($p < 0.01$) during a 1-h preference test [32].

3.5.4. Sucralose

Zhang et al. [45] reported that supplementation of 150 mg/kg sucralose significantly increased ADG and ADFI ($p < 0.05$) compared to the supplementation of 75, 225, 300, and 1500 mg/kg sucralose, as well as a control diet, during a 28-day feeding period. However, supplementation at levels of 75, 150, 225, and 300 mg/kg did not significantly affect the gain-to-feed ratio during the same period [45]. A fitted quadratic model revealed that the optimal dosage of sucralose to maximize ADFI was 137.8 mg/kg during days 0 to 14, 145.8 mg/kg during days 15 to 28, and 141.8 mg/kg over the entire 28-day feeding period [45]. Additionally, supplementation with 150 mg/kg sucralose significantly increased feed consumption and preference percentage ($p < 0.05$) compared to the control diet on days 1, 4, and 7, and throughout the entire 28-day feeding period [45]. Preference was also elicited in pigs when sucralose was supplemented in water at concentrations of 0.062 and 0.125 g/L [44].

3.5.5. NHDC + Saccharin

Sterk et al. [33] observed that ADG, ADFI, and the gain-to-feed ratio, as well as the overall time-related development of feed intake and Kaplan-Meier curves for latency time, were not significantly affected by NHDC + saccharin 3D and C-150 supplementation in 26-day-old weanling pigs during a 19-day feeding period. Additionally, supplementation with NHDC + saccharin 3D and C-150 had no significant effect on feed intake characteristics during the first 4 days, days 5 to 11, and days 12 to 19 of the feeding period. However, NHDC + saccharin 3D supplementation significantly increased feed intake ($p < 0.05$) on days 8 and 10, and NHDC + saccharin C-150 tended to increase feed intake ($p = 0.074$) on day 16 compared to the control diet. The average duration of a feeder visit during the first 4 days post-weaning was 23% higher for pigs supplemented with NHDC + saccharin C-150 compared to the control diet. Moreover, the proportion of feeder visits that included feed consumption was significantly higher for NHDC + saccharin C-150 supplemented pigs during days 5 to 11 and days 12 to 19 of the feeding period compared to the control diet [33].

3.5.6. Other Sweeteners

According to two-choice drinking preference tests, alitame (0.15 and 0.3 g/L), acesulfame-K (0.01, 0.05, and 0.35 g/L), and dulcin (0.6, 1.2, and 2.4 g/L) elicited a preference in 2- to 4-month-old pigs [44]. However, monellin, thaumatin, NHDC, P-4000, perillartine, aspartame, and cyclamate did not elicit a preference in the same age group of pigs [44].

3.6. Clinical Indicators (Diarrhea, Immune Response, and Biochemical/Oxidative Parameters)

3.6.1. Neotame, Saccharin, and Neotame + Saccharin

Lee et al. [38] reported that a 0.02% neotame + saccharin blend significantly reduced ($p < 0.05$) total blood cholesterol compared to both the control and a 0.03% blend after 14 days. However, the 0.03% blend significantly increased ($p < 0.05$) blood triglycerides compared to the 0.02% blend [38]. Additionally, supplementation with 50 and 500 mg/kg neotame had no significant effect on hematological parameters, serum biochemical parameters, and organ index of pigs throughout a 35-day feeding period [31]. Similarly, supplementation of a saccharin-based sweetener had no significant effect on scours patterns in 28-day old pigs during a 10-day feeding period [43].

3.6.2. Rebaudioside A, NHDC + Saccharin, and Sucralose

Supplementation with 100 to 300 mg/kg rebaudioside A significantly reduced the incidence of diarrhea in 28-day-old pigs during a 42-day feeding period, with a broken line regression analysis identifying 191 mg/kg as the optimal dose to minimize diarrhea incidence [30]. In a 19-day feeding trial, supplementation of NHDC + saccharin 3D and C-150 to 26-day-old weanling pigs had no effect on fecal consistency during the first 5 days postweaning. However, from days 5 to 19, both supplements increased the percentage of firm feces compared to the control diet [33]. In addition, supplementation of 150 and 1500 mg/kg sucralose had no effect on hematological parameters, serum biochemical parameters, or organ index [45].

3.6.3. Stevia

Liu et al. [39] reported that supplementing 100 mg/kg stevia to 21-day-old pigs significantly increased serum catalase (CAT) and liver total superoxide dismutase (T-SOD) activities compared to higher doses and the control diet. Additionally, 100 and 200 mg/kg stevia reduced the rate of diarrhea and malondialdehyde (MDA) content, while higher doses (200 to 800 mg/kg) increased MDA content and enhanced total antioxidant capacity (T-AOC) and glutathione peroxidase (GSH-PX) activity. However, stevia supplementation had no significant effect on serum and liver T-AOC levels, CAT activity, or liver T-SOD at any dosage [39].

Xiong et al. [40] observed that 100 mg/kg stevia significantly reduced serum MDA content compared to higher doses and the control diet. Higher doses (600 to 800 mg/kg) also reduced MDA content, but 800 mg/kg increased triglyceride levels. Stevia supplementation (200 to 800 mg/kg) increased serum T-SOD during both 42- and 75-day feeding periods and resulted in a linear and quadratic increase in serum triglyceride, high-density lipoprotein, albumin, T-SOD, and CAT levels. However, stevia had no significant effect on several other serum markers, including glucose, total protein, cholesterol, and enzyme levels [40].

Wang et al. [30] determined that the optimal stevia dose to reduce diarrhea in 28-day-old pigs during a 42-day feeding period was 251 mg/kg.

3.7. Intestinal Development (Taste Receptor/Digestive Enzyme Activity)

3.7.1. Acesulfame K, Aspartame, Cyclamate, Stevia, and Sucralose

Daly et al. [37] reported that supplementing 28-day-old pigs with 10 mM acesulfame K or 2 mM sucralose in drinking water during a 3-day trial significantly increased ($p < 0.05$) the expression and activity of intestinal sodium glucose cotransporter 1 (SGLT1) and activated taste receptors T1R2 and T1R3. In contrast, supplementation with 1 mM aspartame or cyclamate had no effect on T1R2/T1R3 activation or SGLT1 expression and activity [37]. Liu et al. [39] observed that supplementing 100, 200, or 400 mg/kg stevia to 21-day-old pigs over a 42-day feeding period tended to reduce ($p < 0.05$) trypsin, lipase, and amylase activity in the duodenum, but did not significantly affect small intestine morphology or digestive enzyme activity. Similarly, Zhang et al. [45] noted that 150 and 1500 mg/kg sucralose had no significant effect on tissue histopathology compared to a control diet.

3.7.2. Saccharin, NHDC, and NHDC + Saccharin

Moran et al. [34] reported that dietary supplementation with NHDC + saccharin (150 mg/kg body weight) resulted in a 2-fold increase ($p = 0.001$) in SGLT1 mRNA expression and a 1.8-fold increase ($p = 0.002$) in glucose transport. Additionally, supplementation of drinking water with either saccharin (0.25 mM), NHDC (0.02 mM), or a combination of saccharin and NHDC increased SGLT1 mRNA expression in the mid-small intestine by 1.8-fold ($p = 0.003$), 1.6-fold ($p = 0.016$), and 2-fold ($p = 0.001$), respectively. These increases were correlated with rises in SGLT1 protein abundance—1.9-fold ($p = 0.037$),

1.8-fold ($p = 0.040$), and 1.6-fold ($p = 0.035$)—and corresponding increases in glucose uptake. However, no changes in villus height or crypt depth were observed in the intestines following supplementation with NHDC + saccharin or the sweeteners [34].

3.8. Gut Microbiota

Stevia, Neotame, Saccharin + Neotame, and NHDC + Saccharin

The dietary supplementation of 400, 600, and 800 mg/kg stevia significantly reduced ($p < 0.05$) Chao1 and observed indexes and tended to decrease ($p = 0.083$) the Shannon index during a 75-day feeding period [40]. However, no significant effects on the Chao1, observed-species, Shannon, and Simpson indexes of the colon microbes were observed when 100 to 800 mg/kg stevia was supplemented compared to a control diet [39]. Additionally, no significant differences to the gut microbial structure and Kruckal–Wallis rank sum test results were observed [40]. The supplementation of 400 mg/kg stevia during a 42-day feeding period increased the abundances of the genera Coxiella, Prevotella, Subdoligranulum, Akkermansia, and Roseburia in the intestines of 21-day-old pigs [39]. The supplementation of 400 mg/kg Stevia during a 42-day feeding period also tended to increase the relative abundances of the families Lachnospiraceae ($p < 0.067$) and Coriobacteriaceae ($p < 0.085$) and significantly increased ($p < 0.05$) the relative abundance of the family Prevotellaceae and the genera Roseburia and Prevotella in the colon of 21-day-old pigs compared to a control diet [39]. The dietary supplementation of 0.02% Neotame and a 0.02% saccharin + neotame blend during a 14-day feeding period significantly increased ($p < 0.05$) fecal Lactobacillus abundance compared to a control diet [38].

The dietary supplementation of 0.015% NHDC + saccharin to 28-day-old pigs during a 14-day feeding period significantly increased ($p < 0.05$) abundance of Helicobacteraceae [22], Lactobacullus [35], Lactobacilaceae [36], and lactic acid concentrations [35,36] within the small intestinal mucosa, microbiota, and cecum respectively. However, a significant reduction ($p < 0.05$) in the relative abundance of Campylobacteraceae [22], Veillonellaceae, and Ruminococcaceae [36] in the small intestine and cecum, respectively, was observed. No significant differences were found in the quantitative analysis of total 16s rRNA gene copies in the duodenum and jejunum when 0.015% NHDC + saccharin was supplemented compared to a control diet [22].

3.9. Organ Development and Meat Quality

Neotame and Stevia

Normal histological structures of the liver and kidney were observed with both 50 and 500 mg/kg Neotame treatments during a 35-day feeding period [31]. The supplementation of 100 mg/kg Stevia significantly increased ($p < 0.05$) hot carcass weight and gastric index and tended to increase ($p = 0.066$) carcass circumference during a 75-day feeding period [40]. As the Stevia supplementation dosage increased from 100 to 800 mg/kg, the score of carcass appearance increased linearly; however, no significant effects were observed on organ index, meat pH and color, drip loss, shear force, marbling score, or the content of intramuscular fat, moisture, myofiber diameter, density of the longissimus thoracis, score of smell, flavor, abnormal flavor, chewiness, juiciness, and turbidness of soup [40]. The supplementation of 1.67 g/kg stevioside for 14 days was completely converted into steriol (853 ± 48 µg/g dry weight) in the feces of treatment pigs; however, no stevioside or steriol was detected in the blood [29].

4. Discussion

This systematic review aimed to evaluate the effects of various sweeteners on growth performance, feed intake, and gut health in pigs. The majority of studies focused on growth performance, with several reporting significant improvements in ADG and feed efficiency. Studies on gut health were more limited, but some demonstrated that sweeteners modulated gut microbiota by increasing beneficial bacteria. While the review found consistent evidence supporting growth performance benefits, the limited data on other outcomes, such as immune responses and gut health, preclude definitive conclusions on the broader effects of sweeteners in pig nutrition.

The current findings reveal that supplementation of various sweeteners can exert notable effects on the growth performance and feed preferences of pigs. These effects are influenced by both dosage and the specific type of sweetener used, and several results align with the existing literature while others suggest more complex interactions. Neotame supplementation improved both ADG and ADFI during the early stages of the feeding period [38]. This result is consistent with Zhu et al. [31], who demonstrated a dose-dependent effect of neotame on pig growth rates. These findings suggest that neotame may provide an initial boost in feed intake, possibly by stimulating taste receptors and enhancing palatability. However, neotame's intense sweetness limits its maximum tolerable dose, as excessive amounts can have negative consequences. Optimal growth rates and feed intake were achieved with diets containing approximately 20 to 30 mg/kg, when different doses up to 50 mg/kg were tested. This is consistent with previous work by Mayhew et al. [46], who reported that rats preferred a basal diet over a diet containing high concentrations (150 mg/kg or more) of neotame. The reduction in feed intake was attributed to the palatability of the diet rather than any toxicological effects. Another possible explanation is that neotame, at high concentrations, might interfere with the natural flavor of the basal diet, leading to reduced consumption, as pigs may avoid the overpowering taste. Interestingly, the negative interaction between neotame and saccharin observed in [38], where the combination led to reduced ADFI compared to neotame alone, raises important questions about the compatibility of sweeteners. Saccharin inclusion in water was consumed significantly less than the control diet [41]. This effect has been noted in other sweeteners like saccharin, where high doses led to decreased feed intake due to similar palatability issues [47]. Moreover, research by Roura and Fu [48] demonstrated that the interaction between taste receptors and sweeteners can affect feed intake, further supporting the idea that compatibility between sweeteners is crucial. Beyond these effects on growth and feed intake, neotame supplementation also appears to influence gut health. A 0.02% neotame or neotame–saccharin blend significantly increased fecal Lactobacillus abundance [38], suggesting potential benefits for gut microbiota. Lactobacillus is associated with improved digestion and nutrient absorption, indicating that neotame may enhance not only palatability but also overall nutrient utilization. These findings suggest a dual role for neotame in improving both feed intake and gut health, though the optimal dosing remains crucial to avoid palatability issues at higher concentrations.

In contrast to neotame, stevioside and rebaudioside A offer more predictable and sustained improvements in growth performance and feed efficiency. Notably, these compounds enhance the growth performance of weaned piglets not only by improving feed palatability but also by potentially reducing diarrhea incidence [30]. The anti-diarrheal effects are thought to be linked to bactericidal properties, particularly against pathogenic bacteria such as Escherichia coli [49]. Therefore, stevioside and rebaudioside A may present a dual benefit in swine production: promoting growth while simultaneously supporting gut health by reducing the risk of infections. In addition to their sweetening properties, both sweeteners have been shown to offer additional therapeutic benefits, including anti-hyperglycemic, anti-hypertensive, anti-inflammatory, anti-diarrheal, and immunomodulatory effects [50]. These multifunctional properties suggest a broader potential for these compounds in animal health management. Supporting evidence from studies in poultry highlights the potential of stevioside as an immunomodulator. For instance, Daneshyar et al. [51] found that 130 mg/kg of stevioside supplementation not only increased the body weight of broiler chickens but also suppressed pro-inflammatory responses following stimulation of the innate immune system. Similarly, Wu et al. [25] reported a linear increase in serum IgG and IgA levels in broilers fed with stevioside, indicating enhanced immune function. These findings imply that stevioside supplementation could be particularly beneficial for young and susceptible animals, such as weaned piglets. Furthermore, Atteh et al. [20] observed that a diet containing 130 mg/kg of stevioside improved body weight and feed conversion ratio in broilers during the first two weeks. Additionally, stevioside altered the short-chain fatty acid profile in the ceca, promoting beneficial microbial changes, such as increases in Bifidobacteria and reductions in Escherichia coli [25]. These microbial changes could further contribute to the observed improvements in growth performance, particularly in young animals. Although there is limited research on the effects of rebaudioside A on gut microbiota in pigs, supplementation of rebaudioside A in mice increased the diversity of fecal Lactobacilli, which is associated with improved gut health [52]. In pigs, increased populations of Lactobacilli have been linked to improved nutrient absorption, better immune function, and overall growth performance [53], making the modulation of gut microbiota a key mechanism through which rebaudioside A positively affects growth. This modulation of the gut microbiota appears to have broader health benefits, particularly in improving gut integrity.

Stevia supplementation has demonstrated clear dose-dependent effects on growth performance, oxidative stress, and gut health in pigs, though its impact can vary based on supplementation duration and animal age. Xiong et al. [40] reported that 100 mg/kg of stevia increased body weight by day 35, although effects on ADG and feed efficiency varied depending on the length of the study. In contrast, other studies [32,42] found no significant impact on ADG or feed efficiency over shorter feeding periods. This variability suggests that stevia's benefits may become more pronounced with extended supplementation durations, highlighting the importance of study design and animal age when evaluating its effects. In terms of oxidative stress, Liu et al. [39] found that 100 mg/kg of stevia increased antioxidant enzymes such as CAT and SOD, while reducing oxidative stress markers like MDA. Higher doses (200 to 800 mg/kg) further boosted total antioxidant capacity but led to increased serum triglycerides at 800 mg/kg, indicating possible trade-offs at higher levels. These effects were supported by research in rats, which showed that stevia supplementation offers protective benefits against diseases such as ulcerative colitis, hyperuricemia, diabetes mellitus, and acute liver injury, primarily due to its antioxidant properties [54]. Stevia's antioxidant effects are attributed to its high polyphenol content, including phenolic acids and flavonoids, which neutralize reactive oxygen species (ROS) by stabilizing them and preventing cellular damage. Additionally, stevia enhances the activity of key antioxidant enzymes like SOD, CAT, and glutathione peroxidase (GPx), further reducing oxidative stress and inflammation. This dual mechanism supports better cellular health and may help mitigate oxidative damage in animals. Stevia supplementation also had notable effects on gut microbial composition, although its impact on microbial diversity was mixed. Xiong et al. [40] found that higher doses of stevia (400, 600, and 800 mg/kg) reduced microbial diversity during a 75-day feeding period, while Liu et al. [39] observed no significant changes in diversity across various doses. However, stevia at 400 mg/kg positively influenced the abundance of beneficial gut microbes, including Lachnospiraceae, Coriobacteriaceae, and Prevotellaceae, suggesting that moderate doses of stevia can enhance the gut microbial profile even if overall diversity remains unaffected. Similar effects were observed in recent poultry studies, where stevia supplementation modulated intestinal microbial composition and improved production performance, egg nutrition, gut health, and immune capabilities in laying hens [55,56]. In broilers, stevia enhanced intestinal functionality, increased microbial diversity, and promoted the growth of beneficial bacterial genera [57]. These findings suggest that stevia's role in supporting gut health and microbial balance may extend across species, further highlighting its potential as a dietary supplement for improving both animal performance and gut integrity.

Sucralose and NHDC + saccharin have shown potential to influence growth performance, though their effects vary depending on the dose. Sucralose supplementation at 150 mg/kg significantly improved ADG and ADFI [45], suggesting that moderate doses can enhance feed intake and growth in pigs. However, the lack of significant changes in the gain-to-feed ratio indicates that, while sucralose may boost intake, it does not necessarily improve feed efficiency. In contrast, NHDC + saccharin showed less consistent results. Sterk et al. [33] found no significant impact on ADG or ADFI during a 19-day period, although NHDC + saccharin increased feed intake on certain days and extended feeder visits. This suggests that NHDC + saccharin may improve feeding behavior without directly influencing overall growth performance. Similar trends were observed in ruminants, where NHDC + saccharin supplementation showed only a tendency to increase feed intake [15] and exhibited a slight tendency to increase ADG during a 56-day receiving period [19]. The improvements in feed intake seen with both sucralose and NHDC + saccharin suggest that these sweeteners can enhance palatability, which is particularly beneficial for weaning pigs, a period when appetite is often reduced. However, the inconsistent effects on growth efficiency raise questions about their long-term value in improving overall feed conversion. Sucralose and NHDC + saccharin have demonstrated benefits in nutrient absorption and gut health, primarily by enhancing the activity of the SGLT1 and activating taste receptors, potentially improving glucose absorption and energy utilization in pigs [34,37]. However, neither sweetener significantly affected intestinal morphology, indicating their role is focused on nutrient transport rather than structural development. NHDC + saccharin showed a more pronounced effect on gut microbiota than sucralose, increasing the abundance of beneficial bacteria such as Lactobacillus and reducing harmful populations like Campylobacteraceae [22]. This suggests that NHDC + saccharin may improve gut health and immune function, potentially leading to better overall performance. In contrast, the effect of sucralose on gut microbiota remains unclear, with limited data available, indicating the need for further research. A comprehensive review suggests short-term changes in the microbiota with sucralose consumption [58], though more long-term research is needed to fully understand its effects. For instance, inconsistent findings have shown that sucralose reduced obesity in humans by decreasing the Firmicutes/Bacteroidetes ratio and increasing Actinobacteria. However, in mice and rats, sucralose was found to induce obesity by also reducing the Firmicutes/Bacteroidetes ratio [59]. These conflicting results highlight the need for further studies to clarify sucralose's impact on gut health across different species and durations of use. Clinically, sucralose showed no adverse effects on health markers [45], while NHDC + saccharin improved post-weaning fecal firmness, indicating added gut health benefits during the transition period for young pigs [33]. Overall, both sweeteners are safe at tested doses, with NHDC + saccharin offering additional potential for enhancing gut health by positively modulating microbial populations and supporting immune function.

5. Strengths and Limitations

The strengths of this systematic review lie in its comprehensive evaluation of the effects of various sweeteners on growth performance, feed intake, and gut health in pigs. The review effectively highlights consistent improvements in ADG and feed efficiency, demonstrating the potential of sweeteners to enhance pig growth, particularly during critical periods like weaning when appetite is often reduced. Additionally, it emphasizes the benefits of sweeteners in modulating gut microbiota, potentially improving nutrient absorption and supporting immune function. However, several limitations were identified. There is a lack of research on the effects of combining different sweeteners, where interactions may reduce feed intake or alter other outcomes, raising concerns about compatibility and the need for further investigation. Furthermore, there are limited data on the long-term effects of sweeteners, particularly regarding gut health and immune responses, making it difficult to form comprehensive conclusions. The variability in effects on feed efficiency and gut microbiota across studies also suggests a need for more exploration of underlying pathways. While the review discusses potential mechanisms, such as the activation of taste receptors and modulation of gut bacteria, more detailed mechanistic studies are required to fully understand how sweeteners influence growth, metabolism, and health outcomes in pigs.

6. Conclusions

This systematic review highlights the potential of various sweeteners to improve growth performance, feed intake, and gut health in pigs. Many studies reported significant improvements in ADG and feed efficiency, suggesting that sweeteners can be effective in promoting pig growth, particularly during critical periods such as weaning when appetite is reduced. Additionally, certain sweeteners demonstrated the ability to modulate gut microbiota by increasing beneficial bacteria, potentially enhancing nutrient absorption and supporting immune function. However, despite the promising results, limitations were identified. Research on gut health and immune responses remains limited, and the long-term effects of sweetener use in pig diets require further investigation. Additionally, the review raises concerns about the interactions between different sweeteners, as some combinations resulted in reduced feed intake. This highlights the need for more studies exploring the compatibility and mechanisms of action of sweeteners. Furthermore, the variability in findings across studies suggests that dosage, type of sweetener, and study design play crucial roles in determining their effectiveness. Overall, while sweeteners show potential in enhancing pig nutrition, further research is needed to fully understand their broader impacts and optimize their use in swine production systems.

Author Contributions: Conceptualization, methodology, software, validation, resources, visualization, writing—original draft preparation, M.R.J. and K.K.; supervision, project administration, writing—review and editing, K.K. All authors have read and agreed to the published version of the manuscript.

Funding: This project is partially funded by Michigan State University AgBioResearch (Proposal ID #378, East Lansing, MI, USA) and USDA-NIFA Hatch (Project #MICL12073, Washington, DC, USA).

Institutional Review Board Statement: The protocol of this systematic review was reviewed, approved, and registered at the International Prospective Register of Systematic Reviews (PROSPERO) (ID: CRD42024518080). Available from: https://www.crd.york.ac.uk/prospero/display_record.php?ID=CRD42024518080 (accessed on 30 September 2024).

Informed Consent Statement: Not applicable.

Acknowledgments: The authors would like to acknowledge and thank Yoo Jung (Erika) Oh (Department of Communication, Michigan State University) for their invaluable guidance and support throughout the systematic review process.

Conflicts of Interest: The authors declare no conflicts of interest.

References

1. Carocho, M.; Morales, P.; Ferreira, I.C.F.R. Sweeteners as Food Additives in the XXI Century: A Review of What Is Known, and What Is to Come. *Food Chem. Toxicol.* **2017**, *107*, 302–317. [CrossRef] [PubMed]
2. Chen, J.; Lei, Y.; Zhang, Y.; He, S.; Liu, L.; Dong, X. Beyond Sweetness: The High-Intensity Sweeteners and Farm Animals. *Anim. Feed Sci. Technol.* **2020**, *267*, 114571. [CrossRef]
3. Conz, A.; Salmona, M.; Diomede, L. Effect of Non-Nutritive Sweeteners on the Gut Microbiota. *Nutrients* **2023**, *15*, 1869. [CrossRef] [PubMed]
4. Walters, D.E.; Cyclamate, H.P.S. *The Sweetener Book*; Gale Walters Publishing: Wageningen, The Netherlands, 2013; pp. 60–62. Available online: https://library.wur.nl/WebQuery/titel/2043303 (accessed on 29 September 2024).
5. Yalamanchi, S.; Srinath, R.; Dobs, A. Acesulfame-K. In *Encyclopedia of Food and Health*; Caballero, B., Finglas, P.M., Toldrá, F., Eds.; Academic Press: Oxford, UK, 2016; pp. 1–5, ISBN 978-0-12-384953-3.
6. Basson, A.R.; Rodriguez-Palacios, A.; Cominelli, F. Artificial Sweeteners: History and New Concepts on Inflammation. *Front. Nutr.* **2021**, *8*, 746247. [CrossRef]
7. European Parlament Official Journal of the European Union L346/2006. Available online: https://eur-lex.europa.eu/legal-content/EN/TXT/HTML/?uri=OJ:L:2006:346:FULL (accessed on 11 June 2024).
8. Food and Drug Administation HHS Title 21. Available online: https://www.ecfr.gov/current/title-21 (accessed on 11 June 2024).
9. Aldinger, S.M.; Speer, V.C.; Hays, V.W.; Catron, D.V. Effect of Saccharin on Consumption of Starter Rations by Baby Pigs. *J. Anim. Sci.* **1959**, *18*, 1350–1355. [CrossRef]
10. Goatcher, W.D.; Church, D.C. Taste Responses in Ruminants. I. Reactions of Sheep to Sugars, Saccharin, Ethanol and Salts. *J. Anim. Sci.* **1970**, *30*, 777–783. [CrossRef]

11. Hellekant, G.; Hård af Segerstad, C.; Roberts, T.W. Sweet Taste in the Calf: III. Behavioral Responses to Sweeteners. *Physiol. Behav.* **1994**, *56*, 555–562. [CrossRef]
12. Moran, A.W.; Al-Rammahi, M.; Zhang, C.; Bravo, D.; Calsamiglia, S.; Shirazi-Beechey, S.P. Sweet Taste Receptor Expression in Ruminant Intestine and Its Activation by Artificial Sweeteners to Regulate Glucose Absorption. *J. Dairy Sci.* **2014**, *97*, 4955–4972. [CrossRef]
13. Koester, L.R.; Anderson, C.J.; Cortes, B.W.; Lyte, M.; Schmitz-Esser, S. Influence of the Artificial Sodium Saccharin Sweetener Sucram® on the Microbial Community Composition in the Rumen Content and Attached to the Rumen Epithelium in Dairy Cattle: A Pilot Study. *BioRxiv* **2020**, 05.22.110429. [CrossRef]
14. Koester, L.R.; Hayman, K.; Anderson, C.J.; Tibbs-Cortes, B.W.; Daniels, K.M.; Seggerman, F.M.; Gorden, P.J.; Lyte, M.; Schmitz-Esser, S. Influence of a Sodium-Saccharin Sweetener on the Rumen Content and Rumen Epithelium Microbiota in Dairy Cattle during Heat Stress. *J. Anim. Sci.* **2022**, *101*, skac403. [CrossRef]
15. Ponce, C.H.; Brown, M.S.; Silva, J.S.; Schlegel, P.; Rounds, W.; Hallford, D.M. Effects of a Dietary Sweetener on Growth Performance and Health of Stressed Beef Calves and on Diet Digestibility and Plasma and Urinary Metabolite Concentrations of Healthy Calves. *J. Anim. Sci.* **2014**, *92*, 1630–1638. [CrossRef] [PubMed]
16. Siurana, A.; Wall, E.; Rodrguez, M.; Castillejos, L.; Ferret, A.; Calsamiglia, S. The Effect of Dietary Supplementation of Artificial Sweetener on Performance of Milk-Fed Calves. *J. Anim. Sci.* **2014**, *92*, 816.
17. Jiang, J.; Liu, S.; Jamal, T.; Ding, T.; Qi, L.; Lv, Z.; Yu, D.; Shi, F. Effects of Dietary Sweeteners Supplementation on Growth Performance, Serum Biochemicals, and Jejunal Physiological Functions of Broiler Chickens. *Poult. Sci.* **2020**, *99*, 3948–3958. [CrossRef] [PubMed]
18. Molina-Barrios, R.M.; Avilés-Trejo, C.R.; Puentes-Mercado, M.E.; Cedillo-Cobián, J.R.; Hernández-Chavez, J.F. Effect of Dietary Stevia-Based Sweetener on Body Weight and Humoral Immune Response of Broiler Chickens. *Vet. World* **2021**, *14*, 913–917. [CrossRef]
19. McMeniman, J.P.; Rivera, J.D.; Schlegel, P.; Rounds, W.; Galyean, M.L. Effects of an Artificial Sweetener on Health, Performance, and Dietary Preference of Feedlot Cattle. *J. Anim. Sci.* **2006**, *84*, 2491–2500. [CrossRef]
20. Atteh, J.O.; Onagbesan, O.M.; Tona, K.; Decuypere, E.; Geuns, J.M.C.; Buyse, J. Evaluation of Supplementary Stevia (*Stevia rebaudiana, Bertoni*) Leaves and Stevioside in Broiler Diets: Effects on Feed Intake, Nutrient Metabolism, Blood Parameters and Growth Performance. *J. Anim. Physiol. Anim. Nutr.* **2008**, *92*, 640–649. [CrossRef]
21. Sigalet, D.L.; Wallace, L.; De Heuval, E.; Sharkey, K.A. The Effects of Glucagon-like Peptide 2 on Enteric Neurons in Intestinal Inflammation. *Neurogastroenterol. Motil.* **2010**, *22*, 1318-e350. [CrossRef]
22. Kelly, J.; Daly, K.; Moran, A.W.; Ryan, S.; Bravo, D.; Shirazi-Beechey, S.P. Composition and Diversity of Mucosa-associated Microbiota along the Entire Length of the Pig Gastrointestinal Tract; Dietary Influences. *Environ. Microbiol.* **2017**, *19*, 1425–1438. [CrossRef]
23. Connor, E.E.; Evock-Clover, C.M.; Wall, E.H.; Baldwin, R.L.; Santin-Duran, M.; Elsasser, T.H.; Bravo, D.M. Glucagon-like Peptide 2 and Its Beneficial Effects on Gut Function and Health in Production Animals. *Domest. Anim. Endocrinol.* **2016**, *56*, S56–S65. [CrossRef]
24. Sun, D.; Liu, L.; Mao, S.; Zhu, W.; Liu, J. Aspartame Supplementation in Starter Accelerates Small Intestinal Epithelial Cell Cycle and Stimulates Secretion of Glucagon-like Peptide-2 in Pre-Weaned Lambs. *J. Anim. Physiol. Anim. Nutr.* **2019**, *103*, 1338–1350. [CrossRef]
25. Wu, X.; Yang, P.; Sifa, D.; Wen, Z. Effect of Dietary Stevioside Supplementation on Growth Performance, Nutrient Digestibility, Serum Parameters, and Intestinal Microflora in Broilers. *Food Funct.* **2019**, *10*, 2340–2346. [CrossRef] [PubMed]
26. Kim, K.; Jansen, M. Evaluating the Effects of Non-Nutritive Sweeteners on Pigs: A Systematic Review. PROSPERO, 2024. Available online: https://www.crd.york.ac.uk/prospero/display_record.php?ID=CRD42024518080 (accessed on 29 September 2024).
27. Schardt, C.; Adams, M.B.; Owens, T.; Keitz, S.; Fontelo, P. Utilization of the PICO Framework to Improve Searching PubMed for Clinical Questions. *BMC Med. Inform. Decis. Mak.* **2007**, *7*, 16. [CrossRef] [PubMed]
28. Covidence Systematic Review Software, Veritas Health Innovation, Melbourne, Australia, 2024. Available online: www.covidence.org (accessed on 29 September 2024).
29. Geuns, J.M.C.; Augustijns, P.; Mols, R.; Buyse, J.G.; Driessen, B. Metabolism of Stevioside in Pigs and Intestinal Absorption Characteristics of Stevioside, Rebaudioside A and Steviol. *Food Chem. Toxicol.* **2003**, *41*, 1599–1607. [CrossRef] [PubMed]
30. Wang, L.S.; Shi, Z.; Shi, B.M.; Shan, A.S. Effects of Dietary Stevioside/Rebaudioside A on the Growth Performance and Diarrhea Incidence of Weaned Piglets. *Anim. Feed Sci. Technol.* **2014**, *187*, 104–109. [CrossRef]
31. Zhu, L.; Wang, G.; Dong, B.; Peng, C.C.; Tian, Y.Y.; Gong, L.M. Effects of Sweetener Neotame on Diet Preference, Performance and Hematological and Biochemical Parameters of Weaned Piglets. *Anim. Feed Sci. Technol.* **2016**, *214*, 86–94. [CrossRef]
32. Clouard, C.; Val-Laillet, D. Impact of Sensory Feed Additives on Feed Intake, Feed Preferences, and Growth of Female Piglets during the Early Postweaning Period1. *J. Anim. Sci.* **2014**, *92*, 2133–2140. [CrossRef]
33. Sterk, A.; Schlegel, P.; Mul, A.J.; Ubbink-Blanksma, M.; Bruininx, E.M.A.M. Effects of Sweeteners on Individual Feed Intake Characteristics and Performance in Group-Housed Weanling Pigs1. *J. Anim. Sci.* **2008**, *86*, 2990–2997. [CrossRef]

34. Moran, A.W.; Al-Rammahi, M.A.; Arora, D.K.; Batchelor, D.J.; Coulter, E.A.; Daly, K.; Ionescu, C.; Bravo, D.; Shirazi-Beechey, S.P. Expression of Na^+/Glucose Co-Transporter 1 (SGLT1) Is Enhanced by Supplementation of the Diet of Weaning Piglets with Artificial Sweeteners. *Br. J. Nutr.* **2010**, *104*, 637–646. [CrossRef]
35. Daly, K.; Darby, A.C.; Hall, N.; Nau, A.; Bravo, D.; Shirazi-Beechey, S.P. Dietary Supplementation with Lactose or Artificial Sweetener Enhances Swine Gut *Lactobacillus* Population Abundance. *Br. J. Nutr.* **2014**, *111*, S30–S35. [CrossRef]
36. Daly, K.; Darby, A.C.; Hall, N.; Wilkinson, M.C.; Pongchaikul, P.; Bravo, D.; Shirazi-Beechey, S.P. Bacterial Sensing Underlies Artificial Sweetener-induced Growth of Gut *Lactobacillus*. *Environ. Microbiol.* **2016**, *18*, 2159–2171. [CrossRef]
37. Daly, K.; Moran, A.W.; Al-Rammahi, M.; Weatherburn, D.; Shirazi-Beechey, S.P. Non-Nutritive Sweetener Activation of the Pig Sweet Taste Receptor T1R2-T1R3 In Vitro Mirrors Sweetener Stimulation of the Gut-Expressed Receptor In Vivo. *Biochem. Biophys. Res. Commun.* **2021**, *542*, 54–58. [CrossRef] [PubMed]
38. Lee, C.H.; Yun, W.; Lee, J.H.; Kwak, W.G.; Oh, H.J.; An, J.S.; Liu, S.D.; Cho, J.H. Effects of Artificial Sweeteners on Feed Palatability and Performance in Weaned Pigs. *Can. J. Anim. Sci.* **2019**, *99*, 307–314. [CrossRef]
39. Liu, S.; Xiong, Y.; Cao, S.; Wen, X.; Xiao, H.; Li, Y.; Chi, L.; He, D.; Jiang, Z.; Wang, L. Dietary Stevia Residue Extract Supplementation Improves Antioxidant Capacity and Intestinal Microbial Composition of Weaned Piglets. *Antioxidants* **2022**, *11*, 2016. [CrossRef] [PubMed]
40. Xiong, Y.; Liu, S.; Xiao, H.; Wu, Q.; Chi, L.; Zhu, L.; Fang, L.; Li, Y.; Jiang, Z.; Wang, L. Dietary Stevia Residue Extract Supplementation Improves the Performance and Antioxidative Capacity of Growing–Finishing Pigs. *J. Sci. Food Agric.* **2022**, *102*, 4724–4735. [CrossRef]
41. Clouard, C.; Chataignier, M.; Meunier-Salaün, M.-C.; Val-Laillet, D. Flavour Preference Acquired via a Beverage-Induced Conditioning and Its Transposition to Solid Food: Sucrose but Not Maltodextrin or Saccharin Induced Significant Flavour Preferences in Pigs. *Appl. Anim. Behav. Sci.* **2012**, *136*, 26–36. [CrossRef]
42. Munro, P.J.; Lirette, A.; Anderson, D.M.; Ju, H.Y. Effects of a New Sweetener, Stevia, on Performance of Newly Weaned Pigs. *Can. J. Anim. Sci.* **2000**, *80*, 529–531. [CrossRef]
43. Maenz, D.D.; Patience, J.F.; Wolynetz, M.S. Effect of Water Sweetener on the Performance of Newly Weaned Pigs Offered Medicated and Unmedicated Feed. *Can. J. Anim. Sci.* **1993**, *73*, 669–672. [CrossRef]
44. Glaser, D.; Wanner, M.; Tinti, J.M.; Nofre, C. Gustatory Responses of Pigs to Various Natural and Artificial Compounds Known to Be Sweet in Man. *Food Chem.* **2000**, *68*, 375–385. [CrossRef]
45. Zhang, W.; Heng, J.; Kim, S.W.; Chen, F.; Deng, Z.; Zhang, S.; Guan, W. Dietary Enzymatically-Treated *Artemisia annua* L. Supplementation Could Alleviate Oxidative Injury and Improve Reproductive Performance of Sows Reared under High Ambient Temperature. *J. Therm. Biol.* **2020**, *94*, 102751. [CrossRef]
46. Mayhew, D.A.; Phil Comer, C.; Wayne Stargel, W. Food Consumption and Body Weight Changes with Neotame, a New Sweetener with Intense Taste: Differentiating Effects of Palatability from Toxicity in Dietary Safety Studies. *Regul. Toxicol. Pharmacol.* **2003**, *38*, 124–143. [CrossRef]
47. Kennedy, J.M.; Baldwin, B.A. Taste Preferences in Pigs for Nutritive and Non-Nutritive Sweet Solutions. *Anim. Behav.* **1972**, *20*, 706–718. [CrossRef] [PubMed]
48. Roura, E.; Fu, M. Taste, Nutrient Sensing and Feed Intake in Pigs (130 Years of Research: Then, Now and Future). *Anim. Feed Sci. Technol.* **2017**, *233*, 3–12. [CrossRef]
49. Tomita, T.; Sato, N.; Arai, T.; Shiraishi, H.; Sato, M.; Takeuchi, M.; Kamio, Y. Bactericidal Activity of a Fermented Hot-Water Extract from *Stevia rebaudiana* Bertoni towards Enterohemorrhagic *Escherichia coli* O157:H7 and Other Food-Borne Pathogenic Bacteria. *Microbiol. Immunol.* **1997**, *41*, 1005–1009. [CrossRef] [PubMed]
50. Chatsudthipong, V.; Muanprasat, C. Stevioside and Related Compounds: Therapeutic Benefits beyond Sweetness. *Pharmacol. Ther.* **2009**, *121*, 41–54. [CrossRef]
51. Daneshyar, M.; Geuns, J.M.C.; Willemsen, H.; Ansari, Z.; Darras, V.M.; Buyse, J.G.; Everaert, N. Evaluation of Dietary Stevioside Supplementation on Anti-human Serum Albumin Immunoglobulin G, Alpha-1-glycoprotein, Body Weight and Thyroid Hormones in Broiler Chickens. *J. Anim. Physiol. Anim. Nutr.* **2012**, *96*, 627–633. [CrossRef]
52. Li, S.; Chen, T.; Dong, S.; Xiong, Y.; Wei, H.; Xu, F. The Effects of Rebaudioside A on Microbial Diversity in Mouse Intestine. *Food Sci. Technol. Res.* **2014**, *20*, 459–467. [CrossRef]
53. Liao, S.F.; Nyachoti, M. Using Probiotics to Improve Swine Gut Health and Nutrient Utilization. *Anim. Nutr.* **2017**, *3*, 331–343. [CrossRef]
54. Papaefthimiou, M.; Kontou, P.I.; Bagos, P.G.; Braliou, G.G. Antioxidant Activity of Leaf Extracts from Stevia Rebaudiana Bertoni Exerts Attenuating Effect on Diseased Experimental Rats: A Systematic Review and Meta-Analysis. *Nutrients* **2023**, *15*, 3325. [CrossRef]
55. Tang, M.; Zhao, J.; Wu, Y.; Yu, C.; Peng, C.; Liu, H.; Cui, Y.; Lan, W.; Lin, Y.; Kong, X.; et al. Improving Gut Functions and Egg Nutrition with Stevia Residue in Laying Hens. *Poult. Sci.* **2024**, *103*, 103324. [CrossRef]
56. Wen, K.; Zhang, K.; Gao, W.; Bai, S.; Wang, J.; Song, W.; Zeng, Q.; Peng, H.; Lv, L.; Xuan, Y.; et al. Effects of Stevia Extract on Production Performance, Serum Biochemistry, Antioxidant Capacity, and Gut Health of Laying Hens. *Poult. Sci.* **2024**, *103*, 103188. [CrossRef]
57. Medeot, D.B.; Nilson, A.; Miazzo, R.D.; Grosso, V.; Ferrari, W.; Jofré, E.; Soltermann, A.; Peralta, M.F. Stevia as a Natural Additive on Gut Health and Cecal Microbiota in Broilers. *Vet. Anim. Sci.* **2023**, *22*, 100322. [CrossRef] [PubMed]

58. Del Pozo, S.; Gómez-Martínez, S.; Díaz, L.E.; Nova, E.; Urrialde, R.; Marcos, A. Potential Effects of Sucralose and Saccharin on Gut Microbiota: A Review. *Nutrients* **2022**, *14*, 1682. [CrossRef] [PubMed]
59. Chen, C.-Y.; Tien, C.-H.; Chen, Y.-H.; Garrido, D.; Farzi, A.; Herzog, H.; Fan, H.-Y.; Chen, Y.-C. Effect of Sucralose Intake on Human and Mouse/Rat Gut Microbiota Composition: A Systematic Review and Meta-Analysis. *Food Rev. Int.* **2024**, *40*, 1265–1275. [CrossRef]

MDPI AG
Grosspeteranlage 5
4052 Basel
Switzerland
Tel.: +41 61 683 77 34

Animals Editorial Office
E-mail: animals@mdpi.com
www.mdpi.com/journal/animals

www.ingramcontent.com/pod-product-compliance
Lightning Source LLC
Chambersburg PA
CBHW040752250926
51850CB00007BB/1474
9783725865321